Sardul S. Guraya

Biology of Ovarian Follicles in Mammals

With 76 Figures

Springer-Verlag
Berlin Heidelberg New York Tokyo

Professor **Sardul S. Guraya**
ICMR Regional Advanced Research
Center in Reproductive Biology
Department of Zoology
Punjab Agricultural University
Ludhiana, Punjab, India

ISBN 3-540-15022-6 Springer-Verlag Berlin Heidelberg New York Tokyo
ISBN-0-387-15022-6 Springer-Verlag New York Heidelberg Berlin Tokyo

Library of Congress Cataloging in Publication Data. Guraya, Sardul S., 1930–. Biology of
ovarian follicles in mammals. Bibliography: p. Includes index. 1. Ovaries. 2. Graafian follicle. 3.
Mammals–Physiology. I. Title. QL739.2.G87 1985 599′.016 85-2873

© by Springer-Verlag Berlin Heidelberg 1985
Printed in Germany

The use of registered names, etc. in this publication does not imply, even in the absence of a
specific statement, that such names are exempt from the relevant protective laws and regulations
and therefore free for general use.

Product Liability: The publisher can give no guarantee for information about drug dosage and
application thereof contained in this book. In every individual case the respective user must check
its accuracy by consulting other pharmaceutical literature.

Typesetting: H. Hagedorn, Berlin. Printing: Saladruck, Berlin. Bookbinding: B. Helm, Berlin
2131/3020-543210

This Book Is Dedicated to My Parents

Preface

Both functions of the mammalian ovary, the endocrine and (synthesis and secretion of steroid hormones) and exocrine (production of ova), depend upon the presence and cyclic growth of follicles, as the depletion of primordial follicles from the ovary leads to cessation of these functions or female reproduction in mammals, or to postmenopausal period in humans. Actually, various fertility and sterility problems at the ovarian level are related to follicles. Therefore, a thorough understanding of the biology of ovarian follicles in mammals is of fundamental interest to a wide variety of academic and scientific disciplines. Study of their structure, function, and control involves morphology, including ultrastructure, cell biology, physiology, endocrinology, biochemistry, immunology, neurobiology and pharmacology. Zoologists take interest in comparative and evolutionary aspects of biology of ovarian follicles in many different groups of mammals. Agricultural scientists and wildlife biologists need a thorough knowledge of the biology of follicles to control more effectively fecundity in domestic animals and endangered species of mammals. Finally, clinical scientists, toxicologists and physicians want to know the normal and pathological features of ovarian follicles in women, especially in relation to health and the regulation of fertility.

Having in view the great importance of studying various aspects of biology of ovarian follicles, numerous papers and reviews on several aspects of ovarian follicles are published annually in a wide variety of scientific journals. Unfortunately, it is impossible for a scientist, student or physician to go through even a small fraction of these publications. Also, little effort has been made to integrate the vast amount of information available and then to indicate the gaps between the physiological, endocrinological, biochemical, histochemical and morphological parameters of ovarian follicles. A much-needed interdisciplinary approach in the biology of ovarian follicles is very difficult and thus lacking. This book leads therefore to a greater understanding of cellular and molecular aspects of ovarian follicles in relation to their growth and atresia. The purpose of this book is to present a timely, thorough review on various aspects of ovarian follicles in mammals so that the student, scientist, or physician, regardless of discipline, can acquaint himself with the current state of knowledge in this most important compartment of the ovary as well in his own as in other related areas. This book is organized in seven chapters dealing with cellular and molecular aspects of ovarian follicles and besides presenting thorough, up-to-date reviews and extensive bibliographies on various aspects of the biology of ovarian follicles, future research needs related to each chapter are also clearly outlined. Therefore, the various chapters are up-to-date reviews which will serve as an important source for investigators of the biology of ovarian follicles for years to come. Clearly, there exist still great voids in our knowledge of the biology of ovarian follicles. It is hoped that this book will serve as a stimulus for zoologists, reproductive biologists, endocrinologists, cellular, molecular and developmental biologists, animal scientists, gynecologists, obstetricians, etc. to fill in these gaps.

Getting a book such as this by a single author in one volume is a most difficult task, and I am greatly indebted to the following internationally known experts on follicles for critically reviewing and editing the chapters/sections pertaining to their particular fields of specialization: Dr. G. S. Greenwald for Chaps. I, II, III, VI and VII, except for Sects. B and F of Chap. III; Dr. C. A. Shivers and Dr. R. B. L. Gwatkin for Sect. B of Chap. III; Dr. K. P. McNatty for Sect. F of Chap. III; Dr. J. Richards for Chap. VII and Sect. D of Chap. III; Dr. J. van Blerkom for Chap. IV; and Dr. L. Espey for Chap. V. If some lacunae still remain, the author is responsible for these. Thanks are due to Dr. V. R. Parshad for checking the references and helping in the preparation of photomicrographs and subject index, and to S. Inderjit Singh for typing the manuscript. Thanks are also due to authors and copyright holders for permission to republish some of their illustrations, tables etc. Thanks are particularly due to Dr. Dieter Czeschlik and Linda Teppert of the Biology Department, Springer-Verlag, Heidelberg, for their excellent cooperation during the completion and publication of the book. I owe a lot to my wife Surinder and my children, Gurmeet, Harmeet and Rupa for providing constant encouragement and inspiration during the execution of this most difficult job.

Ludhiana, May 1985   Sardul S. Guraya

Contents

Chapter I
Introduction

Follicular growth and maturation in the mammalian ovary represent a process of differentiation which is accompanied by complex morphological, biochemical and molecular transformations of various components of the follicle, such as the oocyte, granulosa and thecal layer. These transformations lead to the formation of eggs from a limited number of follicles depending upon the mammalian species, as most of the follicles in different stages of their growth and differentiation are lost through degeneration (atresia).

During the past few years, follicular growth and atresia, ovum maturation and ovulation in the ovaries of various mammalian species including human and subhuman primates have been extensively studied with the modern techniques of transmission and scanning electron microscopy, histochemistry, autoradiography, biochemistry, biophysics, etc. Isolated reviews have been published in recent books either on the morphology (including ultrastructure with transmission and scanning electron microscopy), histochemistry, or some aspects of the biochemistry, endocrinology and physiology of follicle growth, maturation and atresia (see Mossman and Duke 1973, Zuckerman and Weir 1977, Jones 1978, Peters and McNatty 1980, Motta and Hafez 1980, Schwartz and Hunzicker-Dunn 1981, Channing and Segal 1982, Greenwald and Terranova 1983). There are also numerous individual papers on some aspects of the biology of the follicle in the mammalian ovary (see Biggers and Schuetz 1972, Midgley and Sadler 1979, Channing et al. 1979 a). But no attempt has been made previously to summarize and integrate the vast amount of recent information on different aspects of the cell and molecular biology of the ovarian follicle in mammals in the form of a book. The purpose of this book is, therefore, to summarize and integrate recent information obtained with modern techniques, in order to provide a deeper insight into the cell and molecular biology of normal development, maturation and ovulation of ovarian follicles of mammals as well as their atresia. Such integrated knowledge is essential for a better understanding of the consequences of the influence of chemical agents, including hormones, prostaglandins and other chemical components, to which the follicles and ova of mammals, including human and subhuman primates, are being subjected at various stages of their growth. Such a summary may help to explore the possibilities of improving fertility in farm animals and of controlling human fertility, as well as of coping with the anovulatory condition in humans. Unfortunately, reproduction occurs in an environment increasingly contaminated with many xenobiotic (biologically foreign) compounds (Mattison and Nightingale 1981). Actually, we need a more thorough knowledge of the mechanisms of those factors at the subcellular and molecular levels, and also of their adverse effects on various components of the follicle. Such an account of the subcellular and molecular aspects of follicular growth and atresia, ovulation and ovum maturation in humans, subhuman primates, and other mammalian species, which was not available previously, is also needed to provide a rational interpretation of alterations caused by aging and of prolonged action, even in weak doses, of different types of radiations on the biology of follicles.

Morphological, histochemical, biochemical and endocrinological changes in the follicular fluid, the membrana granulosa and the theca interna and the surrounding stroma of follicles in humans and other mammals, which are associated with preovulatory swelling and ovulation and are described also in some previous review articles published over the past 11 years (see Blandau 1966, Hertig and Barton 1973, Lipner 1973, Espey 1974, 1978 a, b, 1980, Guraya 1974 a, 1979 c), will be discussed in such a way as to place more emphasis on recent data. There has been considerable controversy about the site(s) and regulation of steroid production in the mammalian follicle. Therefore, the functional meaning of correlations of steroidogenic activity with individual cell types or with cell morphology and biochemistry, and regulation of steroid biosynthesis at the molecular level, will be discussed in detail. In this book, recent results on the ovum maturation in vivo and in vitro will also be summarized and integrated in detail as very divergent views have been expressed in this regard. The zona pellucida of mammalian eggs has been extensively studied for its origin, structure, physico-chemical properties, antigenicity, functions, etc., and will be discussed in detail. Such a discussion will be very useful because interest in the zona pellucida as for its being a possible site for manipulating fertility has also greatly increased during recent years. The multidisciplinary approach followed in this book will definitely help us in understanding better the comparative aspects of cell and molecular specializations of oogenesis and eggs in mammals.

Chapter II
Primordial Follicle

Primordial oocytes in the ovary of human and nonhuman primates and other mammalian species remain in an arrested meiotic prophase for a long time in the cortex (Fig. 1) until advanced stages of follicle recruitment when they resume maturation, preliminary to either ovulation or atretic regression (Zamboni 1972, 1974, 1976, 1980, Guraya 1970a, 1974a, Amin et al. 1976, Bjërsing 1978, Dvořák and Tesařík 1980). They are separated from one another by a thick connective tissue stroma (Fig. 1) consisting of fibroblasts, blood vessels and bundles of collagen and reticular fibres. Primordial follicle appears as a relatively simple structure consisting of a centrally located oocyte with a single layer of granulosa cells, demarcated by an all-encompassing basal lamina which separates it from the surrounding stroma (Fig. 2). Follicles remain in this condition of simple structural organization until the time they regress (atresia), or undergo the maturation process that eventually results in ovulation and liberation of mature and fertilizable eggs. This meiotic arrest is believed to be actively maintained by the inhibitory effect of follicle cells on oocyte maturation (Edwards et al. 1977). The primordial follicles lie close to the surface epithelium and

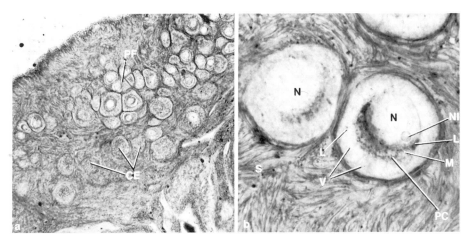

Fig. 1. a Photomicrograph of portion of ovarian cortex of relatively young marmoset *(Oedipo-midas oedipus)* showing primordial follicles *(PF)* and some epithelial cords *(CE)*; **b** Photomicrograph of histochemical preparation of primordial follicles from the ovary of a women of 32 years, showing paranuclear complex *(PC)* of organelles including mitochondria *(M)*, lipid bodies *(L)*, etc. Lipids *(L)* and vacuoles *(V)* are also seen in the peripheral ooplasm. Nucleus *(N)* of primordial oocytes shows large nucleolus *(NI)*. Primordial follicles are surrounded by densely packed stroma *(S)* (From Guraya 1970a)

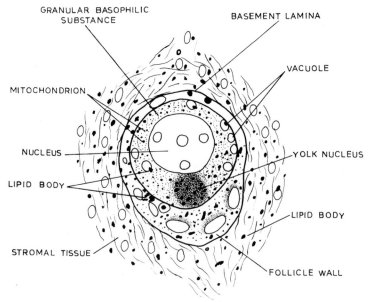

Fig. 2. Diagram illustrating various components of primordial follicle from the cattle ovary. Surrounding stromal tissue lying outer to the basement lamina shows lipid bodies similar to those of oocyte and follicle wall

have no independent blood supply. During the period of sexual maturity primordial follicles and growing follicles, in various stages of their development and differentiation, and atretic follicles lie next to one another (Fig. 3), and their morphological (including ultrastructural) histochemical and biochemical features have been studied.

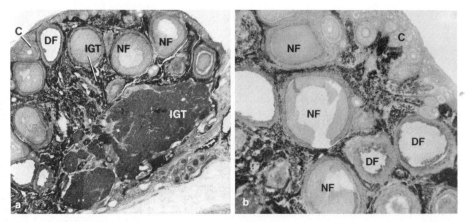

Fig. 3a, b. Photomicrographs of portions of ovaries from marmosets, showing normal follicles *(NF)* of various sizes, and degenerating follicles *(DF)*. In between the follicles is seen the sudanophilic interstitial gland tissue *(IGT)*. Ovarian cortex *(C)* shows primordial follicles as well as pre-antral follicles of various sizes (From Guraya 1968d)

The structural organization of the cytoplasm in primordial oocytes is relatively simple (Zamboni 1974, 1976, 1980, Dvořák and Tesařík 1980). But biochemical and enzymatic studies have revealed that maturational arrest and lack of structural alterations do not indicate a condition of metabolic inertia, as protein and RNA syntheses are active in primordial oocytes (Oakberg 1967, Baker et al. 1969). Various ooplasmic structures form a juxta- or paranuclear complex which is usually designated as Balbiani's vitelline body or yolk nucleus complex (Figs. 1 b and 2) (Guraya 1970a, b, 1979d, Dvořák and Tesařík 1980, Zamboni 1980). Ullmann (1979) has observed that in the bandicoot the conspicuous ultrastructural features of the ooplasm of primordial oocytes are a paranuclear complex, a vesicle-microtubule complex, and an aggregate of tubular cisternae.

A. Balbiani's Vitelline Body

Balbiani's vitelline body consists of various ooplasmic components which show some variations in their development, amount and morphology in different species of mammals. Electron microscope studies of primordial oocytes have revealed a simple large paranuclear aggregate of small, smooth, round or flattened vesicles or saccules interpreted as a large Golgi complex or cytocentrum (Fig. 4) (Hertig and Adams 1967, Hertig 1968, Zamboni 1972, 1980, Hertig and Barton 1973, Guraya 1973a, 1974a, 1979d, Selman and Anderson 1975, Dvořák and Tesařík 1980). The paranuclear complex in the bandicoot is particulate, consisting of five distinct types of bodies, most of which are composed of concentric fibrillar whorls, but others appear homogeneous, granular or crystalline (Ullmann 1979). Embedded among the particles is a group of Golgi-like vesicles. It is believed that vesicles blebbing off the outer nuclear membrane give rise to the early Golgi complex although endoplasmic reticulum could also be the source of the latter (Hertig and Barton 1973). At this stage, the Golgi complex,

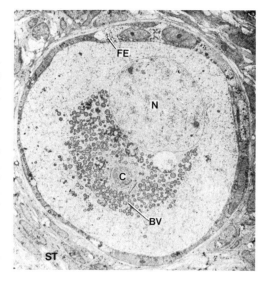

Fig. 4. Electron micrograph of primordial oocyte from the human (34 years old) ovary showing the ultrastructural components of Balbiani's vitelline body *(BV)*. Note cytocentrum *(C)* surrounded by dense fibres, the surrounding halo of endoplasmic reticulum, the dense, vacuolated, compound aggregates, the massed mitochondria and the dispersed vesicular endoplasmic reticulum. See their detailed structure and distribution in Fig. 6. Spherical nucleus *(N)* is also seen. Cells of follicular epithelium *(FE)* are of variable thickness and density. No mitoses are present. Outer to the follicular epithelium are present stromal cells *(ST)* (From Hertig 1968)

consisting of vesicular and short tubular profiles, is the most prominent cytoplasmic
component of the mammalian oocyte. Although the basic organization of the Golgi
complex is essentially identical in the oocytes of different species, some differences in its
detailed structure have also been noticed, which, according to Zamboni (1972), could
be due to different stages of cellular activity, or they may represent actual species
differences. Ribonucleoprotein particles or ribosomes, various profiles of endoplasmic
reticulum (both granular and agranular) and mitochondria are sparsely distributed
around the juxtanuclear Golgi complex to form the Balbiani's vitelline body (Fig. 4)
(Zamboni 1980). The ooplasm of primordial follicles in the dog ovary contains many
large rounded mitochondria, vesicular elements, smooth endoplasmic reticulum and
small Golgi bodies (Tesoriero 1981).

Sparsely distributed mitochondria, having the usual phospholipid-protein composition, show a great diversity in their ultrastructure (Hadek 1965, Zamboni 1972, 1980,
Guraya 1973a, 1974a, Hertig and Barton 1973, Dvořák and Tesařík 1980). They are
usually spheroidal or slightly elongated, and provided with a limited number of cristae
which either arch through the matrix or run parallel to the outer mitochondrial
membranes (Figs. 4 and 5) (see reviews by Hadek 1965, Zamboni 1970, 1972, 1974,
1980, Hertig and Barton 1973, Guraya 1973a, 1974a, Dvořák and Tesařík 1980).
Elongated or filamentous mitochondria with transverse cristae may be seen, but in far
fewer numbers than the round to oval ones. Weakley (1976) has reported variations in
mitochondrial size and ultrastructure during germ cell development in the female
hamster. Mitochondria in oogonia of foetus and the new-born are elongate with
transverse cristae. During pre-dictyate meiotic prophase they become small, rounded,
and electron-dense with pleomorphic cristae. These alterations are largely reversed
when dictyate is reached. There is evidence that mitochondria may divide by fission.
Dvořák and Tesařík (1980) have also reported variations in mitochondrial distribution, size and ultrastructure during the growth and differentiation of the human
oocyte.

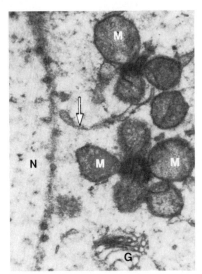

Fig. 5. Electron micrograph of portion of human
oocyte with nucleus *(N)*, mitochondria *(M)* arranged
around dense intermitochondrial substance, the Golgi
complex *(G)* and cisternae of the endoplasmic reticulum *(arrowed)* (From Dvořák and Tesařík 1980)

The elements of endoplasmic reticulum (ER) show several variations in their amount, morphology and distribution in different mammalian species (Hertig and Barton 1973, Zamboni 1980, Dvořák and Tesařík 1980). Rough ER is usually sparse as it appears in the form of short tubules or cisternae with a few ribonucleoprotein particles adhering to the outer surface. Polysomes are inconspicuous (Dvořák and Tesařík 1980). Smooth ER is often vesicular, although cisternae are also seen. Individual mitochondria may be closely associated with flattened and elongated cisternae of the rough ER. This morphological association forms the conspicuous feature of bovine oocytes (Senger and Saacke 1970) where mitochondria are frequently hood- or basket-shaped and contain ER vesicles in the concavity delimited by their appendages. Dvořák and Tesařík (1980) have reported rosette-like arrangement of the mitochondria around the dense, amorphous or fine granular intermitochondrial substance (Fig. 5). The latter is believed to represent stored maternal mRNA in mammalian oocytes.

The lipid bodies of variable size and morphology and consisting of phospholipids usually lie in association with the other cytoplasmic components and in the peripheral ooplasm where they may be associated with pinocytotic vacuoles (Fig. 2) (Guraya 1970a, b, 1974a). According to Dvořák and Tesařík (1980) lipid droplets as well as glycogen are rare in human primordial oocytes. In electron microscope studies, the lipid bodies appear to have been either missed, or described by different names (see Guraya 1970a, b, 1974a). The electron microscope studies have also revealed the presence of a variety of miscellaneous organelles in the quiescent oocytes of different mammalian species. These include dense aggregates or dense bodies, multivesicular bodies, dense lamellar bodies or myelin figures, horseshoe-shaped rods with bulbous ends, lysosomes, etc. (see Hertig and Barton 1973 for their detailed discussion). Microtubules present throughout the oocyte cytoplasm are most prevalent around the circumference of the nuclear envelope (Hertig and Adams 1967).

The vesicle-micro tubule complex in the oocytes of the bandicoot consists of vesicle-like organelles which may be drawn out into tubular extensions, while the bounding membrane may be decorated with granules (Ullmann 1979). Bundles of microtubules ramify between the vesicles, from which they appear to originate. The vesicles contain a matrix similar to the ooplasm. The aggregate of tubular cisternae contains a homogeneous substance, more electron-dense than the surrounding ooplasm. The dense bodies occur in the cytoplasm of both oocyte and follicle cells. These are elongated membrane-bound organelles, circular in cross-section. An electron-dense core is separated from the membrane by a narrow less dense zone. The genesis and morphogenetic significance of these organelles in the bandicoot oocyte need to be determined.

Primordial oocytes of human and chimpanzee also contain masses of stacked or concentric annulate lamellae, and heterogenous spherical bodies (or ultrastructural large compound aggregates or polymorphous complexes) (Baca and Zamboni 1967, Hertig and Adams 1967, Hertig 1968, Barton and Hertig 1972, Zamboni 1972, 1980, Hertig and Barton 1973, Guraya 1974a, Dvořák and Tesařík 1980). Stacks of annulate lamellae are consistently present in the cytoplasm of human oocytes (Fig. 6), sometimes they are arranged concentrically. They are not present in the normal oocytes of other mammalian species investigated (Hertig and Barton 1973). The precise origin and functional significance of these structures are still not known

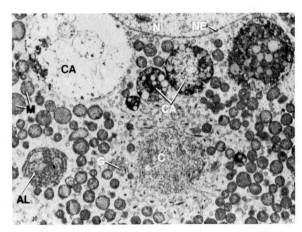

Fig. 6. Higher power view of Balbiani's vitelline body shown in Fig. 4. Note nuclear envelope *(NE)* of oocyte nucleus *(N)*; the prominent dense, vacuolated compound aggregates *(CA)* with and without surrounding membranes; the cytocentrum *(C)* composed of dense granules and closely packed vesicles with peripheral dense fibres; the surrounding halo of smooth endoplasmic reticulum; the scattered Golgi complexes *(G)*; the annulated lamellae *(AL)* cut tangentially; and the clusters of mitochondria *(M)* associated with endoplasmic reticulum (From Hertig 1968)

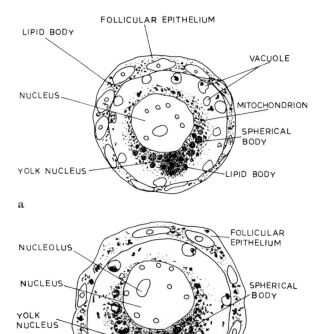

Fig. 7. Diagrams **a** and **b** of primordial follicles from the human ovary showing their various components, especially changes in spherical bodies of primordial oocytes with aging. Diagram **a** is drawn from the ovary of a woman of 18 years, and diagram **b** is from the ovary of a woman of 35 years (Redrawn from Guraya 1970a)

(Zamboni 1972, 1980, Guraya 1974a). Heterogenous spherical bodies consisting of RNA and lipoprotein form the conspicuous feature of oocytes in young women (Fig. 7a) (Guraya 1970a, 1974a). They under go fatty metamorphosis by developing phospholipid-triglyceride granules in women in their middle reproductive years, (Fig. 7b) suggesting that human primordial oocytes possibly become aged (Guraya 1970a, 1974a). According to Dvořák and Tesařík (1980), the secondary lysosomes of human primordial oocytes are represented by compound aggregates (Fig. 6) of Hertig and Adams (1967), described also as spherical heterogenous bodies (Guraya 1974a) or polymorphous complexes (Tesařík and Dvořák 1978a). These structures may or may not be membrane-enclosed. Compound aggregates surrounded by a continuous membrane sometimes show acid phosphatase activity (Tesařík and Dvořák 1978b). The primordial follicle oocytes in the guinea pig ovary show a β-glucuronidase and acid phosphatase reaction (Schmidtler 1980); nonspecific esterase can be seen in the oocytes and granulosa cells of secondary follicles. Electron microscope studies have provided morphological evidence for the transport of material of compound aggregates between the human oocyte and follicle cells (Hertig and Adams 1967).

B. Nucleus

The nucleus of the primordial oocyte is large and of an overall spheroidal shape and is usually eccentrically located (Fig. 1b and 2) (Guraya 1970a, Zamboni 1972, 1980, Hertig and Barton 1973, Dvořák and Tesařík 1980, Tesoriero 1981). The DNA of the primordial oocyte nucleus, which is synthesized just before the onset of meiotic prophase, continues to be present in the growing and mature oocyte in the adult mammals, and this persistence of DNA has been demonstrated with autoradiographic techniques (Baker 1979). After the injection of a radioactive DNA precursor (tritiated thymidine) into an animal during the period of oogenesis, the oocytes in the mature animal still retain the label. The oocyte does not synthesize DNA after the last premeiotic interphase, as evidenced by the inability of oocytes to incorporate tritiated thymidine when it is injected into adult mammals (Peters and Levy 1966, Pedersen 1972).

One or several nucleoli are usually present (Fig. 7), which are prominent and of the reticular type with the finely granular nucleolonema organized into anastomosing strands (Zamboni 1972, 1980, Dvořák and Tesařík 1980, Tesoriero 1981). They consist of RNA and protein. Like the nucleoli, large aggregates of fine granular material forming nucleolus-like bodies also consist of RNA and protein (Guraya 1974a). At diplotene the first meiotic prophase is arrested and the chromosomes become more diffusely distributed. This prolonged diplotene is also called the dictyate stage. The "dictyate" nuclei of rodent oocytes show a uniform dispersion of the chromatin throughout the nucleoplasm (Guraya 1974a, Baker 1979).

Aggregates of finely fibrillar chromatin distributed in wispy strands apparently represent sections through the diplotene configuration of chromosomes (Fig. 8) (Hertig and Adams 1967). The chromosomes in the nucleus of primordial oocytes in primates develop the configuration of lampbrush and act as a site for the synthesis of RNA and proteins (see review by Guraya 1974a). Vazquez-Nin and Echeverria (1976) have also observed that the axial cores in leptotene and the lateral arms in the

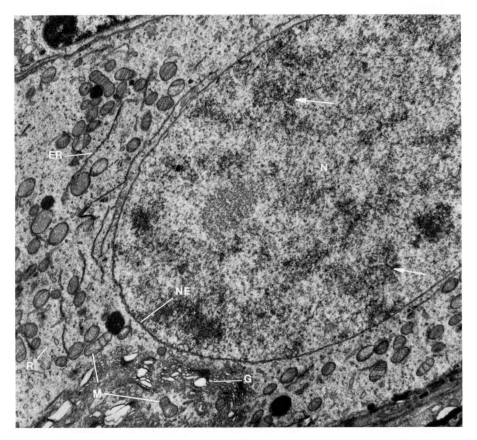

Fig. 8. Electron micrograph of portion of a primordial oocyte from a foetal guinea-pig ovary, showing the aggregates of chromatin *(arrows)* in the nucleus *(N)*. Nuclear envelope *(NE)* shows a few pores. Ooplasm shows the Golgi complex *(G)*, mitochondria *(M)*, and ribosomes *(R)*, of the Balbiani's vitelline body. Elements of endoplasmic reticulum *(ER)* and dense bodies are also seen (From Guraya 1979d)

pachytene synaptonemal complex of rat oocytes in meiotic prophase consist of fibrils. Medial arm and lateral-medial fibrils are formed by filaments of ribonucleoproteins. These filaments form bridges between pairing homologues in the zygotene. In the advanced pachytene stage, the RNA is scarce in these structures. No DNA is detected either in the lateral-medial fibrils or in the medial arm. During diplotene, the synaptonemal complex loses its individuality and the synaptic space becomes wider and irregular. Meanwhile, there are losses of chromatin and a large increase of RNA-containing particles. All these processes lead to the typical interphasic arrangement of nuclear components seen in the dictyate stage. Habibi and Franchi (1978) found that the appearance of nuclear chromatin in hamster oocytes reaching diplotene is changed from that of a compact lampbrush type described in some mammals to that of dictyate in which parts of the genome appear as a highly extended lampbrush and others are condensed. These findings, which have indicated that hamster primordial oocytes also resemble those of other rodents, are also discussed in relation to the reported rise in X-ray sensitivity of oocytes at the time that the chromosomal changes occur.

The development of lampbrush chromosomes in the nucleus of primordial oocytes of mammals are indicative of the synthesis of RNA and protein as demonstrated for the oocytes of nonmammalian vertebrates (Davidson 1976). These ribonucleoproteins have been demonstrated cytochemically in the ooplasm of primordial oocytes (Guraya 1979 d, 1980 a). The transfer of nuclear RNA and protein into the ooplasm may be facilitated by the development of pores or annuli in the nuclear envelope (Dvořák and Tesařík 1980) which is often wavy or undulating. The perinuclear space is also discernible. The outer nuclear membrane sometimes bears adherent ribonucleoprotein particles. Even some cytoplasmic inclusions originate from the nuclear envelope during early growth and differentiation of primordial oocyte (Baker and Franchi 1969, Baker 1970); the process of blebbing involved in their formation may also help in the transfer of informational macromolecules from nucleus into the ooplasm. Baker (1970) suggested that small primordial oocytes utilize some of their RNA as messenger for protein synthesis by ribosomes in the cytoplasm.

C. Follicle Wall

The primordial follicle wall in mammals including human and nonhuman primates forms a crescent-shaped cap of relatively large cells over part of the oocyte (Figs. 1 b, 2, 4 and 7) (Zamboni 1972, 1974, 1976, 1980, Guraya 1970 a, 1974 a, Amin et al. 1976, Dvořák and Tesařík 1980), the remaining cells form a very thin layer around the rest of the oocyte. This single layer of flattened or rectangular granulosa cells rests on a very thin basement membrane which separates the follicle from the surrounding tissue of the ovarian stroma (Fig. 9). Zamboni (1974, 1976, 1980) observed that the walls of primordial follicles in the rhesus and human ovary are often discontinuous due to the presence of irregular gaps between adjacent granulosa cells, which range from small to extensive; the latter may leave large segments of the oocyte perimeter uncovered and directly exposed to the cellular and vascular components of the surrounding stroma from which the oolemma remains separated solely by the interposition of the basal lamina. Often, a naked portion of the oocyte surface may face the thin wall of an adjacent follicle from which it is separated by a few collagen fibrils and attenuated cytoplasm of one or two connective tissue cells. The discontinuities of the follicle wall are believed to constitute a direct route for the transport of extracellular substances to the oocyte surface (Zamboni 1976, 1980). The significance of this route for the metabolic activity of the oocyte during the stage of "quiescence" becomes more clear, especially considering that through these gaps the oocyte surface occasionally comes in contact with the walls of stromal blood vessels.

The structural organization of the granulosa cells in primordial follicles is relatively simple as they are undifferentiated and rest on a basal lamina (Zamboni 1976, 1980, Cran et al. 1979). The nucleus is elongated and highly irregular due to the presence of numerous folds and profound indentations (Fig. 9) (Zamboni 1974, 1976, 1980, Guraya et al. 1974, Cran et al. 1979, Dvořák and Tesařík 1980). There are only a few pores in the nuclear envelope. One or two reticular nucleoli are present in the nucleus (Fig. 9). Mitochondria, lipid bodies, a very small Golgi zone, a few randomly

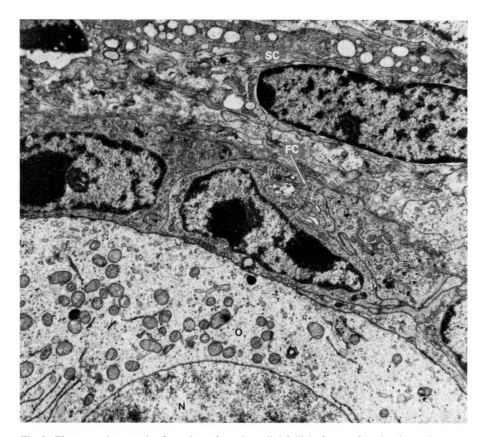

Fig. 9. Electron micrograph of portion of a primordial follicle from a foetal guinea-pig ovary showing follicle cells *(FC)* with a Golgi complex, elements of granular endoplasmic reticulum, mitochondria, and nuclei having heterochromatin masses and nucleoli. The follicle cells with processes are separated from the stromal cells *(SC)* by the basal lamina. The stromal cells *(SC)* show some elements of granular endoplasmic reticulum, free ribosomes, mitochondria, etc. The oocyte shows mitochondria, elements of endoplasmic reticulum, many ribosomes, some dense bodies, and nucleus *(N)* (From Guraya 1977a)

distributed cisternae of granular ER and free ribosomes are found in the cells of the follicle wall; polysomes are numerous. Most of these cytoplasmic structures are distributed in the widened part of follicle cells where the nucleus is also located (Fig. 9). Mitochondria are rod-shaped, elongated and tortuous with numerous transverse cristae and an electron-dense matrix. The Golgi complex is usually placed in the concavity of the nuclear surface facing the oocyte. Ribosomes, small vesicles and lipid droplets make up the major portion of the cytoplasmic components. The numbers of lipid bodies are relatively greater in the rhesus primordial follicle wall (Guraya 1966a, 1970b, 1974a, Zamboni 1974). Frequently, lipid bodies consisting of phospholipid are seen lying across the plasma membrane of the primordial oocyte (Fig. 7). They gradually bulge into the cytoplasm of the oocyte where they are associated with vacuoles probably formed by pinocytosis. The lipid bodies correspond to the small ultrastructural compound aggregates of human primordial follicles, which have also been observed to traverse the oocyte membrane and are often associated with vacuoles in the oocyte cytoplasm (Hertig and Adams 1967, Guraya 1970a). Secondary lysosomes, some of which resemble the compound aggregates in their morphology (Hertig 1968), are similar to those described in the oocyte (Dvořák and Tesařík 1980). The presence of compound aggregates in the follicle wall is typical only of this stage.

At the oocyte–follicle cell junction in the human primordial follicle, there is an aggregation of intensely PAS-positive material which represents the beginning of zona pellucida formation (Guraya 1970a, 1974a). This area is also reactive to tests for adenosine monophosphatase (Hertig and Adams 1967), suggesting active transport on the cell membranes of the oocyte and follicle cells.

For most of its extension, the plasma membrane of the primordial oocyte is smooth and directly apposed to the surrounding granulosa cells (Fig. 9) (Baca and Zamboni 1967, Hertig and Adams 1967, Zamboni 1972, 1974, 1980, Guraya 1974a, Dvořák and Tesařík 1980, Tesoriero 1981). The relationship between granulosa cells and oocytes in primordial follicles is characterized by different patterns of association. The membranes of the two cells may be closely apposed against one another and locally joined by desmosomes characterized by increased thickness and electron opacity of the plasma membranes (Fig. 10) and convergence of cytoplasmic filaments onto the attachment

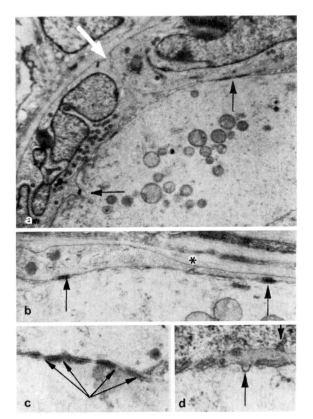

Fig. 10 a–d. Electron micrographs showing follicle cell-oocyte relation in resting follicles of human ovary. **a** Follicle cells lining the wall of a unilaminar follicle. The cells are separated by an intercellular space *(large arrow)* through which the extracellular compartment is in direct communication with the interior of the follicle. The follicle cells are either separated from the oocyte by lacunae partially occupied by microvilli or closely apposed against the oolemma in areas of intercellular contact *(small arrows)*; **b** and **c** Intercellular junctions *(arrows)* between follicle cell and oocyte in unilaminar follicles. *Asterisk* in **b** points to a pinocytotic vesicle; **d** Coated vesicles *(arrows)* on the plasma membranes of oocyte and follicle cell in unilaminar follicle (From Zamboni 1976)

areas, or they may be separated by irregular lacunae (Guraya 1977 a); in this case, the plasma membrane of either cell is thrown into short, interdigitating microvilli and the surface of the granulosa cells may show the presence of cilia. Pinocytotic vesicles are frequently seen on the plasma membrane of either cell.

The associative patterns between adjacent granulosa cells may also vary (Figs. 9 and 10); in places the membranes of neighbouring cells are closely apposed against one another and are frequently associated through interposition of tight intercellular junctions (or desmosomes). In other zones, adjacent granulosa cells may be separated by channels which place the extrafollicular compartment in direct communication with the interior of the follicle (Zamboni 1976, 1980), thus providing pathways for nutritive substance of extrafollicular origin to gain direct access, and become available to the oocytes. This suggestion is supported by the observations of Anderson (1972) who has shown that intravenously or intraperitoneally injected horseradish peroxidase reaches the oocytes within young follicles by transport through the unobstructed clefts which occasionally separate adjacent granulosa cells and granulosa cells from oocytes. Substances of metabolic significance also possibly reach the oocyte after passing through the cytoplasm of the granulosa cells by common methods of transcellular transport (Zamboni 1976); whichever route is followed, it appears that these substances enter the oocyte through a process of membrane diffusion or through pinocytosis or by active transport.

D. Surrounding Stromal Cells

Spindle-shaped cells of the cortex, which are usually designated as fibroblasts, surround the primordial follicles (Figs. 1 b and 2) (Zamboni 1972, 1974, 1980, Guraya 1970a, 1974a). Fibroblasts contain elements of rough endoplasmic reticulum, Golgi body, free ribosomes and mitochondria with simple cristae (Fig. 9). Histochemical techniques reveal RNA and lipid bodies similar to those of granulosa cells and ooplasm. The deeply sudanophilic lipid bodies consist of phospholipids. They correspond to the small compound aggregates of electron microscope studies on the human ovarian cortex (Hertig and Adams 1967).

Chapter III
Follicle Growth

In the prepubertal and fertile periods, the cortices of mammalian ovaries are crowded with follicles of different sizes (Figs. 3 and 11), from primordial follicles described in Chap. II to those showing large distended antra and which are lined with many concentric layers of granulosa cells. This shows that the population of mammalian follicles is in a continuous dynamic state and that at any given time follicles are present which progress from a state of "quiescence" to one characterized by volumetric growth and increased activity of all of its components (Chiras and Greenwald 1980). The early stages of follicular growth (Fig. 11 a–d) are believed to be independent of gonadotrophic support. However, the factors which initiate the selective growth of primordial follicles, are poorly understood (Zuckerman and Baker 1977, Peters and McNatty 1980, Nicosia 1980a). They appear to originate within the ovary. According to Greenwald (1979), a hypothesis proposed by Edwards fits the FIFO (first in, first out) model; i.e. the first formed oocytes are the first ones to be mobilized in the postnatal period. This is an interesting speculation but with no direct evidence – either pro or con. Pulse labelling of the prenatal ovary and subsequent autoradiographic studies at various times postnatally might reveal whether the maturation of oocytes is temporally programmed. Greenwald (1979) also suggested that the recruitment of primordial follicles may represent a random process which may depend upon their proximity to blood vessels, nerves, more advanced follicles or corpora lutea acting as the signal for passage out of the resting pool.

Cahill and Mauleon (1981) suggested that initiation of follicle growth in the sheep ovary involves the passage of follicles from the dormant to the transitory category, and (if it occurs) is independent of gonadotrophins and thence the passage from the transitory category to the growth phase. The second step is dependent on gonadotrophins and the rate of such passage is higher in ewes with a high ovulation rate. The overall mean follicle size of small follicles is larger in ewes with a low ovulation rate; evidently, they have structurally different small follicles or their reserve of small follicles is less depleted than in ewes with a high ovulation rate.

Follicular growth is characterized by increased activity of both granulosa cells and oocyte; shape modifications and numerical increase of the granulosa cells are usually accompanied by volumetric increase of the oocyte (Figs. 11 and 12) (Dvořák and Tesařík 1980, Zamboni 1980) which also frequently resumes, but only exceptionally completes, the first meiotic division. These cellular changes are associated with processes, such as secretion and deposition of zona pellucida and follicular fluid (Fig. 11), which also indicate a state of enhanced cellular activity. Incipient follicle augmentation is heralded by changes in the shape of the granulosa cells (Fig. 11), which become tall and cuboidal, and the appearance of the first mitotic figures.

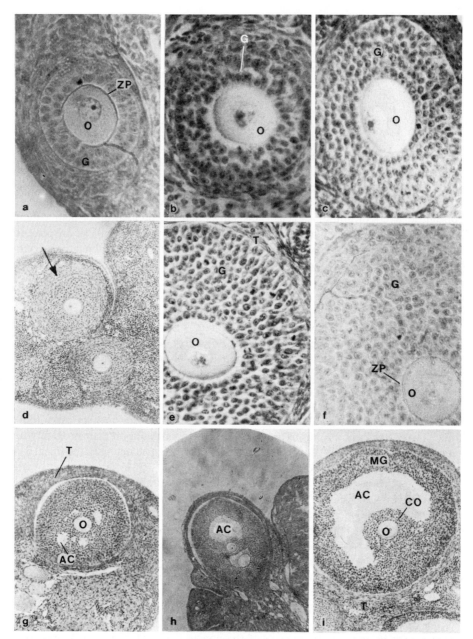

Fig. 11 a–i. Photomicrographs showing changes in the granulosa *(G)*, theca *(T)*, oocyte *(O)* and zona pellucida *(ZP)* during successive stages of follicular growth in the Indian mole rat *(Bandicota bengalensis)*. **a** Stage I follicle with two layers of granulosa cells; **b** Stage I follicle with three layers of granulosa; **c** Stage II follicle with five layers of granulosa cells; **d** and **e** Stage III follicles with six and seven layers of granulosa cells; **f** Stage IV follicle with eight layers of granulosa cells; **g** to **i** The antrum formation in stages V, VI and VII follicles respectively as a result of accumulation of follicular fluid in the antral cavities *(AC)*. *CO* Cumulus oophorus; *MG* membrana granulosa; *T* Theca (Kaur and Guraya 1983)

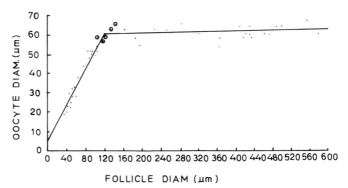

Fig. 12. Relation of the size of the follicle and oocyte in the cyclic nonpregnant mole rat (From Kaur and Guraya 1983)

The first morphological indication of follicular growth is the mitotic activity of the granulosa cells, which coincides with initial growth of the oocyte. Actually, the oocyte and its granulosa begin their growth and maturation together (Fig. 12); both show an increase in size and ultrastructural complexity (Dvořák and Tesařík 1980, Zamboni 1980). The initiation of follicular growth is a continuous process, it starts in infancy and continues during various phases of reproduction including periods of anovulation (Greenwald 1978, Gougeon and Lefévre 1983, Gosden et al. 1983). However, the rate at which follicles leave the nongrowing pool (follicle growth initiation rate) varies with the age of animals: it is highest in the young animal and falls with the attainment of maturity (Richards 1980). Follicles grow sequentially and continue to grow until they become atretic or ovulate.

The results of various studies have suggested that growth of follicles is controlled partly by endocrine events within the ovary (intraovarian and intrafollicular) and partly by those outside it (extra-ovarian) (see references in Nicosia 1980a, Richards 1980). The preantral follicular growth (Fig. 11a–f) appears to be regulated mainly by the follicle-stimulating hormone (FSH) and high local concentrations of oestrogens (Peters and McNatty 1980); FSH seems to play an important role in the structural organization of the follicle envelope. The main hormones within the follicle that influence follicular differentiation include FSH, androgen, oestrogen, luteinizing hormone (LH) and prolactin (Richards 1980, Erickson 1982). Their concentrations in the follicular fluid vary at different stages of follicle development, as will be discussed in Sect. G of this chapter. In hypophysectomized hamsters and rats, replacement therapy with steroids and FSH and LH affects the number and histological organization of small follicles (Greenwald 1979). Similarly, deprivation of pituitary hormones in the mouse also leads to disruption of follicular development. The results of various studies indicate that there are quantitative and qualitative differences in preantral follicles in the absence of pituitary hormones. The concept of pituitary independence in early stages of folliculogenesis appears, therefore, to have little basis in fact.

Lintern-Moore and Moore (1979) described the details of morphological and functional events which occur at the initiation of oocyte and follicle growth in the mouse ovary. Oocytes in the nongrowing pool show RNA synthesis as they contain significant levels of endogenous RNA polymerase activity. With the start of oocyte growth, there is an increase in oocyte nucleolar RNA polymerase activity. Subsequently, increases in nucleolar area and nucleoplasmic RNA synthesis occur in that sequence. As the number of granulosa cells surrounding the nongrowing oocytes increases, there is a transition from flattened (squamous) to cuboidal morphology (see also Zamboni 1974, 1976, 1980). Mitoses are seen only in the follicular envelope of oocytes associated with ten or more granulosa cells in the widest cross-section (Fig. 11). Chiras and Greenwald (1980) have made an analysis of ovarian follicular development and thymidine incorporation in the cyclic hamster. For the sake of convenience, the growth, differentiation and maturation of various components of the growing follicle, which include oocyte, granulosa and theca interna (Fig. 11), will be described separately. The modulations of oocyte-granulosa cell relationship will also be discussed in Sect. C of this chapter.

A. Oocyte

1. Ooplasmic Components

With the initiation of growth in primordial oocyte, the paranuclear complex of organelles forming Balbiani's vitelline body (Figs. 4 and 7) described in Chap. II moves away from the nuclear envelope and its components are meanwhile distributed in the outer ooplasm (Fig. 13) (Zamboni 1972, 1974, 1980, Guraya 1970b, 1973a, 1974a,

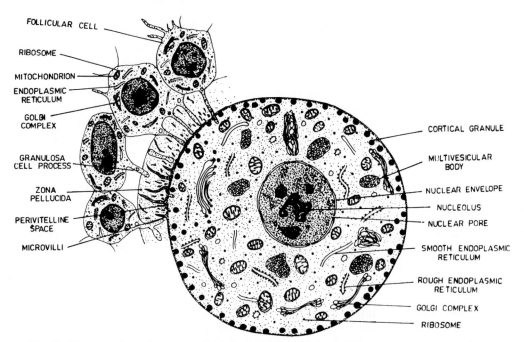

Fig. 13. Diagram of growing mammalian oocyte, illustrating the distribution and structure of its various components. Portion of zona pellucida with its granulosa cell processes and oocytic microvilli is also shown

Dvořák and Tesařík 1980, Tesoriero 1981); simultaneously, the yolk nucleus (or Golgi complex) also fragments (see Weakley et al. 1981). Further oocyte growth shows proliferation of Golgi complexes, mitochondria, free ribosomes and profiles of endoplasmic reticulum and heterogeneous phospholipid bodies of variable size (see also Hertig and Barton 1973, Nowacki 1977). Most of these organelles including heterogeneous phospholipid bodies are randomly scattered in the cortical and perinuclear ooplasm leaving the central areas of the oocyte less populated or nearly devoid of organelles. The oocytes from different species vary greatly in the details of their ultrastructure and cytochemistry. All the Golgi aggregates are eventually localized just beneath the plasma membrane of the oocyte, showing a clear local predilection. Weakley et al. (1981), using thiamine pyrophosphatase and acid phosphatase enzyme techniques and the zinc iodide-osmium tetroxide impregnation technique, have made a cytochemical study of the Golgi apparatus in the developing oocyte of the golden hamster. With all these techniques, there is a falling off of reactivity as oocyte size increases. Within a single oocyte, with techniques for phosphatases, some Golgi bodies are negative while others are positive, suggesting that two or more functional types of this organelles are present within the developing oocyte. These authors have also discussed the possible roles of acid phosphatase and thiamine pyrophosphatase of Golgi apparatus in relation to the growth of oocyte. In the growing oocyte, multivesicular bodies also appear in the ooplasm (Fig. 13) which are believed to correspond to the heterogeneous lipid bodies of histochemical preparations (Guraya 1974a).

Secondary lysosomes in the human oocyte are represented by compound aggregates (Dvořák and Tesařík 1980). However, unlike compound aggregates, which are of autolysosomal form, multivesicular bodies are believed to be of heterolysosomal type (Tesařík and Dvořák 1978b). Secondary lysosomes are for the most part described as phagocytotic vesicles (Gonzáles-Santander and Clavero Nuñez 1973) and multivesicular bodies. The peripheral location of various ooplasmic organelles, including Golgi complexes, suggests that the oocyte is equipped for the absorption and intracellular transport of materials delivered to its surface membrane. Meanwhile, numerous microvilli and micropapillae on the oocyte surface come in close contact with processes of granulosa cells (Fig. 13) facilitating introduction and exchange of extracellular substances as will be discussed later in Sect. C. The diameter of the oocyte increases only slightly after antrum formation (Fig. 12); however, the increase in its volume may be significant.

Systems of cytoplasmic membranes and mitochondria show a great diversity in their development and structure in the growing oocyte of different mammals (Hadek 1965, Norrevang 1969, Szöllösi 1972, Hertig and Barton 1973, Zamboni 1972, 1974, 1980, Guraya 1973a, 1974a, Selman and Anderson 1975, Weakley 1976, Nowacki 1977 Dvořák and Tesařík 1980, Tesoriero 1981, Weakley and James 1982). The physiological significance of these variations is poorly understood in relation to oocyte growth.

The round-to-oval forms of mitochondria with dense matrix predominate and their cristae with irregular configuration are sparse or absent. Cristae usually form arches along the perimeter or may form an inner membrane concentric with the outer one (Fig. 13) (Hertig and Barton 1973, Dvořák and Tesařík 1980). Weakley (1976) observed that maximum mitochondrial size and complexity of cristae are reached just at the beginning of the phase of rapid oocyte growth, and thereafter they decline. As mitochondrial size and number of cristae are decreased in the rapidly enlarging oocyte, the ratio of length to width is increased, as is electron density of the matrix until the formation of an antrum within the follicle. After antrum formation, the mitochondria again become more rounded and cristae are seldom seen. These mitochondrial changes have been correlated with other events occurring during oogenesis. Weakley (1977) demonstrated a coating of electron-dense material on the cytoplasmic surface of outer mitochondrial membranes in medium-sized hamster oocytes. Its possible relationship to the synthesis and transport of mitochondrial protein is suggested. The mitochondria are found in large clusters or rosettes around intermitochondrial substance which is moderately dense and may contain fine, dense granules or a few larger clumps of dense material. The oocytes of ruminants show very unusual mitochondria having a "rodlike" appendage (Senger and Saacke 1970, Fleming and Saacke 1972).

Weakley (1977) observed in the medium-sized hamster oocytes associations between ER, ribosomes and the outer mitochondrial membrane (see also Weakley and James 1982). The close proximity of mitochondria to ER and ribosomes or "intermitochondrial" substance during the oocyte growth suggests a functional relationship between them, which may be involved in protein synthesis or mitochondrial multiplication (Zamboni 1972, Szöllösi 1972, Hertig and Barton 1973, Guraya 1974a). Mitochondria continue to increase in number; that incomplete mitochondria are found continuous with intermitochondrial substance and that membranes resembling cristae are embedded in it suggest that this material perhaps functions in mitochondrial formation or augmentation. But the fission of mitochondria seems to be mainly responsible for their multiplication. Recently Weakley and James (1982) have suggested that frequent contact continuity between ER and mitochondria may be important in relation to lipid metabolism in the hamster oocyte. These authors have suggested that phospholipid and lipoprotein produced in the oocyte ER are delivered in part to developing organelles, such as the mitochondria and Golgi apparatus, and in part to the plasma membrane during the period of rapid oocyte growth.

Granular and smooth ER are present, which show species variations in their amount and distribution (see Hertig and Barton 1973, Dvořák and Tesařík 1980, Tesoriero 1981, Weakley and James 1982). None of the ER in the hamster oocyte appears to be completely rough (Weakley and James 1982). The elements of the ER undergo pronounced development and frequently assume very elongated, tortuous, or concentrically arranged form, limited by ribosome-studded membranes and filled with highly dense material. Weakley and James (1982) have followed the details of differentiation of ER in the developing oocyte of the golden hamster. ER falls into three categories and relationships between these types are described. The changes in the form and distribution of ER are related to developmental processes. Annulate lamellae described in the primordial oocytes of human ovary are rarely seen in its growing oocytes (Dvořák and Tesařík 1980, Zamboni 1980).

Gachechiladze and Togonidze (1976) using histoemzymatic and ultrastructural methods suggested that the oocytes of rat ovaries during the postnatal period are capable of taking part in steroidogenesis (as also proposed by Zamboni 1980), along with the other structures of the ovaries, as they show structural features (smooth reticulum, mitochondria with tubular cristae, etc.) specific to steroid-producing cells. Suzuki et al. (1983) using cytochemical techniques have also demonstrated the presence of Δ^5-3β-hydroxysteroid-dehydrogenase (3β-HSDH) and 17β-HSDH activities in human oocytes, which are implicated in steroidogenesis. They suggest that steroidogenesis, through the adenylate cyclase–cAMP system in human oocytes may play some important role in oocyte maturation, fertilization and early embryonic development. But the present author suggests that these features may be related to the metabolism of steroid hormones by the oocyte rather than to their synthesis.

Besides the usual organelles some workers have also described cytoplasmic filaments, fibrillar arrays, ladder-like structures, lattices or fibrous material in the oocytes of rodents (Fig. 14) (Burkholder et al. 1971, Szöllösi 1972, Zamboni 1970, 1972, Selman and Anderson 1975, King and Tibbitts 1977, Van Blerkom and Motta 1979, Nilsson 1980, Tesoriero 1981). Very divergent views have been expressed about their structure, nature and function (see also Pikó and Clegg 1982, Van Blerkom 1983). Each of the fibrillar arrays consists of 10 to 20 parallel fibrils separated from one another by a distance of 15–20 nm. In cross-section, a hexagonal crystalline pattern is formed. These fibrils increase in number during oocyte meiotic maturation and are abundant in tubal ova (Zamboni 1970). On the basis of morphological evidence suggesting their polysomal origin, it has been recently proposed that they may represent lattices of fibrillar RNA. In the mouse, lattices first develop during the early stages of oocyte growth (Wassarman and Josefowicz 1978), concomitant with active rRNA synthesis, and appear to form by the aggregation of ribosome-like particles

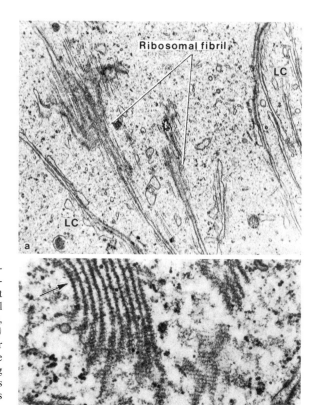

Fig. 14 a, b. Electron micrographs showing bundles of fibrils (or fibrillar arays) in rodent oocytes. **a** Numerous ribosomal fibrils, remnants of cisternae, and "lamellar complexes" *(LC)* in the ooplasm of an older type 5 follicle oocyte from the spiny mouse ovary (From Kang and Anderson 1975); **b** Bundles of fibrils or fibrillar arrays *(arrow)* in the rat oocyte (From Szöllösi 1972)

(Garcia et al. 1979 a, b). On the basis of cytochemical staining characteristics, enzymatic treatments and electron microscopic observations, Burkholder et al. (1971) suggested that the lattices are a storage form of ribosomes.

Garcia et al. (1979 a, b) observed that at the onset of oocyte growth in the cyclic mouse three quarters of its ribosomes occur as singles; these become polysomal ribosomes as growth of the oocyte advances. At the same time, the number of ribosomes increases. Once the major growth period elapses, the number of ribosomes begins to decrease just when lattice-like structures showing periodic organization begin to accumulate in the oocyte cytoplasmic matrix. Evidence, like the particulate organization of these lattices, the size of their particles, its digestion by RNAase and the time of lattice appearance together with other previous data, suggests that near the end of the oocyte growth a great part of the ribosomes are stored in the lattices to be used during early development. Bachvarova and De Leon (1977) using histochemical studies investigated stored and polysomal ribosomes of mouse ova. They suggested that the ratio of nonpolysomal ribosomes to polysomes in mature mouse oocytes is not inconsistent with the electron microscopic evidence on the ratio of particles in the lattices to ribosomes present in free polysomes. Presumably, stored ribosomes become translationally active after fertilization, support protein synthesis during early embryogenesis, and are gradually replaced by ribosomes of embryonic origin (see Van Blerkom 1981 a, b, 1983). Recent studies by Bachvarova et al. (1981) have provided some tentative biochemical evidence so show a ribosomal component in the lattices. Centrifugation of mouse oocyte lysates reveals that at least 75% of the ooplasmic ribosomes can be sedimented at 9000 g or less, suggesting that stored nonpolysomal ribosomes are contained in unusually large structures. Electron microscopic study of this rapidly sedimenting material has revealed that the main components are smooth membrane vesicles, mitochondria, and masses of interlacing fibrils in which are embedded electron-dense particles whose dimensions and RNase sensitivity show a ribosomal nature. However, the fine structure of the material believed to contain stored, nonpolysomal ribosomes bears little resemblance to the highly structured and organized lattices of the intact mouse oocyte, although, as observed by Bachvarova et al. (1981), this may have been the result of the conditions of lysis and isolation. The notion that mouse oocytes accumulate translationally inactive ribosomes in cytoplasmic lattices has been challenged recently by Pikó and Clegg (1982) who, in an extensive study of oocyte and early embryo RNA synthesis and content, have quantitated the amount of RNA and the number of ribosomes present during early development. Both one-dimensional (ID-PAGE) and two-dimensional (2D-PAGE) polyacrylamide gel electrophoretic separation of proteins derived from the detergent-resistant oocyte residue consist primarily of lattices, with protein, resolution by highly sensitive silver staining (Ochs et al. 1981); these studies could not show ribosomal proteins as a major or significant component of oocyte residue (see also van Blerkom 1983). Their observations have made it unlikely that the fibrils (or lattices) are composed of ribosomes or that they contain any significant amounts of RNA. In spite of these advances, the storage of translationally inactive ribosomes in lattice form remains to be proven definitively.

King and Tibbitts (1977) described cytoplasmic lamellar inclusions in the oocytes of primary, secondary and large vesicular follicles in the rodent *Thomomys townsendii*. The numerous cytoplasmic lamellae of small primary oocytes appear as spirals or concentric rings in cross-section. In longitudinal sections, they are arranged as concentric cylindrical sheets of material. Secondary and vesicular oocytes also show a tubular element of the ER associated with the core of the lamellae. Tangential sections through the lamellae also reveal a paracrystalline substructure, regardless of the age of the oocyte. The lamellar inclusions have been reported in the oocytes of other species (Fig. 14a) but several fine structural differences are seen even between these species. *Thomomys* is the only species in which there is a close topographic association between the lamellar inclusions and elements of the ER, indicating some metabolic relationship between them. Sasaki et al. (1979) using glutaraldehyde fixation followed by osmium fixation also reported the presence of lamellar or fibrous inclusions in the oocytes of large Graafian follicles in mouse, rat, Mongolian gerbil and hamster but not in rabbit and pig. These inclusions consist of groupings of individual fibrous elements in the mouse, layers of unilamellar structures in the rat and Mongolian gerbil and layers of bilamellar disks in the hamster. They are distributed almost throughout the cytoplasm, abundantly in the rat and hamster oocytes, but less abundantly in the mouse and Mongolian gerbil. An ultracytochemical examination has revealed that these inclusions contain amino groups, indicating their protein nature. This suggestion is further supported by the fact that the so-called lamellar structures of the unfertilized rat egg ooplasm are filaments constituted of proteins (Baeckelan and Heinen 1980). No RNA is found in them, which indicates that they do not represent a particular organization of polyribosomes. These filaments are considered as protein reserves, possibly of membranous origin, for use during segmentation.

Wassarman and Josefowicz (1978) and Dvořák and Tesařík (1980) using electron microscopy, observed progressive changes in the nucleoli, ribosomes, mitochondria, ER, Golgi complex, and other organelles and inclusions of oocytes of the mouse and human respectively as a function of oocyte size. Growth of the mammalian oocyte appears to involve not just tremendous enlargement of the cell, but extensive alterations in its overall metabolism as reflected in the changing size and ultrastructure of the oocyte and its organelles at various stages of growth (see also Wassarman et al. 1979b).

2. Protein Synthesis

The growth phase of oogenesis is closely accompanied by an increase in the cytoplasmic machinery needed for translation, as well as by the accumulation of protein. The augmentation of the ER and fibrous material or fibrillar arrays during oocyte growth may be related to protein synthesis (Zamboni 1972, 1980). Most studies on protein synthesis have been carried out using autoradiographs of sectioned ovaries after injections of labelled amino acids (see Engel and Zenzes 1976). These results have indicated that protein synthesis occurs during all stages of oocyte growth. With these methods it is possible to distinguish between protein synthesized by the oocyte itself and that synthesized by other cells (granulosa cells) and transported into the oocyte. This problem has been solved by incubating mouse oocytes free from all cellular investments in an in vitro system containing labelled amino acids (Stern et al. 1972, Cross and Brinster 1974). Cross and Brinster (1974), using radioactively labelled leucine, observed the highest uptake for this amino acid in metaphase I oocytes. In regard to the evaluation of protein synthesis during oocyte growth, the value of this type of experiment is likewise restricted. With the autoradiograph procedure it shares the difficulty that the internal precursor pool of amino acids is not yet known in growing and maturing oocytes. Mangia and Canipari (1978) suggested that exogenous protein does not play a "quantitatively relevant role" in oocyte growth and concluded that oocyte growth in mammals is mainly supported by endogenous protein synthesis of the oocyte itself. However, in order to justify this conclusion fully, it is important to find out absolute rather than relative rates of protein synthesis in growing oocytes (i.e. rates of synthesis and turnover, uptake and incorporation and endogenous pool size of radio-labelled precursors). Schultz and Wassarman (1977b) observed that the mouse oocyte is incapable of synthesizing the quantity of protein present at ovulation. This suggests that the mouse oocyte may acquire exogenous protein during its growth phase.

Recently, more information concerning qualitative and quantitative aspects of protein synthesis during mammalian oogenesis has become available. Schultz et al. (1978a), using biochemical techniques, observed that patterns of protein synthesis are very similar in individual oocytes that are at the same stage of meiotic maturation, and the linear increase in protein content of growing mouse oocytes is accompanied by significant qualitative changes in size classes of proteins synthesized. Canipari et al. (1979) found that protein synthesis by mouse oocytes increases linearly with the increase of the cell volume, and such an increase is constant throughout the entire oocyte growth period, as demonstrated by the rate of leucine incorporation into total cell protein. Schultz et al. (1979) found that absolute rate of protein synthesis during mouse oocyte growth increases by about 38-fold, from 1.1 pg/h/oocyte in the pregrowth-phase oocyte to 41.8 pg/h/oocyte for the fully grown oocyte. These authors estimated that preovulatory oocyte can only synthesize 12.8 ng of protein, showing that the fully grown oocyte contains about twice as much protein as it should be capable of synthesizing. The difference between the actual protein content and the protein content based on the calculated rate of synthesis reflects uptake by endocytosis of protein formed elsewhere in the mouse. However, as observed by Kaplan et al. (1982), since protein synthetic rates are measured in the absence of granulosa cells and in medium known to be incapable of supporting mouse oocyte growth, the absolute rate of protein synthesis in vivo may be significantly higher than the in vitro values described by Schultz et al. (1979). The degree to which exogenous protein, if any, is taken up by the growing oocyte from the granulosa needs to be determined more precisely.

Qualitative changes in the pattern of protein synthesis occur during oocyte growth in mammals. Ribosomal proteins are synthesized throughout the period of oocyte growth (LaMarca and Wassarman 1979). According to Schultz et al. (1979), tubulin represents about 1.8% of the total protein synthesized. Some of the proteins appear to be specific to either growing or fully grown oocytes. Recently Kaplan et al. (1982) observed patterns of [^{35}SL]-methionine labelled proteins in mid-growth phase and in fully grown oocytes. Of the 300 spots resolved by 2D-PAGE, several observed in fully grown oocytes are reduced in intensity or absent in growing oocytes. Only a few specific proteins have been found in the patterns obtained from growing mouse oocytes. The cytoskeletal proteins actin, tubulin and putative intermediate filament protein are synthesized during growth and in fully grown oocytes (Van Blerkom 1983). These proteins along with ribosomal proteins, appear as the stable, major bulk proteins of the oocyte. Further studies are needed to determine more precisely the nature and amount of proteins synthesized at different stages of oocyte growth.

3. Lipid Bodies

Phospholipid bodies associated with pinocytotic vacuoles are seen in the peripheral ooplasm adjacent to the plasma membrane (Fig. 15). They are apparently transported from the follicle cells of growing oocyte, which also show similar lipid bodies (Fig. 15) (Guraya 1970b, 1973a, 1974a). Their mechanism of transfer from the corona radiata cells into the oocyte will be discussed later in Sect. C. Heterogeneous phospholipid bodies, which form the most prominent feature of cortical ooplasm in the growing oocyte (Fig. 15), have either been missed in electron microscope studies, or they correspond to the "vesicular conglomerates" or "multivesicular bodies" (Guraya 1974a). Travnik (1977b), using light and electron microscopy, studied the incidence

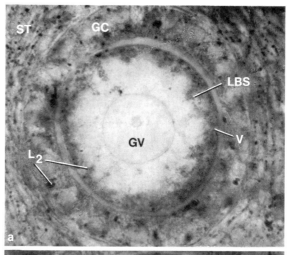

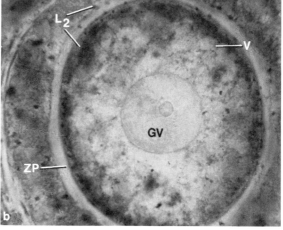

Fig. 15a, b. Photomicrographs of oocytes of cat, illustrating two successive stages in the proliferation of various cytoplasmic components *(LBS)* which are mostly localized in the outer regions of the ooplasm. Also shown are sudanophilic bodies *(L₂)* and sudanophobe vacuoles *(V)*. Similar lipid bodies are also seen in the granulosa cells *(GC)* and surrounding stromal tissue *(ST)*. Some lipid bodies are also seen in the sudanophobe zona pellucida *(ZP)*. The germinal vesicle *(GV)* containing nucleolus does not show sudanophilic lipid bodies (From Guraya 1965)

and localization of lipids in rat oocyte and cleaving ovum. In addition to lipid bodies or fat droplets, there are numerous lipids in the perinuclear and cortical ooplasm which may derive from the lipoproteins of various organelles accumulating during oocyte growth. The other parts of the ooplasm where no lipids are demonstrated contain a large quantity of lamellar structures and very few organelles, indicating that lamellar structures do not possess a demonstrable lipid component.

4. Enzymes

Our knowledge about the enzyme systems of growing oocyte in mammals is still meagre as relatively few enzymes have been studied with the modern techniques of histochemistry and electron-histochemistry. The enzymes, glucose-6-phosphate dehydrogenase (G-6-PDH) and lactate dehydrogenase (LDH), have been reported in growing oocytes (Brinster 1966, Mangia and Epstein 1975, Mangia et al. 1976, Brinkworth and Masters 1978). Their activities are directly proportional to oocyte size up to a diameter of about 85 μm and the enzymes show no further increase in larger oocytes. LDH activity has been demonstrated in the oocytes of rats and mice (Brinkworth and Masters 1978). A masking of its activity is suggested in oocytes and pre-implantation ova. Balabanov and Danilovskii (1976) studied inhibition of exogenous LDH in mammalian ova. These results suggest that enzyme synthesis is completed before the onset of maturational divisions. A histochemical study of a mitochondrial enzyme, succinate dehydrogenase (SDH), has revealed a progressive increase in activity during follicular growth, reaching a peak in pre-ovulatory oocytes (Vivarelli et al. 1976). Schmidtler (1980) has observed an increase in the activities of β-glucuronidase, acid phosphatase and nonspecific esterase during the growth of guinea-pig oocyte.

Tesařík and Dvořák (1978 a, b), using electron-histochemical techniques, demonstrated activity of acid phosphatase in primary lysosomes and some polymorphous complexes of primary oocytes in the ovaries of mature women, which may correspond to heterogeneous phospholipid bodies described by Guraya (1974 a). In guinea-pig oocytes, at every developmental stage, acid phosphatase is histochemically found in cytoplasmic granules (Anderson 1972, Korfsmeier 1979). Ultrastructurally, the reaction product is located in lysosomes and in some cisternae of the ER, but not in cortical granules or in vesicles with a rough ER, filled with a moderately dense homogeneous substance.

The cortical and perinuclear regions of the ooplasm in the secondary and tertiary follicle oocytes of rat show nonspecific esterase (Travnik 1977 a); the greater part of the oocyte organelles are also localized in these zones (Guraya 1970 b). Adenylate cyclase and cAMP phosphodiesterase activities have been studied electron-histochemically in the ovarian and tubal hamster eggs to determine the presence of synthesis and metabolism of cAMP in them (Nimura and Ishida 1981). They first develop on the cytoplasmic membranes of eggs in the secondary follicles, and persist until the eggs become blastocysts. Iyengar et al. (1983) have suggested a possible role for creatine kinase in maintaining the reported high ATP/ADP ratio during oogenesis in the mouse.

5. Cortical Granules

The cortical granules are small, spherical, membrane-limited organelles (Figs. 13 and 16). Their presence was first demonstrated by Austin (1956, 1961) in the eggs of golden hamster. They varied in size from 0.1 to 0.5 µm and are located beneath the plasma membrane or at some distance from it (Figs. 13 and 16). Because of their absence from fertilized eggs, they were considered to play a rather important role in the "zona reaction" (prevention of additional sperm penetration through the zona pellucida after the entry of the first) of the mammalian egg (Braden et al. 1954). Subsequent electron microscope studies on the cortical granules have clearly revealed that they form a general feature of unpenetrated mammalian eggs (Szöllösi 1962, 1967, 1976, Hadek 1963a, b, 1965, 1969, Adams and Hertig 1964, Zamboni and Mastroianni 1966a, Zamboni et al. 1966, Weakley 1966, Fraser et al. 1972, Hertig and Barton 1973, Kang 1974, Kang et al. 1979, Dvořák and Tesařík 1980, Gulyas 1980, Guraya 1982a), and showed their presence in the unfertilized tubal eggs of rabbits, guinea pigs, rats, mice, golden hamsters, coypus and pigs. The cortical granules measure 160 to 350 µm in diameter and are generally localized at or near the plasma membrane of the egg, although some are also distributed throughout the ooplasm (Gulyas 1980). Their

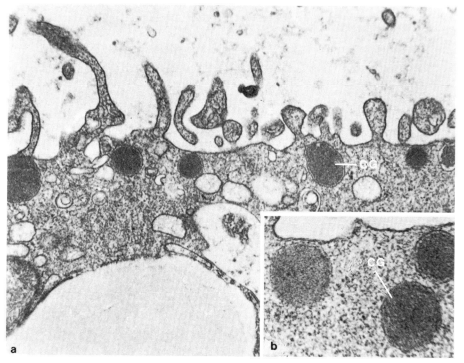

Fig. 16a, b. Electron micrographs of portions of bovine oocytes, showing localization and morphology of cortical granules *(CG)*. Cortical granules either align along the cell membrane or move within a few hundred Angström distance of the oocyte plasma membrane (From Szöllösi et al. 1978)

presence has also been demonstrated in ovarian oocytes of humans (Tardini et al. 1961, Baca and Zamboni 1967, Guraya 1969a), rhesus monkey (Hope 1965, Guraya 1967d, Zamboni 1974), marmoset (Guraya 1967d) and dog (Tesoriero 1981). The cortical granules disappear after fertilization. Since the discovery of cortical granules in mammalian eggs, much work has been carried out on their role in fertilization. Prevention of polyspermy, the zona reaction, and changes in cell surface constitute major roles that have been assigned to cortical granules and their contents (Gulyas 1980, Guraya 1982a).

a) Origin

Divergent views have been expressed about the origin of cortical granules. Adams and Hertig (1964) observed that the cortical granules develop in morphological association with the so-called vesicular aggregates which are distributed in cortical cytoplasm of the growing oocyte in the guinea-pig. The increase in the number of cortical granules in the mouse is closely accompanied by a corresponding increase in the areas occupied by the vesicular aggregates. Cortical granules and vesicular aggregates increase in vivo and in vitro matured eggs.

The proximity of cortical granules and the Golgi complex has been observed during the early observations on cortical granules of several species (Adams and Hertig 1964, Weakley 1966). Such a proximity has also been observed in the dog oocytes (Tesoriero 1981). But the results of all recent electron microscope studies have clearly demonstrated the origin of cortical granules within the peripherally located, multiple, widely scattered Golgi complexes of growing follicle (Szöllösi 1967, Baca and Zamboni 1967, Zamboni 1970, 1980, Kang 1974, Selman and Anderson 1975, Kang et al. 1979, Gulyas 1980, Guraya 1982a), whose vesicles and tubules are first filled with a dense material and then coalesce to form larger vacuoles (Fig. 17), as also clearly demonstrated in thick sections of rat preantral follicles stained by chromic acid-phosphotungstic acid (Kang et al. 1979).

The cortical granules first appear in association with the hypertrophied Golgi complexes, at a stage when the Golgi elements are shifting towards a subcortical region of the egg (Fig. 17). In the hamster and the rat, the membrane-bound vesicles lie near the concave surface of the Golgi complex (Fig. 17a). The contents of these granules consist of homogeneous material showing medium electron opacity (Szöllösi 1967). The limiting membrane of the small vesicles is generally scalloped, indicating the possibility of their recent separation from the Golgi complex. Similar vesicles have been noticed to coalesce into larger ones, which are in turn enclosed by a common limiting membrane (Baca and Zamboni 1967, Zamboni 1970). The progressive accumulation and condensation of the luminal contents of the fusing vesicles result in the formation of the cortical granules (Fig. 17). The large cortical granules are, at first, placed circumferentially at the periphery of a Golgi complex (Fig. 17) (Weakley 1966, Zamboni and Mastroianni 1966b, Baca and Zamboni 1967, Szöllösi 1967, Norberg 1972), then the granules separate from the Golgi units and are shifted towards the surface to form small aggregations.

The oocytes in pre-ovulatory follicles of bovines, hamster, humans, ovines, and rabbit show only a few cortical granules adjacent to the plasma membrane (Baca and Zamboni 1967, Szöllösi 1967, 1976, 1978, Fleming and Saacke 1972, Szöllösi et al. 1978). Instead, the cortical granules may form either aggregations or are scattered

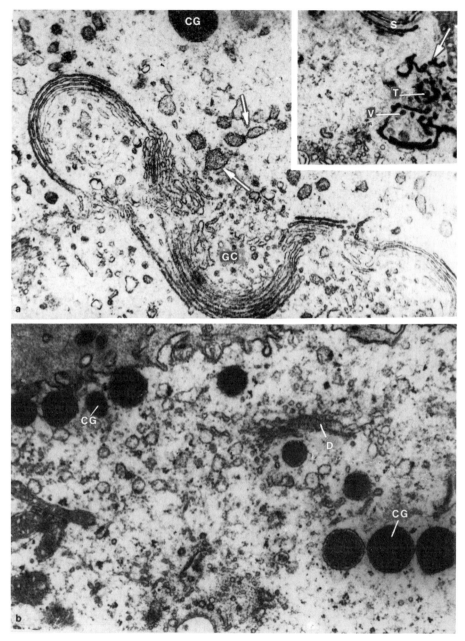

Fig. 17a, b. Electron micrographs showing the origin of cortical granules *(CG)* in the Golgi complex *(GC)* or dictyosomes *(D)* of developing oocyte. **a** Golgi complex *(GC)* within an oocyte of a bilaminar follicle of hamster showing presumptive cortical granules *(arrows)*. The *inset* is a Golgi complex containing thiamine pyrophosphatase in some of its saccules *(S)*, interconnecting tubules *(T)*, and associated vesicles *(V)*. *Arrow* a vesicle in the process of being pinched off from an interconnecting tubule (From Selman and Anderson 1975); **b** Several cortical granules *(CG)* of different sizes lie near dictyosome *(D)*, apparently during formative phase in the rabbit oocyte (From Szöllösi 1976)

individually in the subcortical ooplasm. Szöllösi and co-workers (Szöllösi 1975 a, Szöllösi et al. 1978), carrying out a series of closely timed studies, observed in the rabbit, in which LH is released by the second hour after copulatory stimulus (Goodman and Neill 1975), that the gap junctions between oocyte and corona cell processes are interrupted 5–5½ h after ovulation. Cortical granules are arranged to form a continuous layer beneath the plasma membrane only 6–6½ h after mating, that is, after the gap junctions between oocytes and corona cell processes have disappeared, as will be discussed in detail in Chap. IV. The peripheral shifting of cortical granules has also been noticed in cultured follicular oocytes of rabbit, bovines and mouse in the presence of FSH, LH or HCG, respectively (Nicosia and Mikhail 1975, Szöllösi et al. 1978). These results have indicated that corona cells may have an inhibitory role on cytoplasmic cortical events of the oocyte and that LH sets in motion a number of cellular events essential for cortical maturation (Szöllösi et al. 1978).

Besides the Golgi complex, two other organelles have also been involved in the formation of cortical granules. Several multivesicular bodies and other vesicular complexes are also developed and increased in number corresponding to the hyperplasia of the Golgi complex (Adams and Hertig 1964, Szöllösi 1967). Dense cortical granule-like vesicles are also observed to lie deep within the maturing oocyte of the dog ovary and often are enclosed within the lamellar yolk space (Tesoriero 1981). Granules within this space undergo changes in size, matrix configuration, and vacuolization. These changes are believed to be indicative of a mechanism whereby material is added to the lipid yolk bodies. Adams and Hertig (1964) suggested that the cortical granules in the oocytes of guinea-pig are formed by coalescence of small, dense vesicles with multivesicular bodies. After reinterpreting the results of Adams and Hertig (1964) several workers have attributed the origin of cortical granules to the Golgi bodies in the guinea-pig (Szöllösi 1967, Selman and Anderson 1975). Dense granules, which cannot be distinguished from cortical granules, have also been seen within vesicles of hamster oocytes (Szöllösi 1967). The development of granules within multivesicular bodies remains obscure. Moreover, multivesicular bodies give a positive reaction for acid phosphatase (Stastna 1974), while intact cortical granules remain acid phosphatase-negative (Anderson 1972). These results do not provide any convincing evidence for the origin of cortical granules within multivesicular bodies.

Selman and Anderson (1975), using electron microscopy, observed that cortical granules are formed by the activity of both the Golgi complex and rough endoplasmic reticulum. Numerous small electron-dense vesicles appear to get pinched off from the saccules and interconnected tubules of the Golgi complex, and fuse with one another. These subsequently fuse with vesicles formed from the rough ER to produce mature cortical granules. Similarly to the maturation process of secretory granules in other cell types, there is a condensation of secretory material in these vesicles during the formation of the mature cortical granules (Jamiesson and Palade 1967). The proposed role of ER in the formation of cortical granules is in agreement with the classical concept that protein synthesized in ER is transferred to the Golgi complex, wherein it is further processed and combined with polysaccharides prior to release from the cell in a packed form. Further studies should be carried out, applying available biochemical techniques, to extend and confirm this hypothesis for the formation of cortical granules in the oocytes of mammals. Kang and Anderson (1975) found round membrane-bounded cortical granules in the spiny mouse oocyte, which consist of electron-dense material and first appear in the cortical ooplasm of type 3 follicles that show well-developed Golgi complexes. Later, as soon as a complete layer of zona pellucida is constituted, mature cortical granules show a sparse distribution in the cortex of the oocyte. In type 5 follicle oocyte more cortical granules are produced. They lie in small groups. The coalescence of small vesicles containing electron-dense material, which leads to the formation of cortical granules, is present at the vicinity of Golgi complexes and the "lamellar complexes".

The exact site of synthesis of substances forming cortical granules is not yet known, but they start their development at different stages of oocyte growth in different species of mammals. In the rat and mouse, the first cortical granules appear in those unilaminar, small follicles in which the zona pellucida is only partially developed (Szöllösi 1967, 1976, Odor and Blandau 1969, Kang 1974, Kang et al. 1979). In the guinea-pig, the cortical granules are first seen at the onset of the growth of the follicular cells (Adams and Hertig 1964, Weakley 1966). But in many other mammals, such as humans, monkey, hamster, rabbit, cortical granules are seen first in oocytes of multilayered follicles (Hope 1965, Baca and Zamboni 1967, Szöllösi 1967, Krauskopf 1968a, Zamboni 1974, 1980, Selman and Anderson 1975). Although detailed studies of cortical granule formation in relation to follicular development have been carried out in only a few species, cortical granules have been shown at the time of ovulation in all species studied to date. In the rabbit, the formation of cortical granules continues even after ovulation, as their number appears to increase in aged eggs (Hadek 1963a). The cortical granules in the ova of the rabbit are round to elliptical in shape, measure 0.08–0.2 μm in diameter and show mostly homogeneous density (Hadek 1963a, b); some of the granules have a ligher cortical area and a denser medulla.

The cortical granules constitute a more or less continuous monolayer beneath the plasma membrane of unpenetrated tubal eggs. The formation of cortical granules is believed to be a continuous process until ovulation, at which time it stops. But in aging rabbit eggs there is continued formation of cortical granules (Hadek 1963a, Longo 1974a, Oh and Brackett 1975). In the mouse, in oviduct a full complement of cortical granules is formed just before fertilization (Zamboni 1970). Numerous Golgi elements are said to be prominent and actively involved in cortical granule formation between the times of ovulation and sperm penetration. However, the quantitative studies of Nicosia et al. (1977) on cortical granules of unpenetrated tubal mouse eggs are at variance with Zamboni's observations. Freshly collected tubal mouse ova contain 32 granules per 100 μm of plasma membrane and this cortical granule complement does not change when eggs are maintained in culture for 2 h (Nicosia et al. 1977). The basis for this difference needs to be determined.

b) Morphology

The cortical granules of unfertilized tubal eggs form spherical or slightly ovoid organelles bounded by a single membrane (Figs. 13, 16 and 17). They are generally situated within 2 μm of the plasma membrane.However, some cortical granules are found in the deeper regions of the ooplasm (Fig. 16). The cortical granules vary in size between 80 and 600 μm and show considerable species variations. The density of cortical granule matrix also varies not only from species to species, but also within the same egg. The cortical granules of unfertilized eggs of rabbit (Austin 1961, Zamboni and Mastroianni 1966b, Krauskopf 1968b, Gulyas 1974, 1976), humans (Baca and

Zamboni 1967), and monkey (Hope 1956) are generally of uniform appearance, whereas those in mouse and hamster eggs vary from uniformly dark to light to irregularly dark (Austin 1961, Szöllösi 1967, Selman and Anderson 1975, Nicosia et al. 1977). Light and dark granules may represent different stages of maturation, even though both types lie adjacent to the plasmalemma of the mature egg. The heterogeneity of cortical granule contents may also be indicative of some functional differences. Nicosia et al. (1977) observed that after fertilization in vitro, the number of light granules in the mouse eggs is decreased significantly before penetration by the sperm. In the unpenetrated eggs, 62.5% of the cortical granules are dark and the remaining ones are light.

Wabik-Sliz (1979) studied the number of cortical granules in unfertilized oocytes of four inbred strains of mice (CBA/KW, C 57BL/KW, KP and KE) and F_1 hybrids. KE oocytes show significantly fewer cortical granules than the oocytes of the other strains and F_1 hybrids. The largest number of cortical granules is seen in CBA/KW oocytes. The analysis of ovarian and tubal oocytes of KE and CBA/KW strains before and during ovulation has revealed that in freshly ovulated KE oocytes some cortical granules extrude their contents into the perivitelline space. This premature cortical reaction appears to be partly responsible for the low efficency of fertilization that characterizes the KE strain.

c) Distribution

As the cortical granules begin to be formed, their localization in the peripheral ooplasm adjacent to the plasma membrane also starts (Fig. 17b). But their peripheral shifting is not completed until the time of ovulation (Szöllösi 1976). In unpenetrated tubal eggs cortical granules form an irregularly spaced monolayer beneath the plasma membrane. Other cortical granules are present in the subcortical ooplasm (Szöllösi 1967, Gulyas, 1976, 1980, Nicosia et al. 1977). In the calf, cortical granules constitute large local aggregations in the peripheral ooplasm of nearly mature ovarian oocytes (Szöllösi 1976).

The distribution of cortical granules is generally not uniform and their density may also vary in different portions of the ooplasm of the same egg. Species variations in this regard are also reported (Peluso and Butcher 1974, Fléchon et al. 1975, Nicosia et al. 1977). Irregular spacing and polarity of the cortical granules were reported but their significance was unknown until the studies of Nicosia et al. (1977).

A close association exists between cortical granules and follicle cell processes, which penetrate the oocyte in several species, as will be discussed later in Sect. C. These observations suggest that the cortical granules may also function in corona cell withdrawal before they are involved in the prevention of polyspermy at fertilization (Zamboni 1974). The degradation and separation of the corona cell processes from the oolemma has been suggested to be enhanced by the enzyme(s) of cortical granules which are released prematurely. Premature release of cortical granules occurs in the mouse (Nicosia et al. 1977) and human (Zamboni 1980) eggs, but is apparently not involved in the prevention of sperm penetration (Wolf and Nicosia 1978). It is still to be determined whether the premature release of some cortical granules is related to the withdrawal to the corona cell processes. They are usually absent from the cell membrane overlying the second maturation spindle (Yanagimachi and Chang 1961, Szöllösi 1962, 1967, 1976, Stefanini et al. 1969, Zamboni 1970, Norberg 1973, Longo

1974b, Nicosia et al. 1977, Gulyas 1980). The absence of cortical granules in this region of membrane is apparently due to currents in the ooplasm as a result of cytokinetic events (Szöllösi 1976). The absence of cortical granules in areas immediately overlying the second meiotic spindle in the mouse egg has been suggested to result from the absence of Golgi complexes in the immediate vicinity of the meiotic spindle (Nicosia et al. 1977). This may not be true for other species, as the first polar body which is extruded near the meiotic spindle shows cortical granules (Zamboni and Mastroianni 1966a, Baca and Zamboni 1967, Szöllösi 1967, Zamboni et al. 1972, Norberg 1973), thereby showing that cortical granules are present near the meiotic spindle before the extrusion of the first polar body. This supports the suggestion that cortical granules are dislodged from this area by cytoplasmic currents produced during the rotation of the spindle and expulsion of the first polar body. The cortical granules are absent in the second polar body of penetrated eggs (Szöllösi 1967, Zamboni et al. 1976). The cortical granules are not seen in the first polar body of the human egg even after fertilization (Zamboni et al. 1966). Zamboni et al. (1966) have suggested that the presence of cortical cytoplasm in a polar body can be used as another criterion to distinguish the first polar body from the second.

Nicosia et al. (1977) performed cytological analysis of serially sectioned unpenetrated mouse eggs and showed the existence of a marked polarity in the distribution of cortical granules and microvilli in the egg cortex. Hadek (1963a, b) observed that in the freshly shed rabbit egg cortical granules are not evenly placed and there does not appear to be any visible relationship between their location and cortical villi.

Some studies have revealed the presence of actin-like filaments between the plasma membrane and cortical granules (Szöllösi 1967, 1976). Only cortical granule-free zones in the unpenetrated eggs of the mouse show such a layer of microfilaments (Nicosia et al. 1977). However, in the rat the microfilamentous layer is continuous, regardless of the presence of cortical granules (Szöllösi 1967). The actin-like filaments are believed to be involved in the expulsion and dispersion of cortical granule contents (Szöllösi 1976). These microfilaments may also play some role in cytokinesis during extrusion of the polar bodies and zygote cleavage.(Szöllösi 1970, Gulyas 1973, 1980).

d) Chemistry

The glycoprotein composition of mammalian cortical granules was first demonstrated with the light microscope in hamster, human, rhesus monkey and marmoset oocytes by periodic acid Schiff (PAS) and bromophenol blue staining techniques (Yanagimachi and Chang 1961, Guraya 1967d, 1969a). The results of subsequent electron microscope cytochemical studies have confirmed the glycoprotein composition of cortical granules in the oocytes of different species of mammals (Szöllösi 1967, Fléchon 1970, Zamboni 1970, Selman 1974, Selman and Anderson 1975, Kang et al. 1979, Gulyas 1980, Guraya 1982a). Cortical granules are digestable with pronase on thin sections of both Epon- and glycomethocrylate-embedded rabbit and hamster eggs (Fléchon 1970, Selman 1974, Selman and Anderson 1975). Variability in the degree of pronase digestion has revealed the heterogeneous nature of the cortical granule content. The cortical granules appear insensitive to α-amylase extraction (Selman and Anderson 1975). Phosphotungistic acid stains the cortical granules (Fléchon 1970). The cortical granules are also stained with ruthenium red (Szöllösi 1967). After the cortical reaction only the exudate of the granules stains because the intact granules fail to stain positively with these histochemical techniques (Szöllösi 1967, Anderson 1972, Stastna 1974). Thus, these do not appear to be the most reliable methods to analyze the content of cortical granules which may remain inert and do not participate in some of the histochemical reactions (Gulyas 1980).

The precise biochemical nature of the cortical granules in the mammalian egg is still to be determined, as only fragmentary information is available in this regard (Guraya 1982a). There are suggestions that some hydrolytic enzymes are present in cortical granules. Electron microscopic histochemical observations suggested the presence of acid phosphatase activity in intact granules of rabbit (Hadek 1963b, 1969). This observation needs to be confirmed, as acid phosphatase activity has not been observed in the guinea pig (Anderson 1972) or rat oocytes (Stastna 1974). However, a dense precipitate of the Gomori reaction is seen in cortical caverns of rat eggs immediately after the cortical reaction has taken place (Stastna 1974). Since the cortical granules originate from the Golgi units and no digestion of any substance inside the egg has been demonstrated, the cortical granules are believed to represent a special form of primary lysosome, wherein hydrolytic enzymes function only after their dehiscence (Stastna 1974). Gwatkin et al. (1973) have recently demonstrated a trypsin-like protease in the cortical granule material released from hamster and mouse eggs at fertilization (see also Gwatkin 1976, 1977a, Wolf and Hamada 1977), as also reported for echinoderms (Guraya 1982a). Its activity is reversible by trypsin inhibitors. In intact hamster cortical granules this protease is relatively heat-stable. However, when released from the egg, it is extremely heat-sensitive (Gwatkin and Williams 1974).

6. Deutoplasmic Inclusions

Deutoplasmic inclusions in the form of yolk vesicles, lipid yolk droplets and glycogen show several variations in their development in the fully grown ova of different mammalian species (Fig. 18) (Liss 1964, Guraya 1965, 1970b, 1973a, Szöllösi 1972, Korolev and Zavarzina 1976, Tesoriero 1981). Wherever they are formed, the exact mode of their formation still needs to be determined (Tesoriero 1981). Szöllösi (1972) discussed the problems of yolk in the mammalian egg, as also recently described by

---

Fig. 18a–f. Photomicrographs showing variations in the amounts of sudanophilic lipid yolk in the fully grown ova of different mammalian species. Cumulus oophorus *(CO)* or cumulus cells *(CC)* lying outer to the zona pellucida show some sudanophilic lipid bodies *(L)*. **a** Human oocyte shows sparsely scattered phospholipid bodies, yolk nucleus fragments *(YNF)*, granular mitochondria, spherical bodies *(SB)*, and yolk vesicles *(V)*, but no lipid yolk in the ooplasm. Nucleus *(N)* and zona pellucida *(ZP)* are also seen (From Guraya 1972b); **b** Cat oocyte showing the accumulation of highly sudanophilic lipid yolk *(LY)* droplets of various sizes. Small lipid yolk granules are seen in the process of coalescing to form large droplets of yolk (From Guraya 1965); **c** Hamster oocyte showing distribution of various ooplasmic organelles *(arrows)* and *PL* (peripheral layers of components) which feebly colour with Sudan black B. No lipid yolk is seen. Germinal vesicle *(GV)* containing large nucleolus is seen (From Guraya 1969c); **d** Dog oocyte showing the heavy accumulation of highly sudanophilic lipid yolk. Sudanophobic germinal vesicle *(GV)* and zona pellucida *(ZP)* are also seen (From Guraya 1965); **e** Marmoset oocyte showing little sudanophilic lipid yolk in the ooplasm *(O)* (From Guraya 1969b); **f** Oocyte of small Indian mongoose showing moderate amount of highly sudanophilic lipid yolk *(LY)* in the ooplasm *(O)*

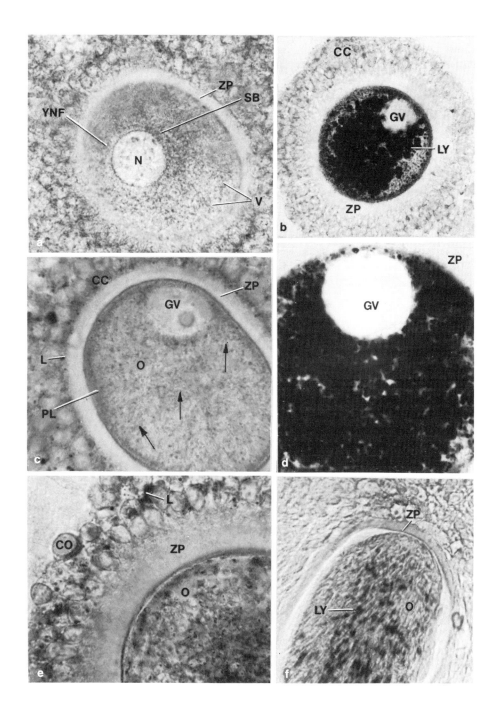

Tesoriero (1981). Yolk vesicles in the oocytes of rat, hamster and primates including humans do not show any appreciable development of yolk material. They have not been much investigated with the electron microscope (see Hadek 1965, Norrevang 1969). However, corresponding yolk vesicles in growing oocytes of some mammals show variable amounts of yolk material in their interior, which consists of a carbohydrate-protein complex (see Guraya 1973a, Norberg 1972, Fleming and Saacke 1972).

Lipid yolk granules composed mainly of phospholipids also show several variations in their amount in the oocytes of different mammals (Fig. 18) (Guraya 1965, 1970b, 1973a, Tesoriero 1981). Korolev and Zavarzina (1976) and Korolev (1979) have made a comparative study of lipids in the oocytes of placental mammals. The ova differ in quantity and manner of lipid distribution. No correlation is discernible between lipid loading of oocytes and general lipid content in ovaries of placental animals (Korolev 1979). Human oocytes, according to the structure and content of yolk, bear a close resemblance to those of hare and rabbit, but differ from oocytes of cat, dog, cow and golden hamster (Korolev and Zavarzina 1976). By thin-layer chromatography on silica gel, a detailed lipid composition has been reported for the rabbit oocytes, with neutral lipid and phospholipid (cephalin, lecithin, sphingomyelin) dominating. In addition, cholesterol is observed. Neutral lipids (triglycerides) accumulate in the ooplasm with the start of atresia (S. S. Guraya, unpublished observations) as will also be discussed in

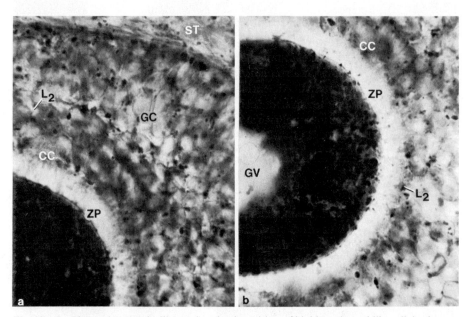

Fig. 19a, b. Photomicrographs illustrating the deposition of highly sudanophilic yolk in the two successive stages of oocyte growth in the dog ovary. The storage of lipid yolk is closely accompanied by the inflow of numerous lipid granules (L_2) into the oocyte from the granulosa cells (GC) as evidenced by their presence along the processes of corona cells (CC) in the sudanophobe zona pellucida (ZP). The phospholipid bodies (L_2) of various sizes and forms are also seen in the granulosa and surrounding stromal tissue (ST) (From Guraya 1965)

Chap. VII. In agreement with the histochemical observations of Guraya (1970b), the total lipid content of the rat oocyte and cleaving ovum is small (Travnik 1977a).

The lipid yolk droplets are formed by the coalescence and condensation of small lipid granules whose precursors are transported from outside the oocyte, as judged from the corresponding infiltration of lipids through the zona pellucida into the oocyte (Fig. 19) (Guraya 1965, 1970b). Fleming and Saacke (1972) have observed a very close association of endoplasmic reticulum and mitochondria with the surface of lipid droplets in the cow oocyte. According to Tesoriero (1981), the lipid yolk first appears in early primary oocytes of dog as aggregated dense bodies that gradually fill the ooplasm as the oocyte matures, thus supporting the histochemical observations of Guraya (1965). However, the site of initial appearance of the yolk is consistently related to a single centriole and often to the lamellae of smooth ER that surround groups of forming yolk bodies.

The exact physiological significance of variations in the development and amount of deutoplasmic inclusions in mammals is still obscure (Tesoriero 1981). Wherever present, they could be important in providing nutrients for the final maturation and fertilization of the oocyte as well as for the development of the very early embryo.

7. Metabolism

Growth of the mammalian oocyte involves not just tremendous enlargement of the cell as a result of accumulation of organelles, proteins, lipids etc., but also extensive alterations in its overall metabolism, as also reflected in the changing ultrastructure and cytochemistry of the oocyte at various stages of growth (see also Wassarman et al. 1979b). Actually, our knowledge is still meagre about the metabolism of growing and maturing oocytes. Eppig (1976) has made a study of mouse oogenesis in vitro with special reference to oocyte isolation and the utilization of exogenous energy source by growing oocytes. The utilization of exogenously administered $[^{14}C]$-labelled energy sources by oocytes in various growth stages has been determined by measurement of evolved $[^{14}Co_2]$. Little or no evolution of $[^{14}Co_2]$ is detected from oocytes of any size incubated in $[^{14}C]$-glucose, lactate or succinate. The production of $[^{14}Co_2]$ from $[^{14}C]$-pyruvate is increased logarithmically when plotted against increase in oocyte volume, with a plateau occurring after the oocytes have reached a volume of 65,000 μm^3 (50 μm diameter). The pattern of energy metabolism for oocyte maturation and early egg cleavage, wherein glucose and lactate are not utilized as efficiently as pyruvate, appears to be established by the earliest stages of oocyte growth. Metabolic changes during different stages of meiosis will be discussed in Chap. IV.

8. Distribution of Organelles and Inclusions in the Egg

Various ooplasmic organelles, RNA and deutoplasmic inclusions accumulated during oocyte growth are rearranged to constitute conspicuous gradients in the ooplasm of the fully grown egg (Fig. 18a) which may be playing some important functions in cell differentiation by influencing nuclear and gene activity (see Guraya 1970b, 1974a). The Golgi complexes, mitochondria, endoplasmic reticulum, and cortical granules are usually placed in the peripheral regions of the ooplasm. The other ooplasmic structures present in this peripheral region include the multivesicular bodies, myelin figures, rods with bulbous tips, lipid droplets and membrane-bound bodies containing dense granules or vesicles (see Hertig and Barton 1973). Ribosomes are usually distributed in the peripheral and perinuclear regions.

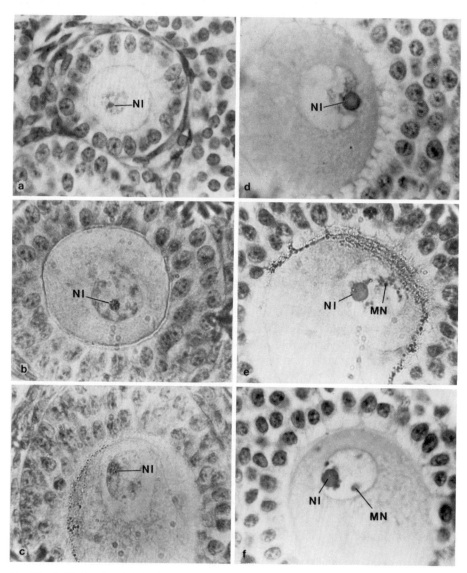

Fig. 20 a–f. Photomicrographs showing morphological changes in oocyte nucleolus *(NI)* during follicular growth in the mole rat. **a** Homogeneously stained nucleolus in the oocyte of a single-layered follicle; **b** and **c** Dense bodies and small vacuoles in the nucleoli of oocytes in stage I follicles; **d** Oocyte of a stage III follicle showing reduction in the number of nucleolar vacuoles; **e** and **f** Note the presence of one large homogeneously and darkly stained principal nucleolus and many small micronucleoli *(MN)* in the oocytes of V and VII stage follicles, respectively (P. Kaur and S. S. Guraya unpublished)

9. Nucleus and Formation and Storage of Informational Molecules

a) Nucleus and Nucleic Acids

The nucleus enlarges to form the germinal vesicle during the growth of the oocyte (Figs. 11 and 20). Its structural components such as nuclear envelope, nucleoli, chromosomes, etc. remain nearly the same as described for the nucleus of the primordial oocyte. The nucleus is located more eccentrically and ultimately it is placed nearly at the oocyte membrane (Fig. 18c). The nuclear envelope becomes generally more irregular, with folds and invaginations. It also shows more pores during the growth of the oocyte, indicating a greater activity of nucleocytoplasmic exchanges. Diplotene chromosomes become more distinct. The nucleolus loses its reticular appearance, a more compact one being formed in humans (Baca and Zamboni 1967, Zamboni 1980), often surrounded by large chromatin aggregates. The nucleolus/nucleoli undergo conspicuous morphological changes during the growth of the oocyte (Fig. 20), which are indicative of some metabolic alterations. In general, the nucleoli of dictyate oocytes are electron-dense, spheroidal bodies with a compacted, fibrillar element. Although nucleoli of such appearance are generally believed to be inactive in transcription, several studies have revealed detectable RNA synthesis up to stage of germinal vesicle breakdown.

While nucleolar fine-structure in growing mammalian oocytes is characteristic of cells actively involved in rRNA synthesis (Davidson 1976), the question of whether or not "amplification" of ribosomal RNA genes (rDNA), described in the amphibian growing oocytes (Davidson 1976), also occurs in growing mammalian oocytes has only been recently answered. Stahl et al. (1975) attributed the enormous amount of ribosomal RNA to the amplification of the ribosomal genes during mammalian oogenesis. Such a mechanism is well established for the oogenesis of invertebrates and lower vertebrates (see Davidson 1976). Two experimental approaches have indicated the occurrence of ribosomal gene amplification in mammalian oocytes. Firstly, the DNA content of the unfertilized mouse egg, which should become diploid after the first meiotic division, appears to be strikingly higher than can be expected for a diploid cell (Reamer 1963, Olds et al. 1973). This is also valid if one assumes that the DNA content of the extruded polar body is included in the measurements, leading to a tetraploid DNA value of unfertilized eggs. Secondly, in contrast to somatic cells, in the diplotene stage micronucleoli of human oocytes have been shown frequently to arise up to 15–20 nucleoli per oocyte (Stahl et al. 1975). Wolgemuth et al. (1980) have also described one or two large primary nucleoli and many additional small, nucleolus-like structures in the oocytes of near-term baboon *(Papio cynocephalus)*, which were predominantly in late diplotene of meiosis. These structures are morphologically identical to micronucleoli of human oocytes during early stages of meiotic prophase. The nucleolus-like bodies of human oocyte nuclei are Feulgen-negative, pyroninophilic, and measure 4–5 µm in diameter (Kurilo 1981). Their extrusion into the ooplasm is observed at all the meiotic prophase I stages from the preleptotene chromosome condensation stage on. The frequency of extrusion of nucleolus-like bodies at the diplotene stage is lower than at the dictyate stage. Quantitative measurements of rDNA in foetal human oocyte nuclei have revealed an approximate twofold increase in rDNA during the early meiotic prophase, with a subsequent increase to approximately fourfold during the pachytene and early diplotene (Wolgemuth et al., 1978). In the baboon, quantitative grain count analysis following rRNA: DNA hybridization in situ suggests the presence of a low level (approximately fourfold) excess of rDNA templates over the predicted 4C number in late diplotene oocytes.

Palombi and Viron (1977) reported the stage-specific modifications of chromatin, nucleolus and extra nucleolar components from the early stages of meiosis to the late prematuration stages. The alterations in the population of perichromatin granules are especially described and compared with the current information on ribonucleoproteins (RNP) production during mammalian oogenesis. The cytochemical approach used has proved particularly helpful in showing a unique richness in RNP-carrying particles in the nucleus of the growing oocyte, and in the identification of new RNP-carrying components at late stages of oogenesis. Zybina and Grishchenko (1977) observed a progressive enlargement of the nucleus and nucleoli up to the 4-layer follicle stage of hamster ova. At this time, numerous micronucleoli (30–40) develop, besides three or four main nucleoli, on the lampbrush chromosomes. There are seen waves of production of micronucleoli – the early and the middle diplonema.

From the morphological and cytochemical characteristics of main nucleoli and micronucleoli, a possible relationship between vacuolization of the nucleolus actively synthesizing RNA and the progressive lack of its granular component has been shown. Crozet et al. (1981) have also suggested that the fibrillo-granular and vacuolated nucleoli present at the onset of antrum formation in pig oocytes are associated with active RNA synthesis as determined by in vitro incorporation of [^{3}H]-uridine. The vacuolated nucleoli are progressively formed by a compact electron-dense fibrillar material. This morphological change is correlated to a significant decrease in nucleolar transcriptional activity. Mirre and Stahl (1978) observed that in the mouse oocyte at pachytene, the newly synthesized nucleolus appears at the junction of the paracentromeric heterochromatin with euchromatic portion of the bivalent containing the nucleolar organizer. The nucleolus is formed by a fibrillar centre penetrated by 5.0-nm-diameter fibres of the nucleolar organizer, surrounded by an electron-dense fibrillar zone and a distal granular component. Following brief [^{3}H]-uridine incorporation labelling is found over the electron-dense fibrillar zone, suggesting that the rDNA located in both the fibrillar centre and the electron-dense fibrillar zone is only transcribed in the latter.

The germinal vesicle of the growing oocyte continues to show lampbrush chromosomes beside the nucleolus/nucleoli (Baker and Franchi 1967, Baaken 1976). Lampbrush chromosomes showing lateral loops and branches are highly extended chromosomes in the diplotene stage of meiosis and have also been described in growing oocytes throughout the animal kingdom (see Davidson and Hough 1972, Davidson 1976). They form sites of marked transcription during oocyte growth. RNA synthesis has been measured autoradiographically by the uptake of labelled nucleotides. These studies have demonstrated that RNA synthesis in very low is small quiescent follicles, but increases during oocyte growth, reaches a peak in fully grown oocytes, and stops with antrum formation (Oakberg 1968, Baker et al. 1969, Bachvarova 1974, Moore et al. 1974). These results are also supported by a qualitative estimation of the activity of the DNA-dependent RNA polymerases in ethanol-acetone-fixed oocytes. Polymerase activity is shown in the nucleolus and nucleoplasm of growing oocytes, but not in the oocytes of mature follicles. The incorporation of [^{3}H]-uridine into the oocyte nucleus or nucleolus has revealed linear correlation with oocyte nuclear and nucleolar enlargement during oocyte growth (Moore and Lintern-Moore 1974).

Zybina et al. (1980) have recently observed that lampbrush type chromosomes are seen at all stages of oocyte growth in the forest dormouse *(Dryomys nitedule)*; highly condensed chromosomes develop at the diplonema stage. Meanwhile, there is seen a poor production of nucleolus-like bodies. This dormouse species is characterized by strongly developed nucleolus and by its intensive formation of small RNA-containing vesicles into the karyoplasm. Karyosphere development is followed. At the start of diakinesis nucleolus-bound chromosomes lose their contact with the nuclear envelope to be arranged on the nucleolar surface. Part of these chromosomes make a Feulgen-positive ring. In late diakinesis all the chromosomes separate from the nuclear envelope to move towards the nucleolus, which becomes the centre of karyosphere formation, having presumably lost its RNA-synthesizing function.

Recent investigations, in which the labelled precursor is injected into the ovarian bursa (Rodman and Bachvarova 1976) or into cultured antral follicles (Wassarman and Letourneau 1976 b, Brower et al. 1981), have shown that RNA synthesis continues in fully grown mouse oocytes until oocyte maturation is resumed. Kaplan et al. (1982) have also arrived at a similar conclusion by measuring rRNA contents of mouse primordial oocytes, three stages of growing oocytes, full-grown oocytes, and ovulated ova with hybridization of RNA samples to excess [^{3}H]-DNA complementary to rRNA. The failure of previous investigations to observe RNA synthesis in oocytes within antral follicles can be attributed either to a reduced uptake of exogenous precursors or to the expansion of the oocytes' endogenous precursors pool accompanying oocyte growth. The bulk of the RNA label in growing oocytes appears to be constituted by rRNA and transfer (t) RNA; the labelled heterogeneous RNA has been estimated to represent between 10% and 15% by polyacrylamide gel electrophoresis (Bachvarova 1974) or 20% by sucrose gradient sedimentation (Jahn et al. 1976). RNA of growing oocytes of all stages is unusually stable. In the mouse, approximately 80% of detectable oocyte transcription is directed to the synthesis of rRNA (and some tRNA), with the remainder belonging to the heterogenous class of RNA (Bachvarova and De Leon 1977). In the mouse, stable, long-lived RNA pool contains both rRNA (Jahn et al. 1976, Brower et al. 1981) and mRNA (Bachvarova 1981, Brower and Schultz 1982 b, Piko and Clegg 1982). Bachvarova (1981) has studied the synthesis, turnover and stability of heterogeneous RNA in growing mouse oocytes. The rate of its synthesis was lower than expected for lampbrush chromosomes with a high rate of transcriptional activity, suggesting that very active lampbrush chromosomes do not exist at any stage in the meiotic prophase of mouse oocytes.

From the results of various studies, it has been suggested that RNA synthesis (as measured by [^{3}H]-uridine incorporation in vivo) is low in the oocyte of the small follicle, increases markedly as oocyte growth starts and reaches its peak just before antrum formation. After the oocyte has ended its rapid growth phase, RNA synthesis falls and is low in large antral follicles (Oakberg 1967, 1968, Moore et al. 1974, Wassarman and Letourneau 1976 b, Moore and Lintern-Moore 1974, 1978, Wolgemuth-Jarashow and Jagiello 1979, Brower et al. 1981). The time course of RNA synthesis has also been followed with autoradiographic methods in vitro (Bloom and Mukherjee 1972). RNA synthesis is totally absent in mature oocytes that have reached metaphase of the second meiotic division (Baker 1969, Engel and Zenzes 1976). From these autoradiographic studies, definite conclusions in regard to the rate of RNA synthesis during mammalian oogenesis cannot be reached, as the size and the state of the internal precursor pool is not known at different stages of oocyte development. But having in view the striking similarity in the morphological and qualitative pattern of oocyte and follicular development in different mammalian species, Moore and Lintern-Moore (1974) assumed the pattern of RNA synthesis to be very similar in oocyte development of all mammalian species. The process by which RNA synthesis is suppressed in maturing oocytes needs to be determined more precisely. However, Moore et al. (1974) claimed that an inhibitor of DNA-dependent RNA polymerase is present in the antral fluid of the mammalian follicle. RNA polymerase is needed for RNA synthesis and is very low in activity in the unfertilized mouse egg (Siracusa 1973, Siracusa and Vivarelli 1975).

Numerous studies have been carried out to determine the duration of RNA synthesis during mammalian oogenesis. Significant incorporation of radio-labelled uridine and adenosine into mouse oocyte RNA occurs between the 19th and the 7th day, but not between the 5th and the day before ovulation (Bachvarova 1974). The phase of marked incorporation takes place during the period of major oocyte growth. Hoage and Cameron (1976), using radioautography, have studied DNA synthesis in the growing oocyte of the mature mouse. The oocytes of follicles that are in the antrum formation stage show significantly higher grain counts. Correspondingly, there appears auto-radiographic visualization of DNA-digestible [³H]-thymidine incorporation into the juxtanuclear region of these oocytes. The scheduled disappearance of this juxtanuclear oocyte DNA and its label during later oocyte growth suggests a degradation or dispersion of the labelled DNA prior to ovulation. Autoradiographic studies of transcription in mature mouse (Rodman and Bachvarova 1976, Wassarman and Letourneau 1976 b) and pig (Motlik et al. 1978 b) have demonstrated transcription up to a few hours preceding the breakdown of the germinal vesicle. But complex morphological associations between germinal and somatic cells of the ovary and the small number of mature oocytes produced have restricted experimental approaches to observations of labelled RNA precursors incorporated into the oocyte in vivo, either by autoradiography or by biochemical analysis of label distribution in oocytes (Moore and Lintern-Moore 1978). In addition, close anatomical associations between oocyte and granulosa cells in developing follicle place a further constraint on interpretation in regard to the actual origin of new transcription.

The duration of RNA synthesis has been determined by means of an in-situ assay for RNA polymerase activity (Moore and Lintern-Moore 1978). Both nucleolar (type I: rRNA synthesis) and nucleoplasmic (type II, heterogeneous nuclear RNA synthesis) RNA polymerase activities are present throughout oocyte growth. Polymerase activity increases progressively during oocyte growth to reach a peak in multilaminar, type -5 a mouse follicles and is decreased to a comparatively low level in oocytes of comparatively large antral follicles. Wassarman and Letourneau (1976 b) also obtained similar results from autoradiographs of ovarian sections obtained after intraperitoneal injection of labelled uridine into adult mice, except that significant levels of uridine incorporation in oocytes of large antral follicles were also observed in in vitro cultures. Kaplan et al. (1982) have shown, by hybridizing RNA samples to excess [³H]-DNA complementary to rRNA, that rRNA is synthesized at a constant rate over the first 9 days of oocyte growth in the mouse and about one and a half times faster in the last 5 days. The maximum value of 0.3 ng per oocyte was attained by about 14 days of growth in oocytes 59 µm in diameter, well below the maximum diameter of 77 µm for full-grown oocytes. The results of various studies have suggested that new RNA transcripts synthesized during the later phases of oocyte growth, and possibly up to nuclear dissolution, are involved in the regulation of resumed meiosis, the maturation of the cytoplasm and perhaps the initial stages of postfertilization embryogenesis (McGaughey and van Blerkom 1977, Young 1977, Schultz et al. 1978 a, van Blerkom and McGaughey 1978 a, b, van Blerkom 1979) as will also be discussed in Chap. IV.

It is becoming increasingly clear for the growing oocyte that a large proportion of RNP produced by the germinal vesicle, especially by its nucleoli and lampbrush chromosomes, is stored in the cytoplasm and its organelles (Burkholder et al. 1971, Guraya 1974a, Engel and Franke 1975, Engel and Zenzes 1976). It has also been shown that the RNA of growing oocytes of all stages is highly stable. At least 80% of the labelled RNA found 2 days after label administration is retained until ovulation 10–20 days later (Bachvarova 1974, Jahn et al. 1976). Recent studies have revealed that approximately 20% of the total population of ribosomes present in mature mouse oocytes is present in polysomes, with the remaining 80% representing a large pool of free ribosomes that are apparently inactive in translation (Bachvarova and DeLeon 1977). But the mammalian oocytes at ovulation show a general scarcity of cytoplasmic ribosomes and polysomes (von Blerkom and Motta 1979). Therefore, the sites of storage become a rather central issue in those mammals that appear to accumulate a population of translationally inactive ribosomes during oocyte growth. As already discussed in Sect. A, the densely packed arrays of proteinaceous superstructures forming lattices or ladderlike structures in the oocytes of rodents are believed to consist of ribosomes embedded in a proteinaceous matrix. If these lattices are indeed stored ribosomes, then it must be considered why they continue to persist until rather advanced stages of pre-implantation development when the embryo is actively involved in the synthesis of rRNA, forming both ribosomes and polysomes (van Blerkom and Manes 1977).

Wassarman and Mrozak (1981) investigated the synthesis and intracellular migration of histone H_4 during oogenesis in the mouse. H_4 is synthesized at all oogenetic stages studied and accounts for 0.07%, 0.05% and 0.04% of total protein synthesis in growing oocytes, fully grown oocytes and unfertilized eggs, respectively. During oocyte maturation the absolute H_4 synthesis rate is decreased by about 40% in comparison to a 23% decrease in the rate of total protein synthesis during the same period. These observations have suggested that enough histone is synthesized during oogenesis in the mouse to support two to three cell divisions. It is concentrated in the germinal vesicle (nucleus), whereas total protein and tubulin are not. Nearly 50% of the H_4 synthesized during a 5 h period is located in the oocyte's germinal vesicle, as compared to 1.9% and 0.9% for total protein and tubulin respectively.

The total amount of RNA in the rabbit egg has been determined as 20 ng of RNA per egg (Manes 1969). In the mouse, Reamer (1963) observed 1.75 ng of RNA per egg. However, a more recent measurements by Olds et al. (1973) has resulted in a much lower value at 0.55 ng of RNA per ovum. Various RNA species have been observed to be present in the mammalian oocyte or unfertilized egg, namely (28 S + 18 S) ribosomal RNA (Bachvarova 1974, Young et al. 1973). 5 S-ribosomal RNA and tRNA (Bachvarova 1974), and messenger RNA (Bachvarova 1974, Schultz 1975). The major portion of RNA in the egg is rRNA. The absolute quantities of (28 S + 18 S) rRNA and tRNA plus 5 S-rRNA in the mouse egg have been demonstrated by Bachvarova (1974) as 0.2 ng and 0.16 ng per egg, respectively. Young et al. (1973) determined a value of 0.4 ng of rRNA in the mouse egg under different experimental conditions. Brower et al. (1981) revealed the presence of putative rRNA precursor, ribosomal (28 and 18 S) RNA, transfer plus 5 S RNA and heterodisperse poly (A)-containing RNA during growth of the mouse oocyte in vitro. A significant fraction of the radiolabelled RNA species was quite large (40 S).

Most of the observations dealing with the presence of messenger RNA (mRNA) in the unfertilized egg made till now are indirect in nature (reviewed by Wolf and Engel 1972, Church and Schultz 1974, Epstein 1975, Engel and Franke 1975, Engel and Zenzes 1976). Bachvarova (1974) used a more direct approach. RNA in ovulated mouse eggs has been labelled by in vivo exposure of growing oocytes for radioactive precursors. The study of RNA by polyacrylamide-SDS electrophoresis has indicated a significant amount of labelled heterogeneous RNA in the mouse egg, which is believed to form the precursor of polysomal mRNA. Schultz (1975) provided evidence for the presence of poly(A)-containing RNA in the rabbit egg. The studies were carried out by hybridizing [³H]-labelled polyuridylic acid and unlabelled RNA of unfertilized and fertilized eggs. The proportion of poly (A)-containing RNA in the rabbit egg has been estimated as 0.25% of the total RNA amount, which is similar to values described for mammalian somatic cells. Mature mouse oocytes contain an "unusually high" amount of poly (A)-containing RNA (putative mRNA) which represents approximately 8% of the total RNA in the oocyte (DeLeon and Bachavora 1978). Brower et al. (1981) observed that about 40%–50% of the radio-labelled RNA during growth of the mouse oocyte in vitro behaved as poly (A)-containing RNA. This value remained fairly constant during the period of oocyte growth in which oocyte diameter increased from 35 to 55 µm. After a 5-h labelling, the percentage of radio-labelled poly (A)-containing RNA in either the fully grown dictyate oocyte, metaphase II oocyte or 1-cell embryo was 20%. Poly (A)-containing RNA has been observed to occur in the nuclear, ribosomal and subribosomal fractions but the major portion of the adenylated RNA has been demonstrated in the ribosomal fraction, showing that these form maternal mRNA's associated with polysomes. Burkholder et al. (1971) suggested that the activation of these particles and the onset of translational activity appear to be caused by partial proteolysis during postfertilization development in mammals. The electron microscope studies have demonstrated lattice-like structures in the mammalian oocytes, which develop during oocyte growth and completely disappear during early cleavage stages (Fig. 14). These ooplasmic components have been suggested to represent maternally derived inactive ribosome-mRNA complexes, which are activated by proteolytic enzymes after fertilization. Therefore, it is likely that RNA synthesized during oocyte growth is utilized at least during very early embryogenesis (Mintz 1964, Epstein 1975). It is currently believed that polyadenylated, oocyte mRNA may be utilized in the support of early postfertilization translation (Davidson 1976).

b) Proteins

The protein content of the unfertilized rabbit egg has been estimated as 100 ng (Brinster 1971 a, b) and as 20 ng or 27 ng in the ovum of the mouse (Lowenstein and Cohen 1964, Brinster 1967 a). The protein content in the rat egg has been observed to be similar to that in the mouse, namely 21 ng per egg (W. Engel and M. T. Zenzes, unpublished observations). Amenta and Cavallotti (1980) have demonstrated by immunohistochemistry a myosin-like protein in the rabbit oocyte. Kaplan et al. (1982) have suggested that cytoskeletal proteins as well as putative LDH are synthesized in growing and full-grown oocytes of the mouse and accumulate to form a significant portion of bulk egg protein.

The enzyme activity of ovarian oocytes has been quantified in relatively fewer studies. Various enzymes have been reported in the unfertilized eggs of mammals (Table 1). Hamster eggs contain amino-peptidase and elastase-like activities but no detectable trypsin-like activity. LDH has been most extensively studied in the ova of mammals. In the mouse ovum, it forms about 5% of the total protein content and its activity is 20 times higher than in any other mouse tissue (Brinster 1965). But a comparison between the LDH activity in mature oocytes of different species demonstrates great variation. The laboratory mouse shows highest activity, the human and the rabbit contain the lowest values observed so far (Brinster 1967 b, 1968). In most somatic tissues LDH consists of five isozymes resulting from random association of the two LDH polypeptide chains α and β to a tetrameric molecule; in the mouse ovum only

Table 1. Enzymes demonstrated as present in the mammalian egg (From Engel and Zenzes 1976)

Lactate dehydrogenase
Glucose-6-phosphate dehydrogenase
NAD-malate dehydrogenase
Hexokinase
Fructose-1, 6-diphosphate aldolase
Adenin-phosphoribosyl-transferase
Glycogen synthetase
Glutamic dehydrogenase
Aspartate aminotransferase
NADP-isocitrate dehydrogenase
Phosphofructokinase
6-phosphogluconate dehydrogenase
Uridine kinase
Phosphoglucose isomerase
Guanine deaminase
Pyruvate carboxylase
NADP-malate dehydrogenase
Hypoxanthine-guanine-phosphoribosyltransferase
RNA-polymerase
NAD-glycohydrolase
Triose phosphate isomerase
Glyceraldehyde-3-phosphate
Dehydrogenase
Phosphoglyceromutase
Phosphoglycerokinase
Creatine kinase

the single isozyme band LDHI (α 4) can be demonstrated (Rapola and Koskimies 1967, Auerbach and Brinster 1967). The polypeptide chains α and β are believed to be the products of two separate LDH genes in the genome. In other rodent species, the LDH activity of their oocytes also consists exclusively of LDHI, whereas in species of the orders Lagomorpha, Carnivora, Artiodactyla, and man, LDH isozymes consisting of both α and β-subunits occur in the oocytes (Engel and Krentz 1973, Engel et al. 1975). The mammalian oocyte and egg, besides the active enzymes, also contain inactive enzyme molecules (Engel and Franke 1975).

B. Zona Pellucida

In the mammalian ovary the zona pellucida develops as a thick, acellular, gelatinous, glycoprotein layer between the surface of growing oocyte and granulosa (or follicle) cells (Fig. 11). Intercellular exchanges between the oocyte and granulosa of growing follicles are mediated through the zona pellucida (Guraya 1974a, Dvořák and Tesařík 1980). As the zona pellucida forms the external investment of the egg, it is of great physiological significance in the initial stages of fertilization, as the sperm must recognize, develop contact, and pass through it before establishing contact with the plasma membrane of the egg (Pikó 1969, Metz 1978, Yanagimachi 1978). Therefore, the zona pellucida of mammalian eggs has been extensively studied for its origin, structure, physicochemical properties, antigenicity, functions, etc. Interest in the zona pellucida's being a possible site for manipulating fertility has also greatly increased during the past few years (Shivers 1976, 1979, Gwatkin 1979, Shivers and Sieg 1980, Tsunoda et al. 1982). This interest has increased, in part, because the zona as a structure is reproduction-specific, several important functional roles in early reproduction having been assigned to it (Gwatkin 1976, 1982), and its existence being terminated at implantation. The various functions assigned to the zona include mechanical protection, osmotic regulation, prevention of polyspermy, species specificity of fertilization and support of blastomeres. But there is doubt about osmotic regulation since large proteins and even viruses can pass across the zona (Glass 1963, Gwatkin 1967).

1. Origin

With the onset of follicular growth the zona pellucida begins to form within the intercellular spaces between the granulosa cells adjacent to the oocyte. The isolated deposits of the material finally coalesce into a homogenous zona in the space between the oocyte and the granulosa cells (Fig. 21). Very divergent views have been expressed about the origin of zona material in the developing follicle. Granulosa cells or oocyte or both are believed to be involved in the formation of zona material (see Stegner 1967, Norrevang 1969, Baker 1970, Hertig and Barton 1973, Kang 1974, Zamboni 1974, 1976, 1980, Guraya 1974a, Nowacki 1977, Kang et al. 1979; see also Bleil and Wassarman 1980a). Patches of zona material are usually first seen in spaces between the cell membrane of the oocyte and junction of two granulosa cells (Zamboni 1974, 1976, Barton and Hertig 1975, Dvořák and Tesařík 1980). They become confluent and finally form a continuous layer around the oocyte with microvilli and granulosa cell processes embedded in its matrix (Fig. 13). The microvilli and follicle cell projections

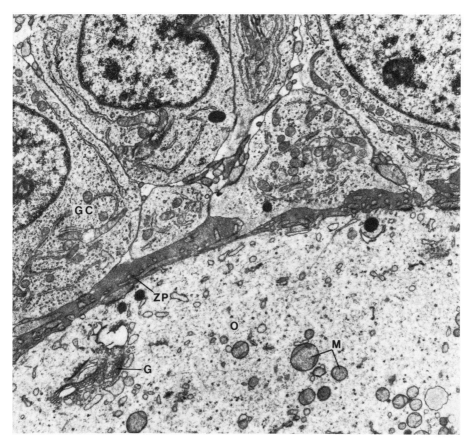

Fig. 21. Electron micrograph of portion of small pre-antral follicle from day 2 postnatal ovary of guinea-pig, showing the formation of zona pellucida *(ZP)* between the follicular epithelium and oocyte *(O)*. The cytoplasm of granulosa cells *(GC)* shows mitochondria of variable size, profiles of granular endoplasmic reticulum and numerous free ribosomes. Note some electron-dense bodies both in granulosa cell cytoplasm and ooplasm. The peripheral ooplasm shows a Golgi complex *(G)*, mitochondria *(M)* of variable size, numerous free ribosomes (electron-dense granules) and profiles of endoplasmic reticulum (From Guraya 1973a)

greatly increase in number as the oocyte and granulosa differentiate, thus increasing the surface area of their cell membranes (Zamboni 1974, 1976, 1980, Dvořák and Tesařík 1980, Apkarian and Curtis 1981).

Guraya et al. (1974) have not observed any special ultrastructural formations in the ooplasm adjacent to the developing islands of zona material in the guinea-pig ovary, which could be correlated to its origin (see also Guraya 1973a). However, a careful examination of granulosa cell cytoplasm, especially of processes, lying in the vicinity of developing zona pellucida, demonstrates the presence of some flocculent and fibrillar material which is apparently similar to that of zona (Fig. 21) (Guraya 1973a). Kang (1974) has also reported the presence of similar material in the cytoplasm of granulosa cell processes and interfollicular cell spaces of granulosa cells, as also discussed by Nowacki (1977). These ultrastructural observations have suggested that the granulosa cells may contribute some material to the formation of the zona pellucida.

The oocyte surface acts as the site for the polymerization and deposition of zona material which consists of a carbohydrate-protein complex (Guraya 1974a, Nowacki 1977), as also reported by Repin and Akimova (1976). According to Kang (1974), the Golgi complexes of the rat oocyte play an important role in the synthesis of the zona pellucida as judged by the intimate association of the Golgi complexes with the oocyte plasmalemma and the presence of mucopolysaccharide in their saccules and vesicles. Kang et al. (1979), using thick sections stained with chromic acid-phosphotungstic acid, have observed that young follicles in the rat ovary show the zona containing nonfibrous mucopolysaccharide-rich material that is transformed into a coarse fibrous structure of mature follicles. The Golgi apparatus of the oocyte appears to sequester the zona-type mucopolysaccharide which is deposited in the extracellular space lying between the oocyte surface and follicle cells.

The variable chemical nature of the zona pellucida in different mammalian species (Kang 1974), as will be discussed later on, suggests that its substances might be derived from different sources (Hertig and Barton 1973, Guraya 1974a). According to Motta and Van Blerkom (1974), the stratified appearance of the zona in the mature follicle suggests a discontinuous secretory process which could be supported not only by granulosa cells, but also by the oocyte itself (Stegner and Wartenberg 1961, Motta et al. 1971). In addition, the same stratified appearance of the zona with the presence of granules on its surface suggests that at least in the mature follicle some new material is produced by the granulosa cells surrounding the zona (Mestwerdt et al. 1977a). The zona granules present on the surface and also within the zona material may originate from the granulosa cells that send their processes towards the oocyte membrane as suggested by Motta and Van Blerkom (1974).

The results of correlative electron-cytochemical and autoradiographic techniques are more significant for determining the respective contributions of granulosa cells and oocyte in the formation of zona pellucida. Haddad and Nagai (1977), using radioautographic technique, have observed that the zona material is synthesized and secreted by the oocyte. Haddad and Nagai (1977), using L-fucose-^{3}H, have also shown that the labelling pattern of the follicular fluid depends on the secretory activity of the granulosa cells. Based on the labelling pattern of large follicles it has been shown that there is very little synthesis of specific glycoproteins for the zona pellucida in the medium-sized follicles. Following the growth of these follicles that have a previously labelled zona pellucida, it has been demonstrated that this extracellular structure is secreted by the oocyte. This suggestion is further supported by the recent studies of Bleil and Wassarman (1980a) who have followed the synthesis of zona pellucida proteins by denuded and follicle-enclosed mouse oocytes cultured in vitro in the presence of either [^{35}S] methionine or [^{3}H] fucose. Evidence has been produced to show that zona pellucida material originates from the oocyte itself, rather than from the surrounding follicle cells. This suggestion is further supported by the recent data obtained with indirect immunofluorescence microscopy using a specific, anti-hamster, zona pellucida, serum (Bousquet et al. 1981). All these observations have established that the oocyte is an important source of the antigenic material of the zona pellucida of hamster and human ova (see also Greve et al. 1982). But the contributions of granulosa cells toward the formation of the zona pellucida still need to be determined more precisely.

2. Structure

The morphological characteristics of zonae have been studied by light as well as by transmission and scanning electron microscopy. These studies have suggested that the zona pellucida of some mammals, such as humans (Stegner and Wartenberg 1961), pigs (Dickmann and Dziuk 1964, Hedrick and Fry 1980), and rabbits (Sacco 1981), consists of more than one layer, as compared to other species such as the mouse (Cholewa-Stewart and Massaro 1972) or the rat (Sacco 1981). Scanning electron microscopy of the zonae of all mammalian species studied to date has revealed that the outer surface of the zonae has a fenestrated lattice-like appearance (Dudkiewicz et al. 1973, Meyenhofer et al. 1977, Jackowski and Dumont 1979, Phillips and Shalgi 1980).

The zona pellucida has long been recognized as a mucopolysaccharide-rich structure (Guraya 1974a) and is seen with the electron microscope as an accumulation of finely textured, filamentous material. When examined with scanning electron microscope, the zona pellucida has an irregular and rough appearance, which results from the presence of numerous, distorted layers of an amorphous material and from the presence of numerous labyrinthine channels and many granules (Motta and Van Blerkom 1974, 1980). Anderson et al. (1978) observed that the zona pellucida in the mouse and rabbit follicle shows numerous fenestrations of varying diameters.

Dudkiewicz et al. (1976), using transmission and scanning electron microscopy, investigated the structure of hamster zona pellucida which shows an intricate network of interconnecting fibres. The latter lead to the formation of a porous outer zona region lying adjacent to a relatively more compact but porous inner zona layer. The canaliculi bounded by the fibres provide via ducts by which processes of granulosa cells may pass through the zona pellucida for interacting with the vitelline surface (see also Anderson and Albertini 1976, Dekel and Phillips 1979). Although eggs treated with zona precipitating antibody (ZPA) show an enhancement of the reticular components of the zona pellucida, ZPA does not occlude the pores or interstices seen prior to treatment. The pores show a decrease from the outer to inner zona regions. The extensive fibrous complexity of the zona surface has also been demonstrated with scanning electron microscopy (SEM) (Fig. 22). In agreement with the observations of Dudkiewicz et al. (1976), von Weymarn et al. (1980a) also observed, in vivo, a consistent development of the zona pellucida surface, i.e. the formation of a fibrous network-like structure interspersed with numerous pores. In normal oocytes with intact germinal vesicle, microvilli are distributed over the entire surface in all age groups of mice. Phillips and Shalgi (1980), using SEM, have studied the structural and sperm-binding properties of the hamster zona pellucida which shows morphologically dissimilar internal and external surfaces. The external surface shows a fenestrated lattice-like appearance, whereas the internal surface shows a regular rough surface. Numerous non-acrosome-reacted spermatozoa and dissociated acrosomes associate with the external surface of zona pellucida but not with its internal surface; acrosome-reacted spermatozoa do not adhere to zona pellucida. These results suggested that the external and internal surfaces of zona pellucida are morphologically dissimilar. These morphological observations are consistent with the lectin-binding properties of the zona, which also reveal the variability of the properties of outer and inner surfaces of the zonae (Nicholson et al. 1975, Dunbar et al. 1980, Dunbar 1980).

In the unfertilized eggs of mouse, the zona pellucida consisting mainly of fibrillar

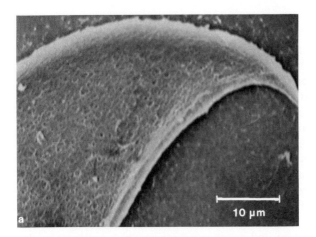

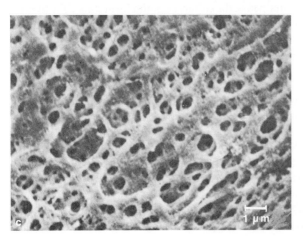

Fig. 22 a–c. Scanning electron micrographs of zona pellucida isolated from the hamster eggs. **a** Outer surface of zona pellucida *on the left* and its inner surface *on the right;* **b** Smooth inner surface of zona pellucida with relatively small apertures; **c** Large openings and channels in the outer surface of zona pellucida (From Meyenholfer et al. 1977)

and granular material shows two layers, a thicker internal layer and a dense external layer (Baranska et al. 1975). But the zona pellucida of the embryo shows an additional third intimal layer of coarse grains inside the internal layer and a fourth fine-grained peripheral layer situated outside the external layer. The intimal layer appears to derive from products of cortical reaction, whereas the peripheral fourth layer is formed and modified during the passage of the embryo through the female genital tract (Dunbar and Shivers 1976). Baranska et al. (1975) have discussed the influence of developmental events and environmental factors on the fine structure of the zona pellucida.

3. Physicochemistry

Studies on the physicochemical characteristics of the zona pellucida are of great significance in determining more precisely the nature of interaction of the egg end sperm during fertilization, as well as to define the relationship of the egg and embryo with the uterine environment prior to implantation. In the fertilization process, sperm penetration through the zona pellucida has been postulated to involve a preliminary sperm binding to the zona pellucida (Gwatkin 1976, 1982, Metz 1978, Yanagimachi 1978). This is followed by a restricted hydrolysis of the zona pellucida with the assistance of sperm enzymes such as acrosin, thereby permitting the sperm access to the egg membrane (see also Guraya 1986). After the fertilizing sperm has gained access to the egg and triggered the cortical reaction, the zona pellucida is chemically and physically altered to inhibit supernumerary sperm penetration (Gwatkin 1976, Metz 1978, Yanagimachi 1978).

The biochemical properties of the zona also being investigated in order to determine the nature of surface receptors, which might determine species specificity for sperm recognition and fertilization, and to define the various proteinic, glycoproteinic and nonproteinic components of the zona pellucida more precisely. Therefore, extensive studies are being carried out for isolating and characterizing zona components and for studying their efficacy as contraceptive vaccines (Gwatkin and Williams 1977, 1978a, Gwatkin et al. 1977, 1980, Gwatkin 1979, 1982, Dunbar and Raynor 1980, Dunbar et al. 1980, 1981, Bleil and Wassarman 1980b, c, Noda et al. 1981, Wood et al. 1981, Sacco 1981). In all these and in the early studies on chemical analysis of zonae (Gould et al. 1971, Inoue 1973, Repin and Akimova 1976, Sacco and Palm 1977) various reagents, such as urea, mercaptoethanol and/or sodium dodecyl sulphate (SDS), have been utilized for solubilization of zonae. The antibodies directed against zona pellucida components inhibit sperm binding to the zona pellucida and fertilization of eggs both in vivo and in vitro, as will be discussed later on.

As the zona pellucida is an amorphous acellular, gelatinous structure consisting of glycoprotein complexes of possibly more than one origin (oocyte and granulosa cells), as already discussed, there is every chance that at the chemical level it must be heterogeneous in nature, consisting of several different chemical components. According to Tadano and Yamada (1978), the zona pellucida of adult mouse oocytes shows positive reactions for complex carbohydrates with 1, 2-glycolytic acidic groups, D-mannoacyl and α-D-glucoacyl residues. These complex carbohydrates are hyaluronic acid, sulfate glycosaminoglycans other than isomeric chondroitin sulfates and neutral glycoproteins. Sialic acid is also present (Sourpart and Strong 1974). Probably it is an artifact. The glycoprotein nature of the zona pellucida is well established (Guraya

1974a, Inoue and Wolf 1974a, Kang 1974, Baranska et al. 1975, Haddad and Nagai 1977, Bleil and Wassarman 1980b, Dunbar et al. 1980, 1981, Gwatkin 1982). All these studies reveal the presence of polysaccharides, mucopolysaccharides and/or proteins or glycoproteins in the zona pellucida of different species of mammals. Physicochemical studies also reveal the presence of disulphide bonds in the zona pellucida, which maintain its structural integrity (Inoue and Wolf 1974a, b). Several disulphide-bond cleaving agents are effective in solubilizing the zona pellucida. Dunbar et al. (1980) studied physicochemical properties and macromolecular composition of the zona pellucida from isolated porcine oocytes. Chemically, the zonae consist mainly of protein (71%) and carbohydrate (19%). No unusual amounts or types of amino acids are found. The monosaccharides present include those typically seen in the animal glycoproteins. Sialic acid in glycosidic linkage, and sulphate and phosphate esters present are believed to be true constituents of the zona pellucida. Its carbohydrate moieties are asymmetrically distributed. Fatty acids (esterified and free) and uronic acids are considered contaminants rather than true constituents. Zona preparations obtained by Gwatkin et al. (1980) show considerably less hexose: 6% vs the 19% demonstrated by Dunbar et al. (1980). Sialic acid and uronic acids are also lower. However, Gwatkin et al. (1980) have demonstrated a close similarity between the composition of bovine and pig zonae. This composition fits a glycoprotein. The rather broad bands obtained on electrophoresis of both the bovine and pig zonae are typical of glycoproteins, where small differences in the number and type of sugar groups can give rise to some variation in electrophoresis mobility.

From solubility test, Dunbar et al. (1980) concluded that the structural integrity of zona pellucida is dependent upon noncovalent interactions between constituent macromolecular components. The zona is effectively solubilized by conditions that will not break covalent bonds. Several observations on the solubility properties of the zona pellucida have been made (Hall 1935, Braden 1952, Gwatkin 1964, Cholewa- Stewart and Massaro 1972, Inoue and Wolf 1974a, b, 1975, Gwatkin and Williams 1978a, Gwatkin 1979, 1982). Dunbar et al. (1980) suggested that physicochemical and immunological characterization of zona solubilized under nondissociating conditions must take into account the fact that zona constituents exist as supramolecular complexes.

Lowenstein and Cohen (1964) found a protein content of 1.8 ng for the zona pellucida of mouse egg. This value represents about 30% of the total dry mass of the zona, the remainder is attributed to carbohydrates. According to Bleil and Wassarman (1980b), each zona pellucida in the mouse contains 4.8 ng of protein, representing 80% or more of the dry weight of the zona pellucida and about 17% of the oocyte's total protein. In contrast to these relatively low values for ovulated eggs, Dunbar et al. (1978) reported a value of 50 ± 10 ng for the protein content of ovarian pig eggs. This relatively high figure, however, may be due to contamination by remnants of follicular cells. Sacco and Palm (1977) reported a protein content of 30 ng for the zona pellucida of porcine egg. This value also represents about 50% of the total dry mass of the pig zona, the rest being contributed by carbohydrates viz. acidic and neutral sugars.

Polyacrylamide gel (PAG) electrophoresis and sodium dodecyl sulphate (SDS) electrophoresis are being used extensively to make a qualitative analysis of solubilized preparations of the zonae pellucidae (see Gwatkin et al. 1980). Gwatkin (1978a) observed three fast-green components in hamster zonae. Bleil and Wassarman (1978) resolved mouse zonae into three glycoproteins. Gould et al. (1971), using mercaptoethanol, but without SDS in their gels, revealed four proteins in the rabbit zona pellucida. Several classes of polypeptides with molecular weight greater than 65,000 have been obtained from SDS-PAG electrophoresis of mouse zona pellucida (Inoue 1973). Repin and Akimova (1976) made electrophoretic studies of rat and mouse zonae pellucidae. Zonae pellucidae from unfertilized rat eggs show an electrophoretic pattern consisting of five protein bands ranging in molecular weights from 20,000 to 82,000. Banding pattern of rat embryo zonae pellucidae revealed six protein bands with molecular weights ranging from 10,000 to 35,000. Similarly, electrophoretic patterns of zonae pellucidae of unfertilized mouse eggs showed five protein bands ranging from 28,000 to 240,000 MW while mouse embryo electrophoretic patterns reveal two protein bands of 100,000 and 140,000 MW.

The recent electrophoretic analysis of as few as five isolated mouse zonae pellucidae treated with diazotized [^{125}I] iodosulfanilic acid revealed the presence of only three radio-labelled proteins, designated ZP 1, ZP 2 and ZP 3 (Bleil and Wassarman 1980 b, c). The same three proteins have also been identified by Coomassie blue staining, when large numbers of isolated zonae pellucidae (approximately 750) are subjected to SDS-PAG electrophoresis. These three proteins migrate as broad bands on SDS-PAG electrophoresis, consistent, with their being glycoproteins having apparent MW of 200,000 (ZP 1), 120,000 (ZP 2) and 83,000 (ZP 3). Results of amino acid analysis and high-resolution two-dimensional electrophoresis of individual proteins have shown that each protein represents a unique polypeptide chain. The proteins ZP 1, ZP 2 and ZP 3 represent about 36%, 47% and 17% respectively of the total protein of the zona pellucida. In the presence of reducing agents which cause dissolution of the zona pellucida, ZP 1 is converted into a species which migrates with an apparent MW of 130,000, suggesting that it exists as an oligomer, stabilized by disulphide bonds, in the unreduced state. Bleil and Wassarman (1980 c) have tested the ability of mouse ZP 1, ZP 2 and ZP 3 glycoproteins to interfere with the binding of sperm to eggs in vitro. These results suggested that ZP 3 possesses the receptor activity responsible for the binding of sperm to zonae pellucidae of unfertilized mouse eggs. Fertilization is believed to cause modification of ZP 3 such that it no longer serves as a receptor for sperm. Zona solutions prepared from ovulated unfertilized hamster eggs contain 17 amino acids, with a relatively high concentration of leucine (Gwatkin 1978 a). On ultracentrifugation, hamster zona solutions behave as a single macromolecule of 8.9×10^6 MW. This molecular weight is too high to allow entry of the native protein into polyacrylamide gels, but after treatment with SDS, mercaptoethanol and urea three subunits have been dissociated (Gwatkin 1978 a).

Gwatkin (1979) observed that SDS-PAG electrophoresis of heat-solubilized bovine zonae pellucidae, in the presence of 1% SDS, 1% mercaptoethanol and 5 M urea, gives three major bands and about ten faint bands. Similar observations have also been made by Sacco and Palm (1977) and Dunbar et al. (1980) who used porcine zonae pellucidae. Analysis of pig zonae pellucidae by SDS-PAG electrophoresis showed at least five Coomassie blue staining bands, of which three are relatively sharp and two are dispersed. Comparison with band pattern obtained on SDS-PAG electrophoresis of follicular fluid revealed that at least one band of dispersed region is specific to zona pellucida extracts and is in the range of MW 120,000–150,000 and also contains periodic acid-Schiff base-positive material. However, the major pig zonae proteins have molecular weights ranging between 40,000 and 70,000 (Dunbar et al. 1980) as compared to the major proteins of rabbit zonae which have molecular weights ranging from 70,000 to 100,000. Gwatkin et al. (1980) described a similar molecular weight range (50,000 to 80,000) for the pig zona protein-staining pattern, and gave three molecular weight ranges for proteins associated with the cow zonae (100,000, 50,000 and 25,000). Dunbar et al. (1981) have demonstrated that both the rabbit and porcine zonae pellucidae are composed of three primary molecular weight species of charged glycoproteins and that the molecular weights, as well as the relative changes of isoelectric point positions of the molecular weight species, are different for the rabbit and porcine zonae. Sacco et al. (1981a) have analyzed [^{125}I]-labelled (Chloramine T method) zonae of pig, human, squirrel monkey, rabbit and mouse. The pig, human, and rabbit zonae contain three major protein families, all showing considerable microheterogeneity, while four protein families are associated with the squirrel monkey, and only two are observed with mouse zonae. The patterns obtained for the radio-labelled pig and rabbit zonae in these studies are similar to those obtained by direct staining of these proteins (Dunbar et al. 1981). All these studies have demonstrated a great variation in proteins of different species as well as the complexity of the chemical composition of zonae proteins.

Menino and Wright (1979) reported that PAG electrophoresis of zonae pellucidae isolated from porcine ovarian oocytes and dissolved in either 6M urea or 2.0% SDS indicates that the zona pellucida consists of a single protein in addition to a carbohydrate component. The structural organization of the protein in porcine zona pellucida is believed to be repeating units of this single protein maintained by noncovalent bonding. Electrophoresis in SDS-B-mercaptoethanol (B-MSH) polyacrylamide gels revealed that the protein consists of four polypeptide or protein subunits with molecular weihgts of 17,000, 57,000, 66,000 and 85,000. These proteins may be cross-linked by disulphide bridges comprising the single protein seen in gels without B-MSH. The subunit proteins may be in a multimeric arrangement supported by covalent bonds which maintain the complex structure. Relative percentages of the subunit proteins demonstrated by scanning densitometry are: 50.3% (17,000 MW), 23.7% (57,000 MW), 18.6% (66,000 MW) and 6% (85,000 MW). The unequal percentages revealed by densitometry may show the number of molecular subunit species within a complex.

Some differences have also been revealed in the chemical composition of zonae pellucidae in different species of mammals by their variable response to different chemical agents, pH, enzymic attack and lectins (Mintz 1962, 1967, Gwatkin 1964, Nicholson et al. 1975, Yanagimachi and Nicholson 1976). Since the zona pellucida of mammalian egg contains carbohydrates (Jacoby 1962, Guraya 1974a, Oakberg and Tyrrell 1975, Lowenstein and Cohen 1964, Soupart and Noyes 1964, Fléchon 1970, Inoue and Wolf 1974a, b, Kang 1974), as also discussed here, a variety of plant lectins that bind to specific saccharide determinants are being used extensively to determine more precisely the process of sperm attachment and binding to zona pellucida (Oikawa et al. 1973, 1974, 1975b). Some of these lectins, which can bind to zona pellucida, are observed to induce egg agglutination and/or cause changes in the ability of the exterior region of the zona to scatter visible light.

Studies with phytoagglutinins help in the analysis of biochemical properties responsible for specificity of sperm attachment, which might be responsible for blocking hybrid fertilization. Recent studies have revealed that agglutinins *Ricinus communis* (RCA I), *Dolichos bifloros* (DBA) and Concanavalin A (Con A) (specific for terminal residues, similar to D-galactose-N-acetyl-D-glucosamine and α-D-mannose or D-glucose respectively) are effective in preventing fertilization, although sperm binding is not prevented. The effective concentrations needed for different lectins are quite variable (Oikawa et al. 1974, 1975 b). The lectin-induced properties are specific because inclusion of the appropriate saccharide inhibitors, such as D-galactose for RCA I and DBA and N-acetyl-D-galactosamine for Wheat Germ Agglutinin (WGA), abolished these effects (Oikawa et al. 1975 b).

Electron microscope studies, using ferritin-lectin conjugates, have revealed that the binding sites for RCA I and WGA are present in the outermost regions of the zona pellucida, while the Con A receptors are distributed sparsely throughout the zona (Nicholson et al. 1975). A comparative study of lectin-binding sites on the zona pellucida of eggs in various stages of their maturation and development has shown that the intensity of binding and the distribution of fluorescent-labelled lectins do not change (Yanagimachi and Nicholson 1976). The fact that the zona-precipitating antibody and the plant lectins produce similar effects on the zona pellucida suggests that antibody and agglutinins react with similar and terminal saccharides or oligo-saccharide groups. Moreover, as these lectins recognize specific saccharide residues, these studies can provide information about the terminal monosaccharides of glycoproteins of the zona pellucida. It is interesting to mention here that some lectins can block fertilization by inducing a cortical reaction (Gwatkin et al. 1976). Furthermore, the resistance of the receptors for spermatozoa to glycosidases argues against a critical role for terminal sugars as receptor determinants.

Solubilization studies on zonae pellucidae are of great help to make the qualitative analysis of different chemical components with regard to their presence and antigenic role in the intact zona pellucida. The solubility of the zona pellucida at different stages of maturing oocytes and of the ovulated and fertilized eggs has been studied with different chemical agents, treatment at different pH and temperatures, and with different enzymes (Sacco and Palm 1977, Gwatkin 1979, 1982, Menino and Wright 1979, Dunbar et al. 1980, 1981, Dunbar and Raynor 1980). Solubilization has been evaluated by light microscopy. Visual criteria of zona dissolution are limited by their subjective nature and do not necessarily correlate with biochemical methods (Dunbar et al. 1980). However, such information is useful as a means of comparing zonae pellucidae of different species. Various studies have demonstrated species variation in zonae, as a variety of conditions affect dissolution differentially.

Proteolytic enzymes cause dissolution of zonae pellucidae. However, the susceptibility of zonae to emzymatic digestion varies greatly among species. Moore and Crayle (1971) investigated the effects of pronase, papain, bromelain, trypsin and α-chymotrypsin on the zona pellucida of fertilized and unfertilized eggs in different animal species. All these enzymes are found to be effective in lysing the zonae pellucidae of unfertilized eggs, while only pronase is able to denude about 34% of fertilized eggs in about 5 min of exposure, and the rest in about 90 min. Trounson and Moore (1974) suggested that ovulation and aging of egg are associated with the alterations in the zona pellucida, which influence its susceptibility to digestion by proteolytic enzymes, as also observed by Longo (1981). The lytic effect of trypsin and pronase on the zona pellucida of eggs from different mammals, such as rabbit (Sacco and Shivers 1973d), goat and cow (Gwatkin 1977a), mouse (Inoue and Wolf 1974a), and hamster and rat (Oikawa et al. 1975a), has also been demonstrated. A preliminary report by Hedrick and Wardrip (1980) has revealed that proteins of the pig zona pellucida characterized by two-dimensional gel electrophoresis are modified by trypsin. These reports are confirmed by recent data which show that pig zonae remain intact (as determined microscopically) after trypsin treatment, but that zonae proteins are chemically changed since a change in protein patterns was visualized by two-dimensional gel electrophoresis (Dunbar and Roberts 1982).

Aonuma et al. (1978) observed that not all mouse ova, which ovulated after the usual superovulation treatment, are susceptible to zona removal by proteases in modified Krebs Ringer bicarbonate medium. This susceptibility decreases as the time between ovulation and recovery is increased. The additional treatment of the mice with PMSG abolishes the decrease in susceptibility and fertilizability. When ova are recovered at 2 h after hCG injection and held in culture for further 12 h, there is no significant decline in either variable. If ova are placed for 2 h in sperm suspension, the dissolution of the zona by protease is completely prevented, but this effect is not produced by sperm-free supernatants of sperm suspension. The loss of dissolution susceptibility in these treatments with sperm suspension is associated with the time of sperm penetration. Rosenfeld and Joshi (1981) have studied the effect of a rat uterine fluid endopeptidase on lysis of the zona pellucida. This endopeptidase causes the lysis of the zona pellucida of unfertilized rat and mouse eggs, but not of fertilized rat and mouse eggs. These results suggest that after fertilization cortical granule discharge may modify the zona pellucida to prevent digestion by the endopeptidase. Since it is now possible to isolate large quantities of zonae, detailed studies can be carried out to study directly the effect of sperm enzyme on zona proteins.

Kaleta and Polak (1978) studied the solubility properties of the zona pellucida of mouse oocytes from inbred strains (KE, KP, C57 and CBA) in different concentrations of chymotrypsin. The time required for zona dissolution of unfertilized eggs depends on the genotype of the ovum and the concentration of enzyme. In a 0.003% solution of chymotrypsin the zonae of CBA eggs are most soluble. In the KE strain after 3 h only 22% of the oocytes is denuded. The C57 and KP strains give intermediate results. Increase in chymotrypsin concentration (0.03%) causes a significant increase in dissolution of the zona in genotypes (except the KP strain). In such conditions the C57, KE and KP strains do not reach the values specific for the CBA strain. There are significant differences between inbred strains in the biochemical structure of zona pellucida. The zonae of ovarian KE oocytes are much more sensitive to the action of chymotrypsin than those recorded from the oviduct after ovulation. This increase in zona resistance of KE tubal ova is probably due to a premature cortical reaction, occurring in this strain at the time of ovulation or shortly after (Guraya 1982a).

Cholewa-Stewart and Massaro (1972) have observed that mouse zona pellucida undergoes a process of "melting" over a narrow range of temperature, near 70°C. The zona of unfertilized eggs melts at a temperature below 70°C, while that of fertilized eggs melts at a temperature above 70°C. Dissolution of the zona pellucida on heating has been reported for rodents (Inoue and Wolf 1974a, b, Gwatkin 1979) and bovines (Gwatkin 1979). But complete dissolution of bovine zona has been contested by Dunbar et al. (1980) who believes that heating is capable of only partially dissolving porcine zonae.

Burne and Psychoyos (1972) observed dissolution of the zona pellucida at low pH (5.06) in 2 h, and at pH 3.36 in 2 min. pH-dependent dissolution of murine zona pellucida surrounding unfertilized eggs and two-cell embryos has also been demonstrated (Inoue and Wolf 1974a, b). It has also been observed that zonae of murine unfertilized eggs are consistently more readily dissolved than those from two-cell embryos in 2-mercaptoethanol, sodium periodate and acidic solutions. But hamster zonae are significantly less sensitive than rat and mouse zonae to pH, and periodate, and dissolution

Recently, zona dissolution has also been achieved by using different chemical agents: SDS urea, triton-x and other detergents (Gwatkin 1978a, b, 1979, 1982, Dunbar et al. 1980, 1981, Menino and Wright 1979, Dunbar and Raynor 1980). The extent of solubilization is a function of pH, ionic strength, temperature and the presence of various solubilizing agents such as detergents and urea (Dunbar et al. 1980). These solubilized preparations are being used now to determine the spectrum of chemical constituents of the zona pellucida. More detailed chemical studies using protein analysis of pig zonae have now been made (Dunbar et al. 1980). Pig zona can be solubilized by conditions which do not break covalent bonds (heat at basic pH). These conditions result in supramacromolecular complexes of zona macromolecules. Dissociation into individual macromolecules can be carried out by the reduction of disulphide bonds in the presence of detergent (Dunbar et al. 1980, 1981).

4. Permeability

For the better understanding of physical properties of the zona pellucida, special attention has been paid to its permeability to different molecules. The permeability properties of the zona are of special importance in regard to sperm penetration and the zona's possible role in immunological protection of the egg from its environment.

Austin and Lovelock (1958) in in vitro studies observed that both Alcian blue and digitonin were able to pass through the zona of hamster, rabbit and rat, but heparin could not. These findings indicated that the zonae pellucidae of these species are permeable to molecular weights of 1,200 or less, but not to those with molecular weights of 16,000 or greater. The zonae pellucidae of eggs and cleavage stages are also permeable to serum proteins (Glass 1963), to Mengo virus (Gwatkin 1967), to such metabolic precursors as [35S]-methionine (Weitlauf and Greenwald 1965, 1967) and perhaps to blastokinins (Krishnan and Daniel 1967) or uteroglobulin (Bier 1968, Kirchner 1972a, b).

By using electron-dense tracers, Anderson (1972) made studies of zona permeability in the intrafollicular oocytes of rats. Unfertilized and fertilized ova and/or pre-implantation stages from a variety of mammalian species (ferret, guinea pig, mouse, mink, rabbit and rat) have also been studied for zona permeability (Enders 1971, Hastings et al. 1972). These findings revealed that zonae pellucidae from various species are permeable to both peroxidase (MW 40,000) and to ferritin (MW 400,000–600,000) after a 10-min exposure, regardless of developmental stage. Similar results have also been obtained by Schlafke and Enders (1973) who investigated the uptake of ferritin and peroxidase by pre-implantation rat embryos in vitro. The zona is penetrated by immunoglobulins (MW 150,000–900,000) 10 min after exposure (Sellens and Jenkinson 1975). Kang et al. (1979) stated that the fibrous zona might not impede movement of macromolecules-like peroxidase (3.4–4 nm diameter) even though it may be an effective barrier to cells and larger macromolecules. These findings together with those obtained by previous investigators (Anderson 1967, 1972, Enders 1971, Hastings et al. 1972) present unequivocal evidence for zona permeability to high molecular weight substances.

5. Receptors for Sperm

Initial studies, using some rodent species, suggested that sperm attachment to the zona was species-specific (Yanagimachi 1977, 1978). These studies have prompted the search for specific "sperm receptors" on the surface of the zona. Therefore, various studies have been made on the distribution and nature of specific sperm receptors, involved in the sperm–egg interaction at the time of fertilization (Gwatkin 1976, 1982, Metz 1976, Yanagimachi 1978, Peterson et al. 1981, Ahuja and Bolwell 1983). The mammalian sperms attach themselves to and penetrate through the zona pellucida before establishing contact with the plasma membrane of the egg. Therefore, the sperm-receptor sites on the surface of the zona pellucida in mammals have been characterized (Dunbar and Shivers 1976, Gwatkin 1977b, Gwatkin and Williams 1977, Ahuja and Bolwell 1983). The universality of the presence of "sperm receptors" is yet to be proven in mammals. For instance, the species specificity of sperm–zona interaction has been observed for some mammals, including mouse, hamster, guinea-pig, and rat (Austin and Braden 1956, Yanagimachi 1977, 1978, Schmell and Gulyas 1980), but less specificity is shown by other mammalian species, such as the rabbit (Bedford 1977, Swenson and Dunbar 1982).

Gwatkin and Williams (1977) observed that capacitated golden hamster spermatozoa bind to the inner as well as to outer surface of the zona pellucida, suggesting that the receptor for spermatozoa may occur throughout this envelope. When solutions of hamster zonae pellucidae are made by heating them in aqueous buffers, receptor activity is retained and is stable up to boiling point. Receptor activity is lost following the reaction, which is a process by which the zona is involved in preventing polyspermy in some mammalian species (see Guraya 1982a). Although this reaction varies among species, it has been attributed to a chemical modification of the zona, most likely as the result of the cortical reaction (Guraya 1982a), although the biochemical basis for this phenomenon needs to be determined more precisely. However, Bleil et al. (1981) have suggested that some of the changes in the biochemical and biological properties of zonae pellucidae, seen following fertilization or activation of mouse eggs, result from modification of the major zona pellucida glycoprotein ZP 2.

The zonae of ovulated hamster eggs are dissolved on heating in an aqueous buffer (Gwatkin and Williams 1977). On exposure of the capacitated sperm to these solutions, the zona material binds to the sperm's plasma membrane (Gwatkin 1978a, 1982) and subsequently they become infertile (Gwatkin and Williams 1977, 1978a). Zona solutions prepared from eggs that have undergone a zona reaction are inactive (Gwatkin and Williams 1977). These observations indicate that the receptor on the zona for capacitated sperm has been solubilized and is still active in the solubilized state.

The sperm-binding property of the zona pellucida is stable to a number of glycosidases and lipases but is inactivated by low concentration of trypsin or chymotrypsin (Hartmann and Gwatkin 1971, Gwatkin et al. 1973) and by a trypsin-like protease produced after the cortical reaction of the egg (Gwatkin et al. 1973, Guraya 1982a). However, sperm attachment and penetration through the zona pellucida of the rabbit are not altered by trypsin or neuroaminidase (Bedford 1974). These findings show that the zona surface receptors which are involved in sperm attachment in some species have critical proteinic components. Saling (1981) has studied the involvement of trypsin-like activity in binding mouse spermatozoa to zonae pellucidae. The trypsin-sensitive receptor for spermatozoa (Hartmann and Gwatkin 1971) is believed to be present throughout the zona pellucida (Ahuja and Bolwell 1983), where it would be expected to maintain gamete association, as the spermatozoon penetrates the zona to reach the vitellus. However, there may be a gradient with the greater concentration of receptor towards the outer surface (Ahuja and Bolwell 1983), and this may account for the apparently greater sensitivity to trypsin of the outer surface binding. O'Rand (1981) has demonstrated inhibition of fertility and sperm–zona binding by antiserum to the rabbit sperm membrane auto-antigen. It will be interesting to mention here that spermatozoa can associate with zona in a nonspecific manner (Swenson and Dunbar 1982).

Now that relatively large quantities of pure zona glycoproteins are available (Gwatkin et al. 1980, Gwatkin 1982), the next step will be to characterize the receptor for sperm by preparing fragments using cyanagen bromide and various proteases. Isolation of the specific receptor for the sperm would not only lead to a better understanding of the molecular mechanism involved in fertilization (Gwatkin 1976), but, if the receptor could be coupled to make it antigenic, might yield an even more effective contraceptive vaccine (Gwatkin 1982).

6. Antigenicity

Immunological techniques are being used extensively to understand the process of fertilization at the molecular level, particularly with studies which pertain to diagnostic tests related to fertility and sterility (Shivers 1976, 1979, Metz 1973, 1978, Shivers and Sieg 1980, Tsunoda et al. 1982). Recently, increasing attention is being paid to antigens present in female tissues, especially in the zona pellucida (Fox and Shivers 1975, Shivers 1974, 1976, Shivers and Sieg 1980, Gwatkin 1982). Immunogenicity of the zona pellucida is now well established and zona-specific antigens appear to be associated with its surface. Therefore, antigen(s) associated with the zona pellucida are being investigated much more extensively in relation to immunological regulation of fertility (Shivers 1976, 1979, Gwatkin 1979, 1982, Tsunoda and Sugie 1979, Trounson et al. 1980, Tsunoda et al. 1981a, b, c, Shivers and Sieg 1980, Verbitskii et al. 1980, Tsunoda et al. 1982, Tsunoda and Sugie 1982, Tsunoda and Whittingham 1982, Sacco et al. 1983). The zona pellucida as a structure is reproduction-specific, functional only in early reproduction, and ceases to exist upon implantation, which makes it an attractive target in immunocontraception. The concept that immunologic procedures might be used in the control of fertility was first postulated from investigations that revealed antizona antibodies change the zona surface in several ways, including zona precipitation, blockage of zona digestion by enzymes, and prevention of sperm attachment to the zona pellucida (Shivers et al. 1972). Tissue-specificity of zona antigens is important because cross-reaction of antizona antibody with antigens in other tissues could lead to interference with nonreproductive processes. But the results of various studies indicate that zona antigens are not present in other tissues (see Shivers 1976, 1979, Shivers and Sieg 1980, Tsunoda et al. 1982). The fact that iodine-labelled antizona antibodies rapidly disappear from all nonovarian body tissues after passive immunization but persist in ovarian tissue for the duration of infertility in immunized females further supports this suggestion. In addition, antibodies produced against other body tissues (serum, uterus, oviduct and follicular fluid) all fail to react with zona. Palm et al. (1979) using radioimmunoassay have tested tissue specificity of porcine zonae pellucidae antigen(s). No zona immunoreactivity was detected in 22 porcine tissues and fluids studied, with the exception of the ovary. These findings strongly indicated that porcine zonae pellucidae contain specific antigen(s). Tsunoda et al. (1981 b) did not confirm the reports of Sacco and Palm (1977) and Dunbar and Rayner (1980), who observed cross-reactions with several other tissues, particularly follicular fluid. It needs to be determined more precisely whether these differences are due to variation in purity of the zona preparations used for immunization or to the differences in the methods used for characterization of zona antigens. Sensitivity of testing procedure may be important. Further studies, including the use of nonspecific antibodies to zona and various other sensitive methods for detecting antigen–antibody reactions, are recommended before the problem of tissue specificity can be solved (Shivers 1979, Shivers and Sieg 1980, Tsunoda et al. 1982). The methodology used in previous studies on the antigenicity of the zona pellucida has been described in several excellent reviews (Gwatkin 1979, 1982, Shivers 1979, Dunbar et al. 1980, 1981, Dunbar and Raynor 1980, Wood et al. 1981, Sacco 1981, Tsunoda et al. 1982, Sacco et al. 1983). Here only the antigenicity of the zona pellucida and its biological effects in relation to fertilization will be discussed.

Antigenicity of the zona pellucida has been shown for several species of mammals. Specific zona antigens have been shown to be present in pre-ovulatory, fertilized and unfertilized ova, and all embryonic stages up to the time when zona is normally shed. Ownby and Shivers (1972) used hamster ovary-raised antiserum in rabbits. In agar-gel-diffusion studies seven antigen – antibody systems were shown, and that among these there are antigen-specific ones to ovary was demonstrated by absorption of the antisera with lyophilized tissues. Further studies have revealed that this antigen is associated with the zona pellucida (see also Garavagno et al. 1974). Antigens have also been shown in the zonae of mouse (Glass and Hansen 1974, Shivers 1974, Jilek and Pavlok 1975). Sacco and Shivers (1973 a, b, c) have observed two antigens specific to rabbit ovary, one of which is associated with zona pellucida. In these studies the absorbed antiserum failed to produce precipitin bands when reacted in double diffusion plates against 11 (in hamster study) and 17 (in rabbit study) other tissues. Also, absorption of antizona antibodies with other body tissues fail to neutralize the antibody. Gonad-specific antigens are also reported in the ovary of the guinea-pig (Porter et al. 1970 a, b). Anti-rat ovary antiserum examined by Tsunoda and Chang (1976 a) is observed to give two to four bands with different tissues of rat, but no band with rat spermatozoa extracts. Studies with anti-hamster ovary antiserum (Tsunoda and Chang 1976 b) demonstrate eight to nine precipitin bands with hamster ovary, three to four bands with rat ovary, and one to two bands with mouse ovarian extracts. Absorption of this antiserum with hamster tissue extracts is observed to remove the cross-reacting bands.

By the use of immunofluorescent techniques it is possible to demonstrate that antibodies against ovary are restricted to certain areas of the ovary, with evidence that these antigens are associated with the zona pellucida (Shivers et al. 1972, Sacco and Shivers 1973 a, b, c, Shivers 1976, 1979, Gwatkin 1979, 1982, Shivers and Sieg 1980, Shivers et al. 1981, Tsunoda et al. 1982, Yurewicz et al. 1983). Studies conducted using anti-pig ovary antiserum and human ovary extracts give several cross-reacting bands in agar-gel diffusion plates. At least one common antigen is present in the zona pellucida, as determined by immunofluorescence (Shivers 1975).

When hamster zona solutions are injected, with Freun's adjuvant, into mice high titers of zona-specific antiserum are produced (Gwatkin et al. 1977). These antisera react with the zonae pellucidae of mice, rhesus monkey and squirrel monkey eggs. Mouse zona is antigenic when antiserum is produced in rabbits against freshly ovulated eggs with cumulus cells and against embryos at two pre-implantation stages, e.g. morula and blastocyst (Glass and Hansen 1974). Immunofluorescence has been used to show that each of these antisera contains antibodies which react with the egg and embryo cytoplasm as well as with the zona pellucida. Using immunohistological tests, it was observed that absorption of each antiserum with mouse serum renders it specific to the zona pellucida, when tested on zona-coated eggs and embryos (Glass and Hansen 1974). In various recent studies, the production of hetero-antiserum to isolated zonae pellucidae has been reported for mouse (Tsunoda and Sugie 1977, Sacco 1979), hamster (Gwatkin et al. 1977), pig (Sacco and Palm 1977, Tsunoda and Sugie 1979 b, Gwatkin et al, 1980, Tsunoda et al. 1981 b), cow (Gwatkin and Williams 1978 a, Gwatkin et al. 1980, Tsunoda et al. 1981 c) and other species (Verbitskii et al. 1980).

Sacco (1977 a) has studied the antigenicity of the human zonae pellucidae. The unabsorbed and absorbed antisera react with isolated human zonae pellucidae, as shown by the formation of precipitation layer on the zona and by the indirect fluorescent antibody technique. Sacco (1977 b) has also studied the problems of antigenic cross-reactivity between human and pig zona pellucida. Each demonstrates a low degree of cross-reactivity when tested by immunodiffusion against ovarian extracts obtained from the other species. This study has also shown that a common antigen is apparently shared by the human and pig zona pellucida (see also Shivers and Dunbar 1977). Cross-reactivity has also been observed between mouse, hamster and rat (Tsunoda and Chang 1976a, c, d), pig, humans, marmoset and chimpanzee (Shivers et al. 1978), mouse, hamster, squirrel monkey and rhesus monkey (Gwatkin et al.1977), cow, rabbit, rhesus monkey, marmosets, dog, hamster, and humans (Gwatkin and Williams 1978 a, Gwatkin 1979), and other mammalian species (Verbitskii et al. 1980). Cross-reaction of rabbit anti-pig zona serum with rabbit, bovine, marmoset, rhesus monkey, and human zonae is also reported (Gwatkin et al. 1980, Gwatkin 1982). Radioimmunoassay studies revealed cross-reactivity between porcine and rabbit zona antigens (Wood and Dunbar 1981). Immunoelectrophoresis studies also demonstrated the immunochemical cross-reactivity between two distinct antigens of rabbit and procine zonae. These two rabbit zona antigens were shown to be immunochemically identical to porcine zona antigens by crossed-line immunoelectrophoresis. Tsunoda et al. (1981 b) demonstrated that passive immunization with goat anti-pig zona serum inhibits fertilization in the cow, sheep, rabbit, rat and mouse, showing cross-reaction. However, with the doses used by these workers, passive immunization with anti-pig zona serum more effectively inhibits fertilization in the rabbit and mouse than in the rat. There is evidence that zonae may contain species-specific as well as non-species-specific antigens, as revealed by the difference in strength of antizona antibody reactions between species (Tsunoda and Chang 1976a, b, c, 1978a, Verbitskii et al. 1980, Tsunoda et al. 1982). Studies by Takai et al. (1980) revealed the presence of at least three different components of human zona pellucida: (1) specific antigen(s) shared by human and porcine zonae, (2) specific antigen(s) proper to human zonae, and (3) nonspecific antigen(s) associated with the blood group substances. The facts that passive immunization with goat anti-rabbit liver and kidney serum does not inhibit fertilization, and that passive immunization with anti-pig zona serum absorbed with liver and kidney inhibits fertilization in the rabbit (Tsunoda et al. 1981 b), indicate that pig and probably other mammalian zonae pellucidae contain tissue-specific, common antigens important in fertilization. Sacco et al. (1983) have evaluated the antigens of the porcine zona pellucida by two-dimensional and line immunoelectrophoretic techniques. Electrophoretic data have revealed that the porcine zona system is antigenically complex, with each zona antiserum tested detecting numerous antigens in the various zona preparations. These electrophoretic studies also suggest that the various zona antisera can distinguish, with different degrees of sensitivity, multiple antigenic determinants on the individual zona macromolecules. These studies also indicate that SDS treatment of zona glycoproteins does alter the antigenicity of the macromolecule, both with respect to the total number and the individual identity of antigens detected.

Gwatkin (1979) reported that bovine zonae solutions prepared from ovarian eggs, although slightly contaminated by follicle cell material give rise to antisera and that these cross-react with human eggs. Column isolation has produced a high molecular weight fraction that contains both sperm receptor and zona antigen, the latter in a concentration adequate to produce infertility in female rabbits (see also Gwatkin et al. 1980).

Nishimoto et al. (1980) reported the presence of autoantibodies to the zona pellucida in the serum of infertile women as also reported by Mori et al. (1978). These auto-antibodies found in serum samples, from several women infertile for unknown reasons, react with pig zonae (Shivers and Dunbar 1977, Sacco 1977a). Shivers and Dunbar (1977) have suggested that the reaction, which is detected by immunofluorescence, may be due to auto-antibodies reacting to zona antigens. Shivers (1979) has discussed the details of auto-antibodies reacting to zonae pellucidae.

The results of various studies as discussed above have clearly indicated that the antigens are present in zonae of developing, preovulatory, unfertilized or fertilized ova, and all embryonic stages up to the time when the zona is normally shed (Dudkiewicz et al. 1975, Shivers and Sieg 1980, Tsunoda et al. 1982). Moreover, it has also been demonstrated that hetero-antibodies against these antigens change the physical and perhaps the chemical nature of the zona surface and thereby interfere with several events in early reproduction (Shivers 1979, Shivers and Sieg 1980, Shivers et al. 1981, Wood et al. 1981, Tsunoda et al. 1982), as will be discussed here.

7. Biological Effects of Antizona Pellucida Antibodies

Various biological effects of zona pellucida antibodies in relation to control of fertility have been demonstrated. They are believed to be powerful inhibitors of fertilization.

Inhibition of sperm attachment to the zona has been shown both in vitro and in vivo. Antizona antibodies appear to cause some physico-chemical changes in the zona surface in several ways including zona precipitation, blockage of zona digestion by enzyme, and prevention of sperm adhesion to the zona surface. Most of the early investigations on the zona approach for control of fertility were carried out with antibodies prepared against homogenates of ovaries or egg masses (see references in Shivers 1979, Shivers and Sieg 1980, Tsunoda et al. 1982). But in recent years, mechanically isolated zonae pellucidae, frozen-thawed preparations and heat-solubilized zona solutions are being used extensively for antizona antibody production (see Gwatkin 1979, 1982, Shivers 1979, Shivers and Sieg 1980, Gwatkin et al. 1980, Wood et al. 1981, Wood and Dunbar 1981, Tsunoda et al. 1982, Yurewicz et al. 1983). Actually, methods have been developed more recently, which allow the large-scale isolation of porcine (Dunbar et al. 1980, Dunbar and Raynor 1980, Wood et al. 1981, Sacco 1981, Yurewicz et al. 1983), cow (Gwatkin et al. 1980), and rabbit zonae pellucidae (Wood et al. 1981). Wood et al. (1981) have designed an apparatus for easily rupturing ovarian follicles en masse, leaving the ovary intact. Eggs used for preparing the zona antigens are being obtained from macerated ovaries or from oviducts after hormonally induced ovulation. The results of all these investigations have suggested that the presence of antizona antibodies in serum can be assessed by the ability of the antibodies (1) to agglutinate zona-coated eggs, (2) to inhibit zona digestion by proteolytic enzymes generally highly effective in dissolving zonae, (3) to localize on zonae in immunofluorescence tests, (4) to change the light-scattering property of zonae by forming a precipitate on the zona surface, (5) to inhibit sperm attachment and penetration of zonae, and (6) to inhibit embryo escape from the zonae at implantation (see Shivers 1979). These tests for the determination of antizona antibodies have been employed successfully with antiserum to ovary and with antiserum to isolated zonae. The most sensitivity test appears to be prevention of sperm attachment to zona and thus inhibition of fertilization, as will be discussed here. Both passive and active immunization, with zonae as the target, have been worked out as methods of regulating fertility in several mammals (Shivers 1979, Shivers and Sieg 1980, Tsunoda et al. 1982). Passive immunization appears to be more effective in fertility regulation. In passive immunization antizona antibodies are produced in a heterologous species of mammals, generally rabbits, and the antiserum is collected and tested in a variety of ways prior to immunization. Active immunization with zona antigens from homologous species is relatively less effective in reducing fertility (see Shivers 1979).

Experiments using the ovary-specific antiserum have been conducted to determine the effects of zona pellucida antibodies on the sperm–egg interactions and fertilization (Shivers 1979, Shivers and Sieg 1980, Tsunoda et al. 1980). Treatment of zona-coated eggs of hamster and rabbit with the absorbed antiserum produce egg agglutination. Exposure of superovulated hamster or rabbit ova to their respective absorbed anti-ovary antiserum, leads to the formation of a precipitation layer on the zona pellucida (Ownby and Shivers 1972, Sacco and Shivers 1973c). This precipitate is found to alter the light-scattering property of zona surface and is believed to be due to formation of the precipitate resulting from antibody reacting with complimentary antigen/s in the zona (see also Gwatkin 1979, Fléchon and Gwatkin 1980, Shivers et al. 1981). The precipitate is seen within a few minutes after incubating the egg in antiserum, and is never seen in the zonae of eggs treated with control serum. The precipitating effects of the antiserum can be neutralized by absorption with ovary- or zona-coated eggs, but not with any somatic tissue tested. Formation of such precipitate has been reported by various workers, with eggs of different mammalian species: golden hamster (Garavagno et al. 1974, Dudkiewicz et al. 1975, 1976, Oikawa and Yanagimachi 1975, Tsunoda and Chang 1976b), mouse (Jilek and Pavlok 1975, Gwatkin et al. 1976), rat (Tsunoda and Chang 1976a), pig (Sacco and Palm 1977), human (Shivers and Dunbar 1977), dog (Shivers et al. 1981), and other mammalian species (Gwatkin 1979, 1982, Shivers 1979, Gwatkin et al. 1980, Shivers and Sieg 1980, Tsunoda et al. 1981 b, 1982). The precipitation is seen near the outer surface of the zona in each of the species of animals tested, suggesting that the active sites of the antigen(s) are present at the external surfaces (Shivers 1979, Shivers and Sieg 1980, Shivers et al. 1981, Tsunoda et al. 1982). The fact that zona-specific antigens and sperm receptors are restricted primarily to zona surface suggests that they may be the same substance (Shivers 1979, Shivers and Sieg 1980). Further evidence for the external localization of antigens is provided by the fact that when the inner surface of the zona is exposed to the antibodies following mechanical rupture of the egg, no precipitate is seen on it (Garavagno et al. 1974). This could be due simply to there being less surface area on the inner surface (fewer pores) to form Ag-Ab matrix. The presence of the antigenic sites at the outer surface of the zona also coincides with the lectin-binding sites (Ownby and Shivers 1972). The uniform labelling in thin sections has suggested that the zona pellucida of cow blastocysts is homogenous antigenically (Fléchon and Gwatkin 1980). Heavy labelling of the inner and outer surfaces of the zona pellucida in thick sections appears to be due to the greater porosity of these regions in which the zona material becomes highly dispersed, or even partly solubilized, thereby permitting the formation of an antigen- antibody matrix.

The precipitation layer is observed to render the zona pellucida resistant to lysis by the proteolytic enzymes trypsin and pronase, which are usually highly effective in dissolving the zona pellucida (Ownby and Shrivers 1972, Shivers 1979). This is interesting observation since acrosin, the alleged zona-penetrating enzyme of the sperm, is trypsin-like in its activity (Gwatkin 1976, Guraya 1986). The enzymatic functions have been postulated as being involved in sperm attachment and penetration through the zona, block to polyspermy and zona shedding (Pikó 1969, Gwatkin 1976, Metz 1978, Yanagimachi 1978, Guraya 1982a). All these studies have indicated that the formation of a precipitate following exposure of zona to antibody is the standard method for determining the presence in serum of antizona antibodies (see also Fléchon and Gwatkin 1980).

The importance of sperm recognition and attachment sites on the surface of zona as being essential to fertilization has been recognized in various recent studies (Gwatkin 1976, Metz 1978, Yanagimachi 1978). The more significant effects of zona-precipitating antibodies appear to be the prevention of the attachment and penetration of sperm through zona (Shivers 1979). When hamster eggs are exposed to the antibodies, and exposed in vitro to capacitated sperms, only a few are seen attached to the zona although many are found to collide with the precipitated surface; they show normal motility. However, the sperms are able to attach to and penetrate through the antibody-treated zona-free eggs. The treatment of hamster (Shivers et al. 1972, Garavagno et al. 1974), and mouse (Jilek and Pavlok 1975) oviducal eggs, in vitro with zona-precipitating antibody prior to insemination with capacitated sperm, inhibits the sperms' attachment to the zona surface and their subsequent penetration into the perivitelline space. Later workers have also observed that passive immunization of female mice, post coitum, with the antibody prevents adhesion of sperm to zona (see details in Shivers 1979, Shivers and Sieg 1980, Tsunoda et al. 1982). Swenson and Dunbar (1982) have suggested that spermatozoa can associate with the zona in a nonspecific, as well as a specific manner. The use of heterologous systems in addition to systems in sperm-binding assays may be useful in distinguishing these two types of interaction.

Gwatkin et al. (1977) observed that when female mice are hyperimmunized they become infertile, but without reimmunization they become fertile again after approximately 3 months, as antibody levels are declined. This provided the first convincing evidence that a complete loss of fertility can be produced by active immunization with isolated zonae. The results of some studies have suggested that active immunization with zona antigens from homologous species is relatively less effective than the passive immunization in reducing fertility (Shivers 1979, Shivers and Sieg 1980). But when the zona are from a heterologous species, active immunization is as effective as passive immunization. This is supported by the fact that active immmunization of female mice with solubilized hamster zonae (Gwatkin et al. 1977), and of female rabbits with solubilized bovine (Gwatkin and Williams 1978a) and pig zonae (Gwatkin et al. 1980), and of female rats with cumulus-free mouse ova (Aitken and Richardson 1981) produces infertility. It also suggests that two species must have cross-reacting zonae antigens. Cross-reactions described between pig and human zonae (Sacco 1977a, Shivers and Dunbar 1977), and between pig, marmoset, chimpanzee and human zonae (Shivers et al. 1978) are significant because active immunization can now be tried in nonhuman primates with pig zonae. Wood et al. (1981) have reported that the effects of active immunization with procine zonae on the fertility of rabbits is dependent upon the means used to solubilize the zona prior to immunization.

More recent studies have further confirmed that the antibody to zona pellucida inhibits fertilization in vitro and/or in vivo in the mouse (Tsunoda and Sugie 1977, 1979a, Tsunoda and Chang 1978a, Sacco 1979, Tsunoda et al. 1981a). The inhibitory effects on fertilization after passive immunization with anti-egg and antizona sera last for at least 25 to 30 days. Reimners et al. (1977) have suggested that passive immunization with rabbit antiserum raised by immunizing with adjuvants alone inhibits pregnancy in the mouse. Tsunoda and Sugie (1979a) have made a quantitative determination of antizona antibody in connection with inhibitory effect on fertilization in the mouse. Tsunoda et al. (1981a) observed that bivalent (IgG) rabbit antimouse zona pellucida antibody completely prevents fertilization in the mouse. Univalent (Fab) zona antibody does not inhibit fertilizability of eggs (see also Ahuja and Tzartos 1981), treatment or injection with antirabbit IgG serum prevents the fertilizability of eggs pretreated with antiserum Fab or eggs from females injected with antiserum Fab. These results suggest that inhibition of fertilization by bivalent mouse zona antibody may be dependent on the possibility that the antisera are directed against the zona as a whole, and not specifically against the receptors for the sperm. This possibility needs to be confirmed in future studies. Gwatkin (1982) has suggested that fertilization is not blocked by a direct interaction of antibody with the sperm receptors, but by the formation of a relatively thick matrix that prevents access of the sperm to the receptor sites. On this basis one would predict that univalent (Fab) antibody will not block fertilization, since a matrix cannot form. Although by utilizing a binding test of F (ab)$_2$ antizona fragments to zona, Sacco (1978) has ruled out the possibility of reaction via the Fc fragments; it is not known whether bivalent Fab lacking the Fc fragment inhibits fertilization as effectively as nondigested IgG. But Tsunoda and Sugie (1982) have produced an antiserum in one rabbit against mouse zonae pellucidae solubilized with 70 mM Na$_2$So$_3$, 1% SDS, and 0.04 mM Cu So$_4$. An IgG of zona antibody completely prevents fertilization both in vitro and in vivo in the mouse. F(ab)$_2$ fragment obtained by pepsin digestion of IgG zona antibody inhibits fertilizability of eggs in vitro, but does not inhibit fertilitation in vivo after passive immunization.

According to Tsunoda et al. (1981 c) passive immunization with antiserum prepared against isolated bovine zonae pellucidae prevents fertilization in the cow. The minimum dose of antiserum (titer 27 by immunoflorescence) needed for completed inhibition is 2 ml/kg of body weight.

A number of other workers have observed that passive immunization with hetero-antiserum, prepared in rabbits against the supernatants of whole ovary homogenates and rendered partly specific for zona pellucida by absorption, can inhibit fertilization both in vitro and in vivo (Ownby and Shivers 1972, Jilek and Pavlok 1975, Oikawa and Yanagimachi 1975, Tsunoda and Chang 1976a, b, c, Yanagimachi et al. 1976, Tsunoda and Sugie 1979b, Shivers 1979, Shivers and Sieg 1980, Tsunoda et al. 1980, 1982). The inhibition of fertilization in all cases is presumably due to the antigenicity of the zona pellucida in the ovary. Tsunoda and Sugie (1979b) have clearly shown that antibody to zona pellucida inhibits fertilization and the inhibition of fertilization in vitro and in vivo is mainly due to a reaction between zona antibody and the zona pellucida. The antifertilization effect of the antibody in all these in vivo and in vitro experiments is probably due to the masking of the essential sperm receptor sites on the zona surface, since fertilization occurs in eggs exposed to the antibody after sperm have attached and passed through the zona surface (Garavagno et al. 1974). In addition, the precipating antibody has no effect on sperm penetration in zona-free eggs, nor can the inhibitory action of the antibody be neutralized by absorbing with somatic tissues.

Antigen(s) of mouse zona pellucida are apparently not strain-specific as antiserum to eggs, and isolated zonae pellucidae of mice can induce a precipitate on the zona of ICR, C3H and CDI mice, and these antisera also inhibit fertilization in vitro and in vivo in these three strains of mice (Tsunoda and Chang 1978b, Tsunoda and Sugie 1979a, Tsunoda et al. 1982).

Jilek and Pavlok (1975), using rabbit antimouse ovary antiserum, rendered specific to zona by absorption, have demonstrated that following passive immunization of mice antibodies associated with zona are detectable on ova aspirated from mature follicles and on ova following ovulation, thus indicating that antizona serum can reach the zona pellucida in situ; and ova obtained from the mated mice are nonfertilizable, thus reaffirming the contraceptive action of antizona antiserum. Similar experiments with antihamster ovary antiserum have also demonstrated block of fertlization of hamster eggs in vitro and in vivo, as the intraperitoneally injected antibody is able to render female mice infertile for approximately three oestrous cycles (Oikawa and Yanagimachi 1975). Yanagimachi et al. (1976) and Tsunoda and Chang (1976a, b) demonstrated that antiserum produced against homogenate of rat ovary has also inhibitory effect on fertility.

Sacco (1977a) demonstrated that the human zona pellucida is also immunogenic and contains specific antigens, since antiserum produced against human ovarian homogenates and rendered specific to the ovary by absorption with human liver and plasma still shows antibody activity. Bovine zona has been observed to be cross-reactive with human zona (Gwatkin 1978a, 1979, 1982). The cross-reaction of zona antibody has also been reported among several species of mammals by using indirect immunofluorescence, as already described. Similar observations have also been made by Mori et al. (1979) for pig ovary. Contraceptive action of absorbed anticanine ovary antiserum has been revealed (Mahi and Yanagimachi 1979). Shivers et al. (1981) observed that pregnancy can be prevented in the dog while a high antizona antibody titer is present. This suggests that vaccination with zona components may eventually be a practical means for controlling reproduction in canines and humans. Trounson et al. (1980) recently made a study of inhibition of sperm binding and fertilization of human ova by antibody to porcine zona pellucida and human sera. The results obtained indicated that hetero-antibody to porcine zona completely inhibits human fertilization in vitro and sera positive for antizona in some women may also inhibit or reduce the possibility of fertilization. These studies suggested that pig zona is a suitable antigen for contraceptive purpose in women, since antipig zona serum reacts with human zonae and pig zona can be easily obtained from slaughter houses (Shivers 1977, Gwatkin 1982). Furthermore, both pig and cattle zona antigens have been purified by various chemical methods (Sacco and Palm 1977, Gwatkin 1979, 1982, Menino and Wright 1979, Dunbar et al. 1980, Gwatkin et al. 1980, Dunbar and Raynor 1980). Keeping this in mind, Tsunoda et al. (1981b) tested the effects of goat antiserum against isolated pig zonae pellucidae on fertilization in vivo in several mammals, such as the pig, cow, sheep, rabbit, rat, and mouse. As revealed by indirect immunofluorescence, antipig zona serum reacts strongly with the zonae of pig, cow, sheep, and rabbit, but the reaction with the zonae of mouse and rat is weak. Passive immunization with antipig zona serum significantly, or completely, inhibited fertilization in all species. However, inhibition of fertilization is more pronounced in the pig, cow, sheep, rabbit, and mouse than in the rat. All of the zonae obtained from the pig, cow, sheep, rat, and mouse after passive immunization with antipig zona serum show strong fluorescence regardless of the incidence of fertilization. These observations indicate that the pig and other mammalian zonae pellucidae tested have tissue-specific antigens. The results of various studies, as discussed here, have indicated that whole antiserum or purified antibody to isolated or heat solubilized zonae pellucidae inhibits fertilization in vitro and/or in vivo in all mammalian species tested so far (see also Aitken et al. 1981). Cross-reaction of antizona antibody with the eggs of subhuman primates also suggests that pig zona material holds promise as a contraceptive vaccine. The pig ovary may also be a superior source of both vaccine and sperm receptor for further study, since it yields four times as many zonae as the calf ovary and, by the procedure outlined by Gwatkin et al. (1980), can be isolated as a single glycoprotein. The information obtained so far has revealed conspicuous differences in the zonae of different species of mammals. Future investigations on the biological roles of zonae as well as studies on the chemical and immunological properties of zonae must take these species variation into consideration. Direct methods such as radioimmunoassay and immunoelectrophoresis, must be used to show the immunological cross-reactivity of the zonae antigens of different species (Sacco et al. 1981 b). Rodent species (rat and mice) are not adequate models for studying the effects of antizona antibodies on human fertility.

Shivers (1976, 1979) and Shivers and Sieg (1980) have suggested that the other possible site for antifertility action using antibodies to the zona is implantation. Several factors are believed to be involved in the embryo's escape from the zona. A hatching process, whereby the embryo escapes from the ruptured zona, and lytic factors (blastocyst and uterine derived) have been reported (Dickman 1969, McLaren 1969, 1970). Antibody binding to the zona appears to inhibit either of the two events from freeing the embryo for implantation. Mouse embryos treated at the two-to-four cell stage to zona-precipitating antibodies do not shed their zonae following in vitro culture even through the embryos apparently develop normally (Shivers 1974). Dudkiewicz et al. (1975) also observed that hamster embryos in the morula and blastocyst stage treated in vitro with zona-precipitating antibody and subsequently transferred to pseudopregnant recipients also do not escape from the zonae. These observations suggested that antibody binding interferes with the process of zona shedding and therefore inhibits implantation. Interference in zona shedding may also explain the reduction in fertility following passive immunization of female mice with sera containing antibodies (see references in Shivers and Sieg 1980). However, Tsunoda and Chang (1978 b) have observed that the treatment of eight-celled mouse embryos with anti-egg or antizona serum slightly inhibits zona shedding in vitro. In all the studies conducted so far, whole serum containing antibody or antibody activity to zona has been used. Such antisera are believed to the contaminated by other proteins. In order to overcome this problem of contamination the effects of purified antibody (IgG) to mechanically isolated or detergent-solubilized zonae pellucidae on the development and implantation of mouse embryos have been studied by Tsunoda and Whittingham (1982) in vitro by culture and utero by transfer of treated embryos. The results of this study have indicated that treatment with purified zona antibody (IgG) does not inhibit the development and implantation of mouse embryos in vitro or in vivo. Labelling of the zona pellucida of cow blastocysts with zona-specific antiserum has suggested that antigenicity is unaffected by abnormal cleavage, in vitro culture, or frozen storage (Fléchon and Gwatkin 1980). There is also apparently no loss of antigenic material from the zona up to the late blastocyst, just before implantation.

From the discussion of various observations made so far it can be concluded that antibodies to the zona act at two sites for inhibiting fertility: one by inhibiting sperm attachment and penetration through the zona provided the egg is exposed to antibody before fertilization, and second, possibly preventing implantation by interfering with zona shedding if fertilization does occur (see Shivers 1979, Shivers and Sieg 1980). But the latter possibility needs to be established in the light of observations made by Tsunoda and Whittingham (1982). A distinct advantage of immunization against the zona, or its components, is that fertility is blocked in an all-or-none fashion, so there is no risk of abortion or stillbirth as the titres fall (Gwatkin 1982).

C. Morphological and Physiological Interrelationships Between the Ganulosa Cells and Growing Oocyte

From an early phase of growth, oocytes are metabolically dependent upon the surrounding granulosa cells. Therefore, the granulosa cells and oocyte during follicular growth develop complex morphological associations (Fig. 13) which are indicative of some important interrelationships. Corresponding to the formation of the zona pellucida during follicle growth, the oocyte develops numerous microvilli mostly clustered around persisting portions of granulosa cell cytoplasm, while the granulosa cells exhibit a few slender cytoplasmic processes which traverse the areas of zona pellucida deposition to anchor on the oocyte surface by means of desmosomes (Fig. 23) (Hertig and Barton 1973, Zamboni 1972, 1974, 1976, 1980, Guraya 1974a, Szöllösi 1975a, b, Albertini and Anderson 1974, Anderson and Albertini 1976, Amsterdam et al. 1976, Nowacki 1977, Gilula 1977, Gilula et al. 1978, Anderson 1979, Dvořák and Tesařík 1980, Heller et al. 1981, Tesoriero 1981). The number and extension of microvilli increase considerably in subsequent stages of follicular enlargement. They are particularly developed in the areas next to the follicle cell processes. Tracer and freeze-fracture electron microscopy of the ovaries of neonatal rat and adult mouse, rat, rabbit and *Macaca mulatta* has demonstrated the presence of desmosomes and gap junctions between granulosa and oocytes, as also demonstrated by Anderson et al. (1978) in the mouse and rabbit, and by Heller et al. (1981) in the mouse (see also Brower and Schultz 1982a). The junctional connections are seen at the

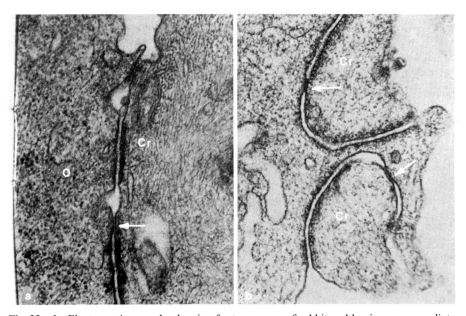

Fig. 23a, b. Electron micrographs showing foot processes of rabbit and bovine corona radiata cells *(Cr)* respectively, which form small gap junctions with the oocyte *(O)* cell membrane *(arrow)*. The adjacent felt-like material may correspond to an intermediate junction (From Szöllösi et al. 1978)

ends of granulosa cell projections which traverse the zona pellucida and terminate upon microvilli and evenly contoured nonmicrovillar regions of the oolemma (Fletcher 1979, Tesoriero 1981). Gap junctions are often seen associated with a macula adherens type of junction. The gap junctions occasionally consist of minute ovoid plaques but more frequently appear as rectilinear single- or multiple-row aggregates of particles on the P-face or pits on the E-face. Fletcher (1979) has suggested that these heterocellular gap junctions are formed during the initial apposition of granulosa and egg. Freeze-fracture has revealed that they occur at least in two forms. One consists of short straight or curved rows of 7- to 8-nm P-face particles. The other is a small macular aggregate of 7- to 9-nm particles – like the gap junctions seen elsewhere in follicles. Szöllösi et al. (1978) have observed that button-like terminals of the corona cells are attached to the oocyte by a further junctional complex which is morphologically of the intermediate junction type (macula adherence) and may serve a structural role (Fig. 23). The gap junctions between follicle cells and oocyte are focal, probably small discs, measuring between 100–200 nm in diameter. That molecular exchange is possible across these junctions has been demonstrated recently by injecting fluorescein into rat oocytes and following its spread, little by little, into the surrounding cumulus cells (Epstein et al. 1976). In all these studies, the functional significance of granulosa cell–oocyte gap junctions is discussed in relation to the regulation of passage of nutrients to the egg, meiosis and luteinization.

At the end of the process of zona pellucida deposition, the oocyte surface shows the presence of numerous regularly distributed microvilli, and of a large number of micropapillae (Pedersen and Seidel 1972). The latter having a high intrinsic electron opacity appear as hemispherical, lens-shaped on the portion of the oolemma which covers the microvilli (Zamboni 1974, 1976, 1980, van Blerkom 1980). The micropapillae, just like the microvilli, increase in number with advancing oocyte maturation and are distributed at random on the oolemma and its microvillosities. Their functional significance in dictyate oocytes is not known. Zamboni (1974, 1980) has suggested that micropapillae play a role in enhancing the adhesiveness of the oocyte plasma membrane and microvilli to the follicle cell processes, as also suggested by Van Blerkom (1980).

In the later stages of oocyte growth, a multitude of complex cytoplasmic processes develop from corona radiata cells, which traverse the zona pellucida to reach the oocyte surface, or even to penetrate for various distances into the ooplasm (Fig. 24c) (see Hope 1965, Zamboni 1974, 1976, 1980, Motta and van Blerkom 1974, Dvořák and Tesařík 1980, Tesoriero 1981, Apkarian and Curtis 1981). In the latter cases, the invaginated portions of the oolemma are characterized by the presence of numerous coated pinocytotic vesicles (Fig. 24c) (see also González-Santander and Clavero Nunég 1973). The processes of corona radiata cells become much more elongated and slender, and assume dentritic features including the tendency to branch repeatedly. These various morphological specializations of cell surface of corona radiata cells and oocyte have been interpreted to indicate an increase in the absorptive surface area during oocyte growth.

Cytoplasmic organelles such as mitochondria and endoplasmic reticulum cisternae are only occasionally seen in the cytoplasm of granulosa cell process; more consistent components include cytoplasmic filaments, lipids and lysosome-like bodies (Fig. 24a, b) (Zamboni 1976, 1980). The filament traverse the whole length of the granulosa cells and are oriented parallel to their long axis. They do not show features typical of myofilaments and thus form cytoskeletal filaments (Zamboni 1974, 1980), which may be needed for cytoplasmic plasticity and contractility. The ultrastructural features of granulosa cell processes, including their irregular profiles and cytoplasmic filaments, suggest that in the living condition they are continuously contracted or expanded, resulting in frequent multiple piercing of the zona pellucida. This suggestion is supported by the observation that the zonae pellucidae in growing follicles generally show numerous defects or areas of rarefaction seen in the form of empty channels whose outlines and orientations are similar to those of the granulosa cell processes.

Amsterdam et al. (1977) observed that large rat oocytes stained with either anti-actin or antismooth muscle myosin show a thin fluorescent band just beneath the zona pellucida, suggesting that actin and myosin are associated with the oolemma. The functional significance of these contractile elements in the oocyte surface remains to be elucidated. They may be playing some role in the extrusion or release of cortical granules during fertilization, as suggested for the contractile elements demonstrated in the egg cortex of different animal species (Guraya 1982a).

Corona radiata cells and other granulosa cells apparently contribute phospholipid bodies, glycogen, amino acids, and possibly some ribonucleoproteins to the growing oocyte (see Baker 1970, Guraya 1965, 1969b, 1970b, 1973a, 1974a, Moor and Smith

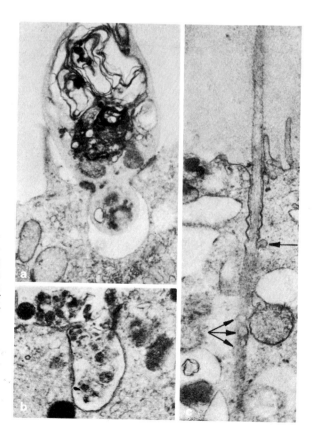

Fig. 24 a and b. Electron micrographs showing portions of follicle cell cytoplasm containing lysosome-like bodies and whorls of membranes (myelin-figures) apparently in the process of being engulfed in the ooplasm of rhesus monkey oocyte by phagocytosis; **c** Follicle cell process penetrating deeply into human oocyte. Note the numerous pinocytotic vesicles *(arrows)* along the invaginated portions of the oocyte membrane (From Zamboni 1976)

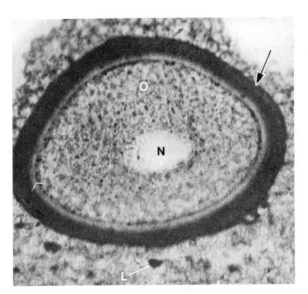

Fig. 25. Photomicrograph showing fully grown oocyte from the antral follicle of the rhesus monkey ovary. Note the accumulation of highly sudanophilic lipids *(arrow)* in the outer portion of the zona pellucida. Cumulus oophorus shows sparsely scattered sudanophilic lipid bodies *(L)* of variable size. Sudanophilic lipid bodies are also seen in the oocyte *(O)*. Such lipids are not seen in the nucleus *(N)* of oocyte (From Guraya 1969 b)

1979, Colonna and Mangia 1983). This suggestion is supported by the fact that such components have been demonstrated in the cumulus cells (Fig. 25). Colonna and Mangia (1983) demonstrated the nutritional role of mouse granulosa cells in antral dictyate mouse oocytes by measuring the transfer of different amino acids through gap junctional channels between somatic and germ cells. When present in the incubation medium at concentrations resembling in vivo conditions, glycine, alanine, proline, serine, tyrosine, glutamic acid and lysine entered cumulus-enclosed oocytes cooperatively, while valine, leucine and phenylalanine did not. However, cooperative uptake of leucine and phenylalanine was observed at higher external precursor concentrations. These results clearly support the suggestion that in vivo antral oocytes depend on surrounding granulosa cells for amino acid uptake which appears to be dependent on precursor concentrations in the coupled cells. According to Cran et al. (1979), in the cumulus of sheep follicle, glycogen is associated with electron-lucent areas, whereas in the membrana granulosa it is invariably associated with membranes. In large antral follicles large membrane-bound bodies are present in the basal cells of the cumulus. At late oestrus numerous lipid droplets are formed in the cumulus and membrana granulosa, which may be indicative of incipient luteinization.

The exact mechanisms of the transfer of various substances from corona cells into the oocyte are still unknown. Standard membrane uptake mechanisms provide the major route of entry of amino acids and small ions into the oocyte. In this regard, diffusion, pinocytosis, phagocytosis and phosphatases are believed to play significant roles, since there is no direct cytoplasmic continuity between granulosa cell processes and oocyte (Fig. 13 and 23) (González-Santander and Clavero Nunéz 1973, Guraya 1974a, Nowacki 1977, Dvořák and Tesařík 1980). Zamboni (1974, 1976, 1980) observed that the cytoplasmic processes of corona cells, which contain lipid droplets and pleomorphic bodies with morphologic features reminiscent of lysosomes, traverse the zona pellucida and terminate in the perivitelline space or penetrate into the ooplasm of the rhesus oocyte (Fig. 24a und b). Lipids appear in the form of prominent, intensely osmiophilic droplets, which are especially abundant in the terminations of the granulosa cell processes at the oocyte membrane, even though they may also be present in the intrazonal segments of these processes. This can be best illustrated in light microscopic preparations of growing oocytes which show a constellation of sudanophilic lipids throughout the thickness of the zonae pellucidae (Fig. 25). Most of the granulosa cell processes containing lipids, which terminate in the perivitelline space, undergo degeneration and are subsequently engulfed by the ooplasm in a manner not dissimilar from phagocytosis (Zamboni 1974, 1976, 1980). Observations on infiltration of lipid granules into the oocyte (Fig. 19), as first reported by Guraya (1965, 1969b, c, 1970b) are thus substantiated by the electron microscope studies of Zamboni (1974). The engulfed lipid-laden portions of the granulosa cell processes subsequently appear to undergo enzymatic degradation within the walls of phagocytic vacuoles (Fig. 24a, b). The nature of enzymes involved in degradation needs to be determined.

The lysosome-like bodies appear as membrane-bound, irregular structures consisting of pleomorphic and variously aggregated material of usually high electron density which frequently includes prominent concentric arrays of membranes; just like the lipid droplets, these bodies are also present in the dilated extremities of the granulosa cell processes at the oocyte surface. The functional significance of these components is unknown at present. They are believed to play a role contributing proteinaceous and lipid material to the oocyte as these bodies are also incorporated in the ooplasm by phagocytosis (Zamboni 1976, 1980). The lysosome-like bodies seen at or near the oocyte surface or in the granulosa cell processes are also suggested to represent catabolites of oocytic origin which have been expelled in the perivitelline space and picked up by the cytoplasmic processes of the granulosa cells; as such, they could represent senescent or otherwise degenerating ooplasmic components (cytosegregosomes) earmarked for extracellular disposal.

These various observations have clearly suggested that the granulosa cell processes function to provide isolated oocytes with nutrients required for their metabolic needs, including their growth as the oocytes become isolated within an expanded follicle cavity and cut off from the circulation in the ovarian stroma by a liquor-filled antrum and a follicle wall consisting of multicellular layers. Cross and Brinster (1974) reported that the presence of granulosa cells significantly alters leucine fluxes in oocytes. The results of this study indicated that granulosa cells enhance entry during the germinal vesicle stage, depress entry at the first metaphase division, and are without effect at metaphase II. The reverse observations were, however, made in experiments using the nonmetabolizable amino acid, cycloleucine, which shares a common entry system with methionine and phenylalanine (Moor and Smith 1979). The involvement of granulosa cells in facilitating the entry of amino acids into oocytes remains to be determined more precisely (Colonna and Mangia 1983). By contrast, granulosa cells play an essential role in providing a means of entry for certain nucleotides and lipid precursors. This is supported by the fact that there is practically no entry of uridine, guanosine, choline and inositol into oocytes devoid of surrounding granulosa cells, this showing an absence of carrier-mediated membrane uptake system for these compounds. The rate of entry is high in the presence of granulosa cells, but only when the structural contact between the somatic cells and oocyte remains intact (Wassarman and Letourneau 1976a, Heller and Schultz 1980, Moor et al 1980).

Szöllösi et al. (1978) have studied permeability of the ovarian follicle with special reference to corona cell–oocyte relationship in mammals. Graafian follicles incubated in toto with electron-dense markers, such as horse-radish peroxidase and lanthanum nitrate are penetrated freely by the tracers through the granulosa cells and the cumulus oophorus, reaching the oocyte. This indicates that no blood-follicle barrier exists. A series of control experiments, using medium conditioned by granulosa cells and co-cultures of denuded oocytes and granulosa cells in which physical contact between the two-cell types is not allowed, have indicated that contact between follicle cells and the large contribution of nutrients apparently provided to the oocyte by the granulosa cells are consistent with the concept that gap junctions-mediated metabolic cooperativity between follicle cells and their enclosed oocytes is vital for mammalian oocyte growth (Heller and Schultz 1980, Heller et al. 1981). Heller et al. (1981) made a biochemical study of metabolic cooperativity between granulosa cells and growing mouse oocytes (see also Heller and Schultz 1980). The various lines of experimentation used have suggested that gap junctions are functional and that in most cases studied 85% of the metabolites present in follicle-enclosed oocytes contain more intracellular radioactivity than do oocytes with no attached gap junctions (denuded oocytes). Treatments known to disrupt gap junctions in other cell types are effective in reversibly uncoupling metabolic cooperativity between gap junctions and oocytes. Metabolic cooperativity between granulosa cells and oocytes has also been suggested in other recent studies (Brower and Schultz 1982 a, Pivko et al. 1982, Colonna and Mangia 1983). Pivko et al. (1982), after studying autoradiographic topography of tritium-labelled fucose incorporation in pig oocytes cultured in vitro, have observed a relatively intense glycoprotein synthesis in oocytes and cumulus cells during resumption of meiosis, at least before germinal vesicle breakdown. Metabolic cooperation may occur as long as oocyte and cumulus cells keep membrane junctions.

The transfer of essential metabolites from granulosa cells to the oocyte, as demonstrated in various studies, underlines the functional dependence of the oocyte on the somatic elements in the follicle for its survival during the prolonged period of meiotic arrest. Permeable junctions between the granulosa cells and oocyte provide the channels through which metabolites enter into the oocyte cytoplasm. These inter-communicating channels develop at an early stage in follicular development and persist despite the deposition of the zona pellucida around the oocyte, as already discussed. Intercellular coupling between the oocyte and granulosa cells remains constant until just before ovulation when it declines, as will be discussed in Chap. IV. This decline is induced primarily by FSH (Moor et al. 1980), is confined to coupling between somatic cells and oocytes, and does not affect somatic–somatic cell coupling (Cran et al. 1981).

The ingested food materials and organelles at some stages of oocyte growth might be digested with the help of hydrolytic enzymes of primary lysosomes derived from the peripherally placed Golgi complexes (Anderson 1972). This suggests that the Golgi complex is associated with the formation of lysosomes that are involved in the intracellular digestion of nutritive material. The lysosomes of Golgi origin may also participate in the necrosis of oocytes and follicle cells (Guraya 1973 b), as will also be discussed in Chap. VIII. The peripheral location of various organelles in the developing oocyte also suggests that this cell becomes equipped for the absorption, utilization and intracellular transport of materials delivered to its surface membrane by the corona cells. Moor and Smith (1979), while studying uptake of [^{14}C]-labelled amino acid by the mammalian oocytes, have observed that the proportion of sheep oocytes with a functional amino acid transport system (i.e. cycloleucine flux; nmol $cm^{-2}h^{-1}$) is highest in pre-ovulatory follicles (97%), and lowest in atretic follicles (59%). Amino acid fluxes in functional germinal vesicle oocytes are similar at all stages of development studied. Amino acid uptake by mouse oocytes is approximately half that measured in sheep oocytes.

Eppig (1977), studying mouse oocyte development in vitro, has used four systems for the culture of oocytes, either alone or in association with granulosa cells; isolated oocyte culture; isolated oocyte-ovarian cell culture; isolated follicle culture and ovarian organ culture. Oocytes grown in isolated follicles culture under defined conditions and in the absence of gonadotropins resemble oocytes grown in vivo in terms of their ultrastructural characteristics, with exception of enlarged mitochondria. In addition, these oocytes show normal functional characteristics in terms of their increased levels of CO_2 evolution from exogenous pyruvate, and the ability of the fully grown oocytes to start meiotic maturation when free from granulosa cells. The results obtained indicate that association of granulosa cells and oocytes is necessary for oocyte growth. Isolated oocytes in culture with ovarian cells fail to grow. Addition of FSH or oestradiol to the cultures fail to promote oocyte growth or delay oocyte degeneration. The results of this study also indicate that under the culture conditions used granulosa cells must be in contact with the oocyte, perhaps by means of specialized cell junctions, for oocyte growth to occur.

In addition to supplying essential nutrients to the growing oocyte, the granulosa cells also play a significant role in supporting protein synthesis within the oocyte. Very conspicuous changes are induced in the patterns of protein synthesis by the removal of granulosa cells from the oocyte. The most conspicuous effect of denudation is a 15-fold decrease in the synthesis of a 43,000 MW band distinguished by co-migration studies as actin (Osborn and Moor 1981). But it needs to be determined whether the actin is directly transferred into the oocyte by endocytosis, or the granulosa cells provide signals for the direct synthesis of actin in the oocyte.

In summary, it can be stated that the follicular epithelium (or granulosa cells) provides the basic metabolic needs of the growing oocyte, regulates the exchange of substances between the extrafollicular and intrafollicular compartments and regulates the internal stability of the environment where oocyte growth and differentiation occur. Thus granulosa cells support the oocyte in a number of ways during the period of meiotic arrest.

D. Follicle Wall

The follicle wall consists of granulosa and thecal layer (Fig. 26), which are separated by the basement membrane (or lamina). During follicle growth both the granulosa and theca cells undergo remarkable developmental changes (Fig. 11) related to the organogenesis of an endocrine/exocrine gland that will synthesize androgenic and oestrogenic hormones and secrete an ovum at ovulation. This complex process of growth and differentiation is closely accompanied by series of sequential intracellular molecular mechanisms which are regulated by a set of hormonal stimuli. Recent studies have revealed that follicle growth proceeds through a process of cell

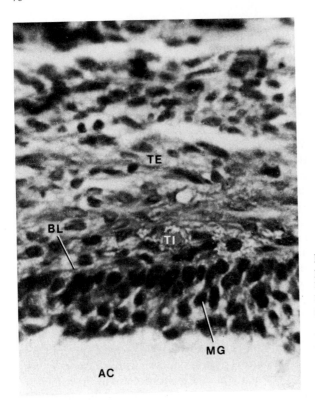

Fig. 26. Photomicrograph of portion of Graafian follicle from the ovary of small Indian mongoose, showing membrana granulosa *(MG)*, basal lamina *(BL)*, theca interna *(TI)* and theca externa *(TE)*. Cellular theca interna contains numerous blood capillaries. A portion of antral cavity *(AC)* is also seen

proliferation and fluid accumulation (Fig. 11), and follicle maturation proceeds through a process of hormone-dependent cytodifferentiation.

Recent biochemical studies have revealed that the follicles can secrete various steroid hormones including progestins (progesterone, and 17 α hydroxyprogesterone), androgens (androstenedione and testosterone) and oestrogens (oestrone and 17 β-oestradiol) which have been identified in fluid obtained directly from follicles or in ovarian venous blood (Peters and McNatty 1980, Greenwald 1980, Murdoch and Dunn 1982), or in vitro experiments (Mills 1979, Armstrong 1980, Janson et al. 1982, Hubbard and Greenwald 1982). In their quality and quantity, these steroid hormones vary with the growth and differentiation of the follicle (Yoshinaga 1978, Bahr et al. 1979, Shahabi et al. 1979, Greenwald 1980, Ermini and Carenza 1980, Erickson 1982, Hubbard and Greenwald 1982). The secretion of hormones by the pre-ovulatory ovarian follicle is a complex developmental process which integrates several modalities of endocrine regulation. It is still not known when the follicle starts secreting oestrogen as there is so little direct evidence on this issue. It is axiomatic that the antral follicle is the principal source of the hormone, at least in nonprimate species (Greenwald 1979). However, Hubbard and Greenwald (1982) have provided the first direct evidence that pre-antral follicles of the cyclic hamster produce oestrogen, albeit in small amounts. But it needs to be determined more precisely whether earlier pre-antral stages secrete oestrogens for either local or distant use. The methods are now available to determine directly whether oestrogen secretion is limited to the antral follicle. If this turns out to be true, the interesting possibility then arises that younger follicles somehow "trap" the oestrogen needed for their further growth and differentiation from more advanced follicles. However, the synthesis and secretion of steroid hormones appear to be regulated by the synergistic action of gonadotrophins (FSH, LH, prolactin) and steroid hormones (androgen, oestrogen) (Schomberg 1979, Bahr et al. 1979, Leung et al. 1979, Armstrong et al. 1979, Young Lai 1979, Mills 1979, Fortune and Armstrong 1979, Nimrod and Lindner 1979, Thanki and Channing 1979, Fortune and Hansel 1979, Anderson et al. 1979, Armstrong 1980, Erickson 1982). It is also possible that steroids may regulate their own synthesis through a local negative feedback system. Most of these studies have also revealed that this regulation is brought about through adenyl cyclase (AC) and cyclic adenosine monophosphate (cAMP) which act as mediators of hormone action (Fig. 27) (Marsh 1975, Clark et al. 1978a, Hunzicker-Dunn et al. 1979a, Bahr et al. 1979, LeMaire 1979, Lamprecht et al. 1979, Kraiem and Lunenfeld 1979, Vaitukaitis and Albertson 1979, Mills 1979, Armstrong 1980, Peters and McNatty 1980, Hubbard and Greenwald 1982). However, little is known about the mechanism by which cAMP mediates the actions of LH in follicles (Hunzicker-Dunn et al. 1980). It has been shown that rat pre-ovulatory follicle contains protein kinase activity whose phosphorylative capacity is increased by cAMP (Lamprecht et al. 1977, Hunzicker-Dunn et al. 1980).

Follicular hormone receptors have also been extensively studied during the past few years (Eshkol et al. 1976, Midgley and Richards 1976, Richards 1978, 1979, 1980, Schreiber 1979, Vaitukaitis and Albertson 1979, Peters and McNatty 1980, Ermini and Carenza 1980, Oxberry and Greenwald 1982). Most protein hormones initiate a biological response in their target cells by initially binding to a membrane-localized receptor, thereby activating AC and raising intracellular levels of cAMP (Fig. 27)

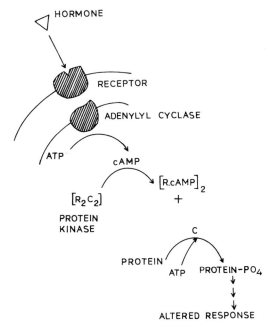

Fig. 27. Receptor-coupled adenylyl cyclase *(AC)* and cAMP-dependent protein kinase *(PK)* enzyme systems in protein hormone action. Hormone-receptor interaction leads to activation of the membrane-localized adenylyl cyclase *(AC)* promoting synthesis of cAMP from ATP. Cyclic AMP mediates many hormone-induced events by regulating the activity of cAMP-dependent protein kinase *(PK)*. Cyclic AMP is shown to activate PK by binding to the regulatory subunit R of the inactive holoenzyme, thereby releasing the active, cAMP-insensitive catalytic subunit *(C)* which catalyzes the transfer of phosphate from ATP to cellular proteins. By catalyzing the phosphorylation of specific cellular protein, PK presumably changes cellular functions in such a way that the known hormonal response ensues (Redrawn from Hunzicker-Dunn et al. 1980)

(Marsh et al. 1973, Hunzicker-Dunn and Birnbaumer 1976a, Hunzicker-Dunn et al. 1980). At present, the only known manner in which cAMP can mediate cellular function is through cAMP-dependent protein kinases (PK). cAMP is known to activate cAMP-dependent PK by binding to the regulatory subunit of the inactive holoenzyme, thereby releasing the active, cAMP-insensitive catalytic subunit which can catalyze the transfer of phosphate from ATP to cellular protein (Langan 1973, Hunzicker-Dunn et al. 1980). Presumably, the phoshorylation of specific cellular protein(s) regulates cellular activity in such a way that the known hormonal response ensues. Hunzicker-Dunn et al. (1979c) have concluded that the activity and intracellular distribution of cAMP-dependent PK in ovarian follicles are subject to regulation by hCG. The apparent reduction in the catalytic activity of PK in nuclear and microsomal fraction occurs in a time-dependent manner after hCG administration under identical tissue manipulations, which strongly suggests that these changes are of physiological significance. Hunzicker-Dunn et al. (1980) have discussed the ovarian responsiveness to gonadotrophic hormones. They have concluded that the response evoked by gonadotrophic hormones at ovarian target cells is related to the levels of circulating gonadotrophins; the presence, sensitivity and activity of the gonado-hormone receptor-coupled AC system; and the activity, state of activation, subcellular distribution, and isoenzyme forms of cAMP-dependent PK. However, the nature of the involvement of alterations of PK in pre-ovulatory and/or postovulatory follicular events is yet to be determined more precisely (Hunzicker-Dunn et al. 1980).

In spite of all these advances in the molecular biology of follicle growth, controversy still continues to exist regarding the cellular sites and mechanisms of steroid hormone synthesis (especially of oestrogen biosynthesis) in the growing follicle of the mammalian ovary. Theca interna and/or granulosa cells, or both cell types, have been believed to constitute the sites of steroidogenesis in the developing follicle (Guraya 1971a, 1974b, 1980a, Bjërsing 1978, Yoshinaga 1978, Schomberg 1979, Armstrong et al. 1979, Young Lai 1979, Fortune and Armstrong 1979, Mills 1979, Dvořák and Tesařík 1980, Hoyer 1980, Armstrong 1980, Erickson 1982). In order to make direct quantitative comparisons of steroidogenesis in vitro by the thecal and granulosa cell types, it may be necessary to establish their metabolic and/or endocrine requirements in greater detail. Because of its life history – the tertiary follicle passes from infancy to maturity to senescence – Greenwald (1979) suggested that the failure to keep this sequence in mind may account for some of the confusion on the steroidogenic capabilities of thecal and granulosa cells. Morphological, histochemical, biochemical and physiological differences have been shown to occur between immature and mature antral follicles, which must be kept in mind while interpreting the results. Furthermore, antral follicles of the same size and age may differ in their steroidal secretions and responses depending on the local hormonal milieu (e.g. the influence of corpus luteum).

For the better understanding of follicular function, it will be essential to work out the life cycle of the antral follicle in much greater detail. Follicular maturation from small antral to pre-ovulatory sized follicles takes up a relatively short time. In rats, showing 4-day oestrous cycles, the final stages of follicular maturation need only 3 days (Hirshfield and Midgley 1978a). Morphological studies have shown that the follicles grow rapidly during this time. This rapid growth of the follicle is characterized by enlargement of and increases in the vascularity of the theca interna layer, granulosa cell proliferation, and an increase in the volume of follicular fluid and coincident increase in the size of the antrum (Fig. 11). The follicles undergoing these final stages of maturation are believed to be selected by the surge of FSH of the previous cycle (Welschen and Dullaart 1976, Hirshfield and Midgley 1978b, Greenwald and Siegel 1982). When follicles reach pre-ovulatory size, a surge of LH plus FSH causes the final spurt of follicular growth, culminating in ovulation of the ripe follicle within 10–12 h in the rat. This final growth spurt is closely accompanied by conspicuous changes in the follicle wall and in the other components.

1. Theca

The formation of the theca layer outer to the basement lamina is little understood. With the initiation of growth in the follicle, cells indistinguishable from fibroblasts are aligned concentrically around the follicle to form the thecal layer (Fig. 11) (Bjersing 1978). Thus, the secondary or preantral follicles are surrounded by the concentric sheath of undifferentiated stromal elements or fibroblasts which show some development of blood vascularity (Gillet et al. 1980). At this stage theca cells contain a few organelles, such as small Golgi complex, rod-shaped mitochondria with transversal cristae, profiles of granular ER and some free ribosomes, which are specific to common undifferentiated stromal cells of the ovary (Guraya 1971a, 1974b, O'Shea 1973, Balboni 1976, Mestwerdt et al. 1977a, b, Bjersing 1978). At this stage they are devoid of mitochondria with tubular cristae, smooth ER, lipid droplets and enzyme activities indicative of steroidogenesis (Guraya 1971a, 1974b, 1980a, Bjersing 1977, Hoyer 1980). Thus they do not show definite morphologic submicrosopic criteria of steroid biosynthesis (Mestwerdt et al. 1977b, Bjersing 1978, Dvořák and Tesařík 1980).

a) Development and Differentiation of Theca Interna (or Thecal Gland) and its Vascularity

With the appearance of the antrum in the granulosa, the theca layer consisting mainly of fibroblasts becomes stratified and thicker (Figs. 11 and 26) and meanwhile some of the surrounding stromal elements, lying adjacent to the granulosa, begin to hypertrophy or "luteinize" to form large glandular-looking cells having abundant cytoplasm with organelles specific to steroidogenic cells (Balboni 1976, Nowacki 1977, Bjersing 1978, Guraya 1980a, Dvořák and Tesařík 1980). These organelles include mitochondria with tubular cristae, widespread Golgi complex, abundant ER of a smooth type, besides the lipid droplets (Fig. 28). Factors regulating the differentiation (or luteinization) of thecal cells at the molecular level are poorly understood. It has been suggested that theca inducer is produced by the granulosa cells (Dubreuil 1957). However, recent studies have suggested that LH plays a specific role in inducing the differentiation of theca interstitial cells (or ovarian androgen-producing cells) (Erickson 1982). This process of theca cell differentiation is modulated by prolactin during follicle growth.

Two parts (theca externa and theca interna) of the layer are generally distinguished in the developing antral follicles, especially in frozen sections coloured with Sudan black B (Fig. 29). The inner part, the theca interna, consists of polyhedral or elongated cells having all the morphological characteristics of steroid-secreting cells. Their nucleus is vesicular, sometimes with a slightly undulating nuclear envelope; one

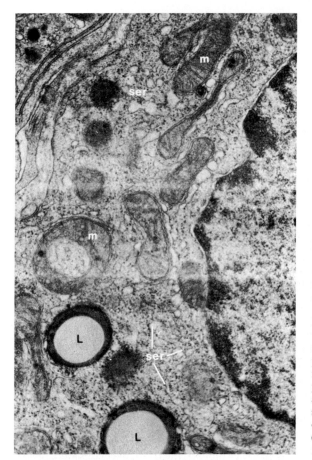

Fig. 28. Electron micrograph of rat theca interna cells from a Graafian follicle. The most prominent organelles are the smooth endoplasmic reticulum *(ser)*, mitochondria *(m)*, and lipid droplets *(L)* with extracted centres. Free ribosomes are numerous, but rough endoplasmic reticulum is uncommon. Cup mitochondria (appearing as an annular profile) and other bizarre mitochondrial shapes frequently seen in ovarian steroid-secreting cells (From Christensen and Gillim 1969)

distinct reticular nucleous is present in it. The cytoplasm contains ovoid mitochondria with tubular cristae occurring besides plate-like ones, numerous profiles of smooth ER, the widespread Golgi complex, and lipid droplets (Fig. 28). Polysomes and lysosomes are also present. The occurrence of granular ER is less conspicuous. The cell membrane is smooth; the cell surface being only slightly irregular in outline. Among the theca interna cells numerous capillaries penetrate, completing the character of an endocrine gland (Gillet et al. 1980).

Fig. 29 a–f. Photomicrographs of histochemical preparations of portions of Graafian follicles having antral cavity *(AC)* from the ovaries of different mammalian species, showing variations in the development of sudanophilic lipids in the hypertrophied theca interna. Such lipids are not seen in the theca externa *(TE)*, and membrana granulosa *(MG)* which, however, show sparsely distributed phospholipid bodies *(L)* of variable size: **a** Showing theca interna *(TI)*, membrana granulosa *(MG)*, antral cavity *(AC)* and cumulus oophorus *(CO)* around the oocyte of human antral follicle; **b** From rabbit ovary; **c** From ovary of small Indian mongoose; **d** From rat ovary (From Guraya 1972a); **e** From two adjacent Graafian follicles of cat ovary (From Guraya 1969d); **f** From hamster ovary; surrounding interstitial gland tissue *(IGT)* is filled with highly sudanophilic lipid bodies (From Guraya 1972a)

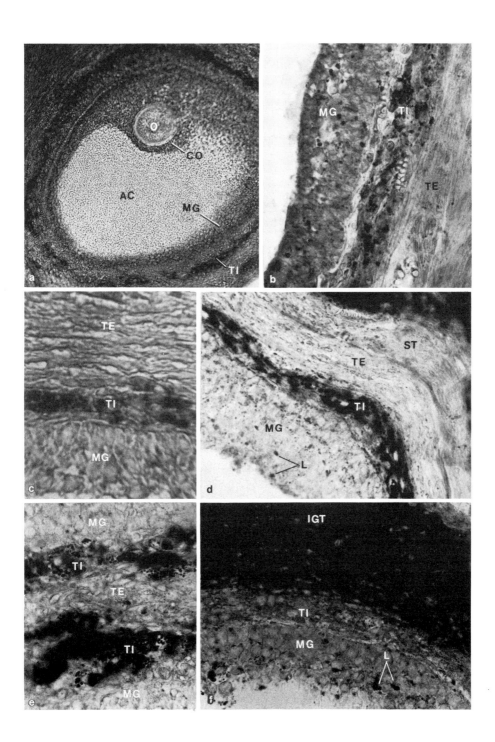

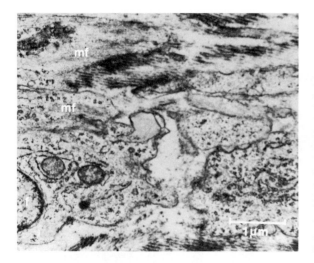

Fig. 30. The centre of this electron micrograph shows an unusual protrusion from a thecal fibroblast *(f)* in a mature rabbit follicle. If this protrusion represents a modified form of the multivesicular structures shown in Fig. 56, then the origin of their contents is obviously rough endoplasmic reticulum. Note the microfilaments *(mf)* in this fibroblast, and in the cell above it. Collagen fibres are also seen (From Espey 1978a)

The outer part of the theca, the theca externa, consists of cells which in their morphology resemble the fibroblasts (Figs. 26 and 30). Martin and Talbot (1981a) observed that in the hamster, the theca externa forms a continuous layer around each follicle and consists of one to three layers of spindle-shaped cells which have the morphological features of either smooth muscle cells or fibroblasts. Fibroblasts contain an abundance of ER, Golgi bodies, free ribosomes and coated vesicles. They can be further distinguished by a paucity of glycogen and microfilaments in the cytoplasm, and lack of caveolae and dense attachment plaques along the plasma membrane. The fibroblasts also show very long, thin, tapered processes extending from both poles. They are generally separated from each other by intracellular spaces containing bundles of collagen fibres. When fibroblasts are placed next to each other,

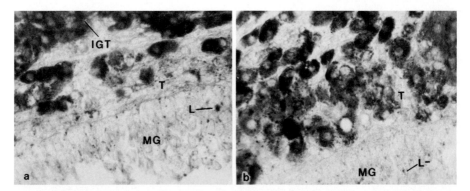

Figs. 31 a, b. Photomicrographs showing two stages in the development, differentiation and maturation of sudanophilic glandular theca cells *(T)* in developing antral follicles of the civet cat ovary. Membrana granulosa *(MG)* shows sparsely distributed lipid bodies *(L)*. Finally, hypertrophied theca cells *(T)* and surrounding interstitial gland cells *(IGT)* show similar sudanophilic lipids (From Guraya 1981)

junctions are rare and then of the desmosome type. The plasma membrane of fibroblasts occasionally contain a thin layer of microfilaments (5 nm diameter) along its inner margin and its external surface lacks a basal lamina. It is interesting that the thecal layer of developing follicle in the civet cat *(Paradoxurus hermaphroditis)* is not differentiated into theca interna and theca externa (Fig. 31) (Guraya 1979 b, 1981). Its hypertrophied cells are formed partly by the hypertrophy (or differentiation) of some fibroblasts and partly by the incorporation of some hypertrophied cells from the surrounding greatly developed interstitial gland tissue (Fig. 31). The hypertrophy of fibroblast-like thecal cells is closely accompanied by the development of sudanophilic lipids (Fig. 31) which consist of diffusely distributed lipoproteins, and some lipid granules consisting mainly of phosholipids; no cholesterol and/or its esters and triglycerides could be detected histochemically. Meanwhile, many hypertrophied interstitial gland cells having comparable cytological and histochemical features are also incorporated into the developing thecal layer (Fig. 31). Finally, it is not possible to make any distinction between both cell types, as they resemble each other in their cytological and histochemical features. As a result of these morphogenetic changes, the highly vascularized thecal layer of Graafian follicles in the civet cat ovary becomes continuous with the surrounding greatly vascularized and hypertrophied interstitial gland tissue (Fig. 31).

Mestwerdt et al. (1977a, b) also observed that structural differentiation of the normal stromal cells around primordial primary and secondary follicles in the human ovary finally leads to the development of definite submicroscopic steroid gland cells in the wall of tertiary follicles. The cell contacts between these hypertrophied theca interna cells of developing pre-ovulatory human follicle consist of septate-like cell contacts and gap junctions (Fukushima 1977, Dvořák and Tesařík 1980, Toshimori and Oura 1982), as also reported for other mammalian species (Bjersing 1978). Anderson (1979) also observed that the cells constituting the theca interna in the ovaries of rodents are associated with each other via gap junctions (see also Fletcher 1979, Amsterdam and Lindner 1979). Elements of smooth ER are frequently associated with gap junctions between thecal cells of large pre-ovulatory follicles. According to Fletcher (1979), in the large follicles, either of the mature or immature rat, the dominant contact between thecal cells resembles the septate variety reported in adrenal cortex.

Hiura and Fujita (1977a) recognized three different cell types around the secondary and Graafian follicles, which include fibroblast-like cells, transitional cells and thecal gland cells (steroid secreting cells). The fibroblast-like cells, like the general stromal cells of the ovary, show well-developed endoplasmic reticulum, moderately developed Golgi apparatus, rod-shaped mitochondria with lamellar cristae, few lipid droplets and free ribosomes in their cytoplasm, which are indicative of protein synthesis. A large ellipsoidal nucleus, abundant lipid droplets, mitochondria with tubular cristae and smooth ER are seen in the hypertrophied theca interna (or thecal gland) cells and are indicative of steroid hormone synthesis. The transitional cell shows a round or oval nucleus, many lipid droplets, mitochondria with tubular or lamellar cristae and a small amount of smooth ER, which support the transformation of stromal cells into steroid-secreting thecal gland cells during follicle growth. After the administration of PMSG or hCG, many fibroblasts continue to be seen in the thecal layer of Graafian follicles. They are believed to have no ability to differentiate into thecal gland cells.

Corresponding to the formation of hypertrophied theca interna (or thecal gland) cells, the vascularity of the thecal layer in the wall of the Graafian follicle is greatly increased (Szöllösi et al. 1978, Ellinwood et al. 1978, Bjërsing 1978, Gillet et al. 1980). Szöllösi et al (1978) observed that the theca interna actually consists of two layers: (1) A vascular layer, several microns thick along the basal lamina of the follicle-containing capillaries, a few endocrine cells, sometimes clustered, and fibroblasts. (2) A glandular layer, composed of endocrine cells and capillaries. The inner vascular wreath always separates the theca interna from the membrana granulosa, and capillaries are never seen to penetrate through the basement membrane (or lamina) into the membrana granulosa until after ovulation (Guraya 1971a, Bjërsing 1978, Gillet et al. 1980). Kanzaki et al. (1981), using a method of resin injection-corrosion casts, have observed that the vasculature of follicles in juvenile and adult rabbits starts as a finger-like ramified capillary network. As the follicle grows, follicular vessels increase remarkably and form a multilayered complex vascular network surrounding the follicles. Observations of circular impressions around the arterioles near the follicle and in the follicular wall suggest the presence of sphincteric control mechanism of blood flow into the follicle.

The mature follicle is also surrounded by a complete lymphatic crown between the theca interna and granulosa (Gillet et al. 1980). This lymphatic network is made up of large meshes. A few connections are present between this network and lymphatics of the theca externa. Numerous connections between the lymphatics of the theca externa and the lymphatics of the stroma have been observed. These changes in the blood and lymphatic vascular system of the follicular wall may be brought about by gonadotrophins, especially LH (Guraya 1974b). Koos and LeMaire (1983) have proposed that developing follicles in the rat ovary produce an angiogenic factor; the probable source of this factor is the granulosa cells.

With the maturation of the tertiary (Graafian) follicles in the human ovary, the number and the activity of hypertrophied epithelioid cells of the theca interna are further increased (Schaar 1976). In the pre-ovulatory period, the entire width of the theca interna consists of epithelioid cells showing a particularly high activity. They are provided with a rich capillary network which reaches the basal lamina of the granulosa layer (Schaar 1976). The ultrastructural changes of theca interna, which are related to their steroidogenic activity, are also further developed. The hypertrophied and well-vascularized cells finally constitute the fully developed thecal gland cells around the Graafian follicle (Mestwerdt et al. 1977a, b, Hiura and Fujita 1977a, b; see other references in Bjërsing 1978). The concentric sheath of undifferentiated, fibroblast-like stromal cells continue to surround the thecal gland cells during the pre-ovulatory period. They also continue to show organelles typical of stromal cells (Bjërsing 1978).

Development and differentiation of theca interna cells at the light and electron microscopic levels vary greatly in follicles of different mammalian species (Fig. 29) (Guraya 1973a, 1974b, Bjërsing 1978). In the marmoset and human ovary, hypertrophied theca interna cells show sparsely distributed lipid granules consisting of phospholipids, suggesting that these cells might be functioning in the secretion of some steroid hormones rather than in the storage of hormone precursors during follicular growth. In the rhesus monkey, the theca interna cells show relatively less cytoplasmic differentiation and contain some lipid granules consisting mainly of phospholipid. The hypertrophied theca interna cells possess the ultrastructural features typical of most-steroid-secreting cells which, however, show some variations in different species of mammals (Bjërsing 1978). These features consist of the presence of tubular smooth ER, mitochondria with predominantly tubular cristae, and lipid droplets (Christensen and Gillim 1969, Guraya 1971a, 1974b, 1976a, b, 1978, Bjërsing and Cajender 1974f, Nowacki 1977, Mestwerdt et al. 1977a, b, Bjërsing 1978). Diffusely distributed lipoproteins are presumed to derive from the abundant agranular ER of ultra-structural studies (Bjërsing 1978).

b) Smooth Muscle Cells

The presence of nonvascular smooth musculature in the mammalian ovary has for a long time been a subject of great controversy (Burden 1973, Espey 1978b). Electron microscopy was first used for this problem by O'Shea (1970b) and Osvaldo-Decima (1970). The reason for this controversy is that smooth muscle cells in the ovary are not as easily demonstrable as in other peripheral organs: they cannot, for example, always be clearly distinguished from fibroblasts; and the transitional forms between these and ovarian smooth muscle-like cells appear to be a characteristic feature (O'Shea 1970a, Burden 1972, Espey 1978b). Various recent transmission electron microscope studies have revealed the presence of cells with the characteristic subcellular features of smooth muscle in the theca externa of various mammalian species (Fig. 32) (Sjöberg et al. 1979, DiDio et al. 1980, Owman et al. 1980, Motta and Familiari 1981, Martin and Talbot 1981a). Martin and Talbot (1981a) have observed that smooth muscle cells in the follicle wall of the hamster are similar to fibroblasts but may be distinguished from them by the following features: (1) The intercellular spaces are small and collagen fibres within these spaces are rare; (2) both desmosomes and gap junctions are common (the latter are most often observed in transverse sections of smooth muscle cells); (3) smooth muscle cells are surrounded by a basal lamina; (4) Caveolae are commonly seen; (5) dense attachment plaques and microfilaments are seen along the plasma membrane; (6) synthetic organelles are accumulated at the nuclear poles. The cytoplasm of smooth muscle cells shows many longitudinal filaments (myofilaments)

Fig. 32a, b. Electron micrographs of smooth muscle-like cells in the theca externa of follicle wall. **a** Smooth muscle-like cell in the theca externa of the bovine follicle wall. The cell is characterized by (longitudinally sectioned) myofilaments *(m)* and dense bodies *(d)*; **b** Theca externa of an ovarian follicle from a guinea-pig, showing a naked adrenergic nerve terminal *(arrow)* containing dense-cored synaptic vesicles characteristic of adrenergic axons, approaches the smooth muscle cells *(mc)* with a neuro-muscular distance of only about 25 mm (From Owman et al. 1980)

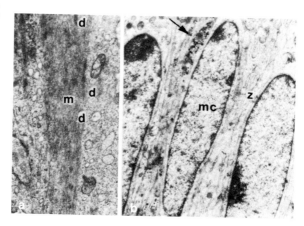

with a diameter of 6–8 nm, free ribosomes in the perinuclear cytoplasm, moderate amount of rough ER, well-developed, lamellar (Golgi) complexes, lipid droplets, and occasionally glycogen particles (20–25 nm in diameter). Mitochondria are generally not numerous, but when present they often show long cristae. Along the internal aspect the plasma membranes of perifollicular myocytes show dense bodies which appear to be associated with myofilaments; they are frequently covered externally by a dense lamina. The plasmalemma shows many invaginations (pinocytotic vesicles) with dense areas of close contact between adjacent cells similar to junctional complexes. Martin and Talbot (1981a) have distinguished three types of cytoplasmic filaments based on their cross-sectional dimensions in the hamster follicle. The most abundant type is the thin actin filament (4.7–6 nm diameter) which lies in bundles throughout the cell. Intermediate or desmin filaments (8.7–12 nm diameter), which are less abundant, occur primarily along the periphery of bundles of thin filaments. Thick filaments (12.5–14 nm diameter) are seen in favourable transverse sections to be partially or completely surrounded by a ring of thin filaments. Both the thin and intermediate filaments are seen in smooth muscle cells prepared by all the fixatives used. In contrast, the thick, presumably myosin filaments are difficult to preserve and seemed to be stabilized best when the cells are fixed at a low pH.

The immunohistochemical demonstration of actin and myosin in cells with a perifollicular localization (Fig. 33) has recently provided strong evidence for the presence in the mammalian ovary of nonvascular contractile elements in amounts sufficient to account for the follicular contractile properties (Amsterdam et al. 1977, Walles et al. 1978, Sjöberg et al. 1979).

Amsterdam et al. (1977) studied the localization of actin and myosin in the rat follicular wall by immunofluorescence, as also confirmed in other recent studies (Owman et al. 1980). Staining with antiserum has shown the presence of several layers of intensely fluorescent elongated cells within the theca externa, forming a coherent band around the large Graafian follicles (Fig. 23), thus supporting the electron microscope data. In smaller follicles, this band of fluorescent cells is incomplete and ovaries of 6-day rats are devoid of strongly fluorescent cells. DiDio et al. (1980) have discussed the significance of the presence of myocytes in the mammalian ovary. Their

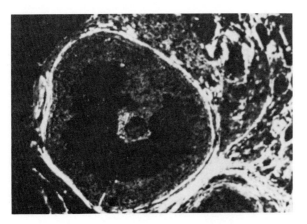

Fig. 33. Immunofluorescence localization of smooth muscle myosin in relation to a large antral follicle in the rat ovary. The follicle is enclosed by a well-defined layer of intensely myosin-containing contractile cells. Similar immunoreactive muscle cells are present in strands within the stroma and in vessel walls (From Sjöberg et al. 1979)

contraction appears to influence several events, such as (1) the detachment of the cumulus oophorus, (2) expulsion of the follicular contents after the rupture or, better said, opening of the apical wall, (3) the related vascular phenomena, and (4) collapse of the follicle and its transformation into a corpus luteum. However, Motta and Familiari (1981) have suggested that if other factors do not simultaneously occur to cause a normal ovulatory process, the contraction of contractile tissue, instead of assisting in the rupture of the follicle, may favour its collapse and subsequent involution during atresia (see also Espey 1978 b).

c) Innervation

The theca externa of the Graafian follicle contains numerous cholinergic nerves as demonstrated by cholinesterase histochemistry following specific inhibition of pseudo-cholinesterase (Fig. 34) (Walles et al. 1976, Burden 1978, Sjöberg et al. 1979, Stefenson et al. 1981, see other references in Owman et al. 1980). Several varicosed nerve terminals enter into the thecal layer to run isolated in an irregular manner; sometimes the fibres are seen to accompany small blood vessels. The arrangement of adrenergic nerve terminals is the same in the protruding part of the follicle, though they are less numerous than in the intra-ovarian portion of the follicular wall. The human Graafian follicle has been investigated with the glyoxylic acid fluorescence method (Lindvall and Björklund 1974). Adrenergic nerve terminals, running in a meridional direction, are seen in the wall of the whole Graafian follicle, except for the most attenuated region of the apex (Owman et al.1975).

Gimbo et al. (1980) have studied ovarian innervation in the camel *(Camelus dromedarius)*. Numerous nerve fibres are present in the perifollicular stroma of primary and secondary follicles. Their number is increased in the theca externa of growing antral follicles. Then the nerve fibres also enter the theca interna radially without going beyond the basement lamina. Fibrils or nerve endings are not seen in the granulosa and ova. Morimoto et al. (1981), using a combination of fluorescence and light microscopic observations, have studied adrenergic innervation around bovine ovarian follicles at various stages. No fluorescent nerve fibres are seen in the membrana granulosa at any stage of follicular development. The sparse nerve fibres seen in the

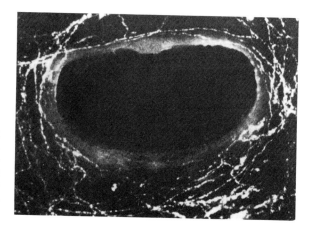

Fig. 34. Adrenergic innervation of the sheep Graafian follicle. The nerve terminals can be followed as they run within the follicle wall, approaching the theca interna (From Sjöberg et al. 1979)

theca interna of Graafian follicles are apparently those accompanying blood vessels and are believed not to play a significant role in the vasomotor control of follicular blood flow. The fluorescent fibres in the theca are increased in density and intensity in the course of follicular development. This increase appears to coincide with increasing contractility of the follicular wall about the time of ovulation. Stefenson et al. (1981) have made a comparative study of the autonomic innervation in 12 mammalian species. Ovaries from cow, sheep, cat and guinea-pig are richly supplied with adrenergic nerves in the cortical stroma, particularly in enclosing follicles. In the follicular wall the nerve terminals are placed in the theca externa where they run parallel to the follicular surface. Numerous adrenergic terminals also surround ovarian blood vessels. The adrenergic innervation is of intermediate density in the human ovary and in the pig, dog, cat and opossum. Ovaries from rabbit, mouse and hamster show a sparse adrenergic nerve supply. The amount of intra-ovarian adrenergic nerves agrees well with the tissue concentration of noradrenaline in the various species. The cholinergic innervation is generally less well developed but shows the same distribution as the adrenergic system around blood vessels and in the ovarian stroma, including follicular walls.

d) Neuromuscular Relationship

In attempts to correlate the distribution of sympathetic nerves with that of the ovarian smooth musculature (Walles et al. 1978), adjacent cryostat sections have first been reacted with glyoxylic acid in order to visualize the adrenergic transmitter (de la Torre and Surgeon 1976), and then treated for immunohistochemical demonstration of actomyosin. It is thus observed that sympathetic axon terminals run in between the contractile cells of the externa. The more detailed relationship between the adrenergic nerves and contractile cells has been studied at the ultrastructural level (Fig. 32) (Sjöberg et al. 1979). Adrenergic terminals are frequently seen close to smooth muscle cells in the follicular theca externa. Occasionally, naked axon varicosities are seen only 20 nm from the membrane of a muscle cell, further supporting the presence of a functioning neuro-muscular relationship (Owman et al. 1975).

Nonadrenergic, probably cholinergic, axons have been observed with the electron microscope, where they can be distinguished from adrenergic tissues (Owman et al. 1980). Naked cholinergic terminals are located at a distance of 150 nm from smooth muscle-like cells, i.e. a distance short enough to represent a functioning neuro-muscular relationship. In addition, close contacts of 20–25 nm are sometimes seen between the two types of axon varicosities, showing an axo-axonal interaction. The presence and characteristics of cholinergic receptors have also been studied in follicle strips on the basis of amine-induced contractile activity (see Owman at al. 1980). All the results discussed above have provided strong evidence for the presence of a well-developed system of smooth muscle cells in the theca externa of the ovarian follicle, which are innervated by autonomic nerve terminals. The results of recent studies, as integrated and discussed by Owman et al. (1980), have also indicated a functioning neuro-muscular relationship in the follicle wall. Burden and Lawrence (1980) have described the origin, morphology and distribution of ovarian innervation, and considered evidence for neuronal modulation of ovarian events. This is probably done by modulating the sensitivity of ovarian compartments to humoral factors.

e) Enzymes

Almost all workers agree that the development of organelles typical of steroidogenic cells in the hypertrophied (or epithelioid) theca interna cells of the developing follicle is closely accompanied by the appearance of enzyme activities, such as steroid dehydrogenases (Fig. 35c) (Guraya 1971a, 1974b, Schaar 1976, Bjersing 1977, 1978, Hoyer 1980, Parshad and Guraya 1983). Δ^5-3β-hydroxysteroid dehydrogenase (3β-HSDH) has been most extensively studied. But all other cells (undifferentiated stromal cells) including the fusiform types in the outer and inner layers of thecal tissue are totally negative for 3β-HSDH activity (Schaar 1976, Hoyer 1980). Some of the enzymes are blieved to be involved in the production of NADPH. The theca interna in the ovaries of the Indian gerbil (*Tatera indica*) shows a progressive increase in the activities of NADH-and NADPH-tetrazolium reductases, 3β- and 17α-HSDH, and succinate and lactate dehydrogenases during follicular growth (Fig. 35) (Parshad and Guraya 1983); meanwhile sudanophilic lipids are also increased in their amount.

Stoklosowa et al. (1979) suggested specific criteria by which one can distinguish theca interna cells from granulosa cells prior to and after the establishment of cultures. Theca interna cells prior to culture show more intense staining with oil Red 0, stronger 3β-HSDH activity and large cell size. They can be distinguished from granulosa cells in culture by their elongated shape, slower and more chaotic growth, weaker enzyme activity and increased oestrogen production. Stadnicka and Stoklosowa (1976) earlier observed that theca interna cells in tissue culture show a high and constant activity of 3β-HSDH and G-6-PDH (glucose-6-phosphate dehydrogenase), but activities of both 17β- and 20α-HSDH are lower and show some fluctuations. Addition of LH to the medium brings about an increase of all dehydrogenases investigated. This is also supported by the presence of LH receptors in the theca interna (Bjersing 1978, Richards 1978, Richards et al. 1978, 1979, Carson et al. 1979). FSH is less effective and little or no receptors for FSH exist in the theca interna cells (Oxyberry and Greenwald 1982). Oestradiol shows inhibiting effects on dehydrogenases. But all these hormones (LH, FSH and oestradiol) increase alkaline and acid phosphatase activities in cells of cultured porcine theca interna (Bjersing 1977, 1978). Brietenecker et al. (1978) have shown the presence of alkaline phosphatase, acid phosphatase, leucine aminopeptidase and LDH in the theca interna of human follicles; 3β-HSDH activity is also present in the theca interna of nonovulatory tertiary follicles and pre-ovulatory Graafian follicles. Elfont et al. (1977) observed an increase in acid phosphatase reaction of thecal steroidogenic cells of antral follicles in the ovary after treatment with PMSG or PMSG-hCG. Acid phosphatase forms an important hydrolytic enzyme of lysosomes. In a number of ovarian tissues it has been shown that lysosomal enzyme activity increases, concomitant with a stimulation of steroid production in response to LH stimulation (see Dimino and Elfont 1980). One role of lysosomes in steroidogenesis may be to interact with lipid inclusions in order to make substrates available for steroidogenesis. The other roles that have been suggested, may (1) serve to reshape ovarian subcellular components for facilitating the differentiation of follicular (thecal) tissue and (2) be related to a mechanism for the secretion of steroids.

Miyamoto et al. (1980) have observed that the theca cells of normal pre-antral follicles in the goat ovary show a very weak activity of NADP-dependent G-6-PDH, which becomes intense in the theca cells of antral follicles, suggesting their role in steroidogenesis. Succinate dehydrogenase (SDH) is sparse in the theca cells of normal

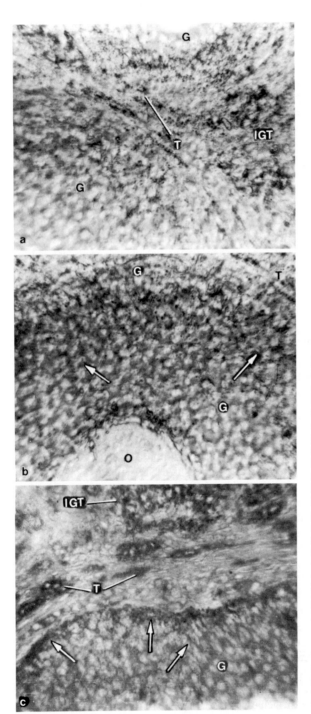

Fig. 35. a Activity of succinic dehydrogenase in the theca cells *(T)* and outer layers of granulosa *(G)* of pre-antral follicles and the interstitial gland tissue *(IGT)* of the ovary of the Indian gerbil *Tatera indica;* **b** Showing moderate activity of lactate dehydrogenase in theca *(T)* and moderate to intense in granulosa *(G)* of the pre-antral follicle of *T. indica.* Outer layers of granulosa contain relatively more activity of the enzyme *(arrows)*. Intense reaction can also be seen around the oocyte and weak reaction within the ooplasm *(O);* **c** Portion of antral pre-ovulatory follicle of rat showed intense reaction for Δ^5-3β-HSD in the thecal gland cells *(T)*, and interstitial gland tissue *(IGT)*. Note moderate to intense reaction in granulosa *(G)* along the basal layer, and rest of granulosa showing weak reaction (V. R. Parshad and S. S. Guraya, unpublished)

growing pre-antral follicles in the goat ovary (Miyamoto et al. 1981). Weak or very weak enzyme activity is also present in normal antral follicles.

Hiura and Fujita (1977 b) carried out an electron microscope cytochemical study of adenylate cyclase (AC) activity in the theca layer of mouse Graafian follicles. Deposits of reaction products are seen on the plasma membrane of the fibroblast, theca cell and transitional cell from the fibroblast-like cell to the theca cell (partially or incompletely differentiated theca cell) after incubation with adenyl-imidodiphosphate as an effective substrate for AC. These cells appear to have receptors on the plasma membrane, and the AC-cAMP system may be important for steroid secretion or collagen fibre production in the theca layer.

f) Steroid Hormone Synthesis

The results of various morphological (including ultrastructural) and histochemical studies have clearly demonstrated the presence of organelles and enzyme systems (specific to steroidogenic tissues) in the hypertrophied theca interna (or thecal gland) cells of large antral ovulatory follicles, as already discussed, indicating that they form an important site for steroid biosynthesis (see Guraya 1971 a, 1973 a, 1974 b, 1980 a, 1982 c, Bohr et al. 1980). Systems for oestrogen synthesis have also been demonstrated in some in vivo and in vitro studies on the theca interna cells, but not in the granulosa of growing antral follicles of some animal species in vivo, as already discussed (Guraya 1971 a, 1974 b, Abel et al. 1975, McClellan et al. 1975, Nowacki 1977, Mestwerdt et al. 1977 a, b, Bjersing 1977, 1978, Young Lai 1979, Schomberg 1979, Stoklosowa et al. 1982). The recent in vitro experimental studies have, however, strongly suggested that the theca of follicles is responsible for most or all androgen secretion and that androgen production is under the control of LH (Armstrong et al. 1979, Fortune and Armstrong 1979, Peters and McNatty 1980, Richards 1979, Schomberg 1979, Armstrong 1980, Makris and Ryan 1980, Terranova et al. 1982, Koninckx et al. 1983); FSH has no effect. But Armstrong et al. (1979) also believed that the theca interna, forming an additional source of oestrogen formation, may assume greater quantitative importance in some species or physiologic states than in others (see also Armstrong 1980, Stoklosowa et al. 1982). Baird (1977 a) also suggested that follicular oestrogen is of thecal origin as in the sheep; small episodic releases of LH are accompanied by episodic secretion of androstenedione and oestradiol-17β (Baird 1977 b), and theca cells have an apparent aromatase activity (Hay and Moor 1975, Channing and Coudert 1976, Stoklosowa et al. 1982). Murdoch and Dunn (1982) have quantified progesterone, testosterone and oestradiol within the thecal layer, granulosa layer and fluid of the pre-ovulatory follicle and in ovarian and jugular venous sera of sheep. These follicular steroid hormones undergo conspicuous changes in their quantity and quality during the pre-ovulatory period. Prior to the pre-ovulatory surge of LH, the total content of oestradiol-17β is elevated within each follicular constituent. The levels are decreased rapidly to minimum values concurrent with the surge of LH.

The action of LH on steroidogenesis in the thecal cells is also supported by the presence of LH/hCG receptors in them (see Richards 1976, Richards et al. 1978, 1979, 1980, Peters and McNatty 1980, Oxberry and Greenwald 1982); the thecae of large antral follicles in the hamster ovary also bind prolactin, showing receptors for it (Oxberry and Greenwald 1982). Oxberry and Greenwald (1982) have observed that (1) hCG binding to the thecae of the large antral ovulatory follicles is slightly reduced after the pro-oestrous gonadotrophin surge and (2) prolactin binding to the thecae of the large antral ovulatory follicles is also decreased during the surge followed by a nearly complete loss of prolactin binding to the thecae after the surge. However, Dunaif et al. (1982) have not found prolactin receptors in the theca of rat follicle. Carson et al. (1979) observed that in contrast to the granulosa cell, binding of gonadotrophins to the theca does not vary significantly, remaining relatively high for LH (hCG) and low for FSH. The role of cAMP as a second messenger in the action of LH on steroid biosynthesis in the thecal cells has also been established (Midgley and Richards 1976, Peters and McNatty 1980, Armstrong 1980). FSH causes no significant increase in cAMP level or steroidogenesis in the theca layer (Zor et al. 1983). In contrast, LH increases the accumulation of cAMP, progesterone and testosterone as well as of prostaglandin E (PGE) in follicular theca. Besides the gonadotrophins, the direct neural (autonomic nervous system) control of thecal steroidogenesis must be kept in mind in the light of various studies carried out in this regard (see Weiss et al. 1982 for references), as the theca has been shown to contain autonomic nerves, as already discussed.

The ability of antral follicles to accumulate oestrogen in follicular fluid paralleling the ultrastructural, histochemical, and biochemical differentiation of theca interna cells in vivo also agrees well with the ability of these cells in vitro to synthezise oestrogen from its labelled precursors (Guraya 1971 a, 1974 b, Young Lai 1972, 1979, Abel et al. 1975, McClellan et al. 1975, Stoklosowa et al. 1982).

Fortune and Hansel (1979) observed that bovine theca interna is also capable of secreting progesterone, that high intrafollicular concentrations of oestradiol may inhibit thecal progesterone production in vivo prior to the LH surge, and that oestradiol concentration may be an important modulating factor in the control of follicular progesterone secretion (see also Armstrong 1980). Leung et al. (1979) have reported an inhibitory action of oestradiol-17β on LH-induced thecal androgen production in vitro, as also reported by Armstrong (1980). However, addition of different concentrations of oestradiol-17β does not affect the ability of LH to stimulate progesterone production by the thecal cells.

Terranova et al. (1982) have observed that theca is the major source of androgen in proestrous follicles of the hamster ovary. But there occurs a shift from androgen to progesterone production by theca in vitro in hamster pre-ovulatory follicles exposed to an endogenous LH/FSH surge. In other words, progesterone does not accumulate in theca incubations prior to the LH/FSH surge. It is possible that either progesterone is produced and rapidly converted to androgen or some other steroid, or that prior to the LH surge the Δ^5 pathway of steroidogenesis is the preferred pathway for androgen production. The reason for the decrease in theca androgen production following LH/FSH stimulation in vivo may be due to the previously reported "inhibitory" action of LH on oestrogen production during the pre-ovulatory period (Armstrong and Dorrington 1977). It appears that hydroxylase-lyase system (converts progesterone to 17α-hydroxyprogesterone to androgen) may be inhibited in the post-LH/FSH surge follicle (specially in the thecal layer). Terranova et al. (1983) have observed that LH surge alters the in vitro responsiveness of hamster pre-antral follicles to LH. Prior to the LH surge, these small follicles are capable of synthesizing androgen and oestradiol. In response to the LH surge, androgen and oestradiol synthesis is inhibited and progesterone synthesis increases. Theca of pre-antral follicles appears to be the major target of the LH surge. Stoklosowa et al. (1982) studied oestrogen and progesterone secretion by isolated cultured porcine thecal and granulosa cells from a pool of the same pro-oestrous large follicles. Both cell types alone or in combination are capable of secreting oestrogen, androgen, and progesterone in variable amounts.

Mestwerdt et al. (1977b) observed that a conspicuous size-increase of the mitochondria in the steroid biosynthetic active cells of theca interna occurs in the pre-ovulatory as well as in the freshly ruptured follicle, indicating some change in steroid metabolism. The alteration in organelles related to steroidogenesis have also been reported for various other mammalian species (Björsing 1978). Domino et al. (1979) observed that mitochondrial preparations of theca from medium follicles (3 to 6 mm in diameter) of porcine ovary show some ability to convert 4-[^{14}C]-cholesterol to 4-[^{14}C]-pregnenolone and 4-[^{14}C]-progesterone. This steroidogenic activity is greatly increased in thecal mitochondria of large follicles (7 to 12 mm). Higher 4-[^{14}C]-progesterone formation occurs in thecal mitochondria than in granulosa mitochondria of large follicles. Gonadotrophic (LH) stimulation is well known to bring about the development of the steroidogenic apparatus within mitochondria of the follicle (Dimino and Campbell 1980). After treatment with LH, mitochondria from follicles show a twofold increase in cholesterol conversion activity after 18 h of incubation and a fivefold increase in activity after 24 h incubation. LH appears to act by increasing cAMP, which then in turn activates protein kinase, and the activated protein kinase catalytic subunit stimulates the cholesterol side-chain cleavage enzyme complex (Dimino and Campbell 1980).

2. Granulosa

a) Granulosa Cell Origin, Multiplication and Morphology

Considerable controversy exists about the origin of granulosa cells. But recent ultrastructural studies indicate that they originate from the undifferentiated, fibroblast-like stromal cells of the developing mammalian ovary (Stegner and Onken 1971, Guraya et al. 1974, Guraya 1977a, b). These somatic cells in the central region of the ovary are of mesonephric origin (Mauléon 1978). Once the granulosa (or follicle) cells are formed around the primordial oocyte, they increase in their number during follicle growth by mitosis, possibly in response to some hormones (oestrogen) or intra-ovarian factors (Hammond et al. 1982, 1983), thus forming unilaminar, bilaminar, trilaminar follicles (Fig. 11) (Björsing 1978). Granulosa cell multiplication is closely accompanied by synthesis and deposition of follicular fluid and zona pellucida (Fig. 11). The granulosa remains avascular throughout its development (Dvořák and Tesařík 1980).

The follicle cells proliferate, follicular fluid is secreted and an antrum is formed (Fig. 11). Mitotic activity within the granulosa is such that a constant thickness is maintained during much of the follicular growth. Cahill and Mauleon (1980) have observed that follicular growth rates, investigated by measuring the mitotic index before and 2 h after colchicine treatment, vary greatly between different animals studied and do not vary significantly between breeds, time of cycle or season. From three layers of granulosa cells until antrum formation the mitotic index increases slowly but then the follicles grow rapidly, reaching maximum growth rate at a follicular

diameter of 0.85 mm; thereafter the mitotic index decreases almost to zero in pre-
ovulatory follicles. Highly significant correlations exist in each ewe between the
number of follicles and the mean mitotic index per class, suggesting the presence of an
intra-ovarian mechanism regulating folliculogenesis. Coulson (1979) has characterized
polyploidy in ovarian granulosa cells. High ploidy values are shown for the first time in
granulose cells from bovine and porcine ovaries. This polyploidy appears to be hightest
in small immature follicles and decreases as the follicles mature. Ploidy changes in
granulosa cells have been suggested to form an essential part of the differentiation
process. The control factors involved need to be determined, but may be hormonal in
nature. Wyche and Noteboom (1977) demonstrated the requirement of a specific
factor for the multiplication of ovarian cells in serum-free medium. This factor isolated
from foetal calf serum appears to be a large glycoprotein of MW between
600,000–700,000. Ovarian growth factor has survival but not mitogenic activity in the
serum-free system. A cooperative interaction between ovarian growth factor and
limiting concentrations of the serum factor stimulates ovarian cell multiplication. By
far the most potent stimulators of granulosa cell replication evaluated in vitro are the
low molecular weight peptides, epidermal growth factor and fibroblast growth factor
(see Hammond et al. 1982, 1983). In addition, requirements for insulin, transferrin, as
well as steroids and thyroxine, have also been defined for optimal maintenance of
granulosa cells under in vitro conditions (Hammond et al. 1982, 1983). Follicular fluid
is found to be comparable to serum in stimulating replication of porcine granulosa cells
in vitro (Hammond et al. 1982). Qualitative and quantitative studies have shown the
presence of somatomedins and somatomedins-binding proteins in the follicular fluid
(Hammond et al. 1982, 1983). In most cases, the concentrations of somatomedins are
higher in the follicular fluid than those found in the serum, and they are much greater
than insulin concentrations. However, a physiological role for those compounds in
vivo has remained uncertain.

The development, arrangement, morphology and number of constituent granulosa
cells vary in comparable follicles of different mammalian species (Mossman and Duke
1973, Bjërsing 1978, van der Merwe 1979, Nicosia 1980a). In the Natal linging bat
(Miniopterus schreibersi natalensis) only one Graafian follicle develops, which shows a
large antrum with the ovum-bearing mass of cells occupying only a relatively small
portion of the antrum (van der Merwe 1979). Kvlividze (1976) studied age-related
characteristics of the number of granulosa cell layers of ovarian follicles in Wister rats
of various ages. In the process of aging, the number of layers in the growing follicles is
decreased after 12 months of age. This decrease coincides with decreased maturation
and growth of follicles and precedes the loss of gonad reproductive function.

For the unilaminar follicle the granulosa cells change from flattened to cuboidal or
columnar, forming a continuous investment around the centrally placed oocyte. Their
nuclei, which are spherical to slightly ovoid, become voluminous, more profoundly
indented, and frequently subdivided into irregular lobes conjoined by tenuous
chromatin bridges (Björkman 1962, Zamboni 1974, 1976, 1980, Guraya 1977a,
Bjërsing 1978). Chromatin distribution begins to become more regular. Conspicuous
are large nucleoli which develop reticular structure. In the multilaminar follicle the
granulosa cells are separated by irregular lacunae which soon become filled with liquor
folliculi (Fig. 11) (Bjërsing 1978, Dvořák and Tesařík 1980). Anderson et al. (1978)
observed that during follicular development in the mouse and rabbit changes in the

granulosa cells are especially noteworthy and include dramatic modifications in the cell shape coincident with antrum formation, as also described for other mammalian species (Bjersing 1978, Zamboni 1980).

Whatever may be the source of follicular fluid, its progressive accumulation induces distension and coalescence of the lacunae, resulting in the formation of the antrum (Fig. 11 g–i). Finally, the granulosa cells are separated into a membrana granulosa lying adjacent to the basement membrane, and several to many concentric layers of granulosa cells forming the cumulus oophorus around the oocyte (Fig. 11 i) (Björkman 1962, Mossman and Duke 1973, Zamboni 1974, 1976, 1980, Hay and Cran 1978, see also Cran et al. 1979). The granulosa cells lining the antrum of Graafian follicles and the cumulus cells surrounding the oocyte are thus derived from the follicle cells of unilaminar follicles. The mitotic rate in the cumulus is greater resulting in its growth (Hay and Cran 1978). This is accompanied by the development of a reticulum of lacunae, few of which have egress to the antrum. In normal follicles, the granulosa and cumulus cells show a similar structure as seen both at the level of the light and electron microscope. Both compartments have two morphological types; basal columnar cells and above these several layers of more rounded cells with larger intercellular spaces. At least in small and medium-sized antral follicles there is generally seen a considerable degree of similarity between the cells comprising the granulosa and the cumulus (Fig. 11). The oocyte remains embedded at one side of the follicle in the cumulus oophorus forming a mound of cells (Fig. 11 i). The cumulus oophorus (or somatic cell mass) appears to be directly involved in the nourishment of the growing oocyte, as discussed in Sect. C. Due to progressive increase in the size of the antrum, the cumulus oophorus and its oocyte become eccentrically displaced. The oocyte with its remaining granulosa cells (the corona radiata) retains its attachment to the follicle wall through the cumulus oophorus (Mossman and Duke 1973). As the amount of follicular fluid increases, the cumulus with the oocyte becomes further displaced to a more eccentrical position within the follicle (Fig. 11 i). Bjersing (1978) has described the detalls of morphological variations of granulosa cells of the membrana granulosa, cumulus oophorus and corona radiata in different species of mammals (see also Zamboni 1980). The immediate relationships between the oocyte and its extracellular environment are not significantly altered, however, until just before ovulation, as will be discussed in Chap. IV.

Granulosa cells during follicular development show distinct morphologic patterns when examined with transmission and scanning electron microscope (Fig. 36) (Motta and van Blerkom 1974, Blersing and Cajander 1974b, Anderson et al. 1978, Bjersing 1978, Dvořák and Tesařík 1980, Zamboni 1980). Most of them are polyhedral and possess regular surfaces. Cells of this type are particularly present near the basal lamina in large follicles and form the largest portion of the membrana granulosa. The mural granulosa cells lying close to the basement lamina become pseudostratified and their nuclei elongated and eccentric (Lipner 1973, Hirschfeld 1979, Zamboni 1980). These morphological alterations coincide with the cessation of cell division in this layer and with the development of LH receptors (Richards and Midgley 1976, Oxberry and Greenwald 1982), as will be discussed later on. The cells facing the follicular cavity or antrum are elongated or flattened. All granulosa cells of large antral follicles have irregular evaginations and surfaces (Fig. 36) which are covered with material having a filamentous/reticular texture. This material, which probably corresponds to pre-

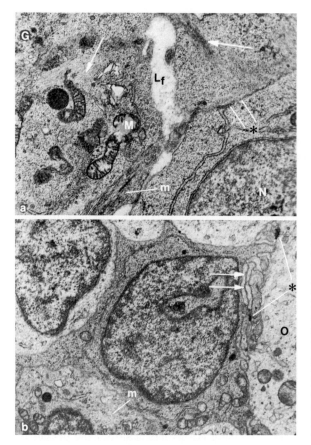

Fig. 36 a, b. Electron micrographs showing morphologies of granulosa cells of a developing follicle. **a** Bundles of microfilaments are present in the cytoplasm and in the larger cellular evaginations; some of them appear in contact with the plasma membranes *(*)*; *(m)* microtubules; *M* mitochondria; *N* nucleus; *G* Golgi complex; *Lf* Liqour folliculi; **b** Microvillous projections (→) towards the oocyte *(O)*; attachment zones and microtubules *(m)* are also occasionally present (From Motta and DiDio 1974)

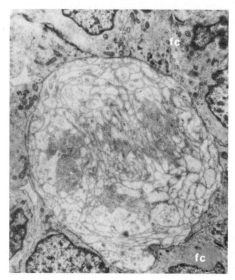

Fig. 37. Electron micrograph showing Call-Exner body in ovarian follicle of rabbit; *fc* follicle cells (From Zamboni 1972)

cipitated follicular fluid, has attached to its flattened cells and cytoplasmic debris and forms a lamina-like structure which limits the interior surface of the antrum of Graafian follicle.

b) Call-Exner Bodies

In the growing follicles of some mammals (e.g. rabbit and human) there are developed Call-Exner bodies (Fig. 37) (Motta 1965, Zamboni 1972, Motta and van Blerkom 1974, Mestwerdt et al. 1977a, Bjërsing 1978, Guraya 1980a, Dvořák and Tesařík 1980). They are surrounded by granulosa cells bound to one another by tight intercellular junctions. The granulosa cells are arranged in a radial fashion around

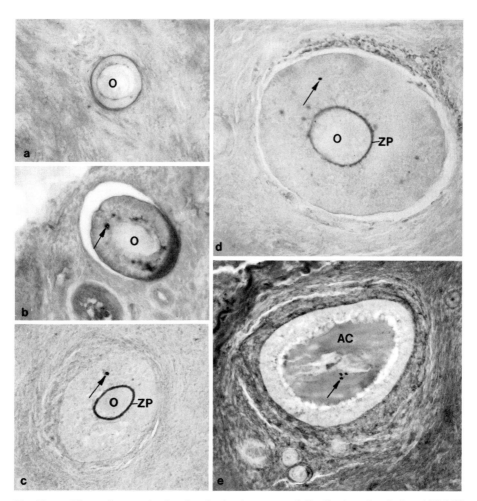

Fig. 38a–e. Photomicrographs showing the development and distribution of periodic acid Schiff (PAS)-positive Call-Exner bodies or droplets *(arrows)* in the granulosa of developing pre-antral follicles and then their final release *(arrow)* in the antral cavity *(AC)* of antral follicle of human ovary. Strongly PAS-positive zona pellucida *(ZP)* is developed between the surface of oocyte *(O)* and granulosa

spherical fluid-filled vesicles (Fig. 37). A basement membrane-like lamina lines the periphery of the cyst running parallel and close to the plasma membrane of the granulosa cells. Scanning electron microscope stereo-views show that the basal lamina limiting the Call-Exner body is composed of a thin membranous layer to which are attached large extensions of granulosa cells.

Histochemical and ultrastructural studies have shown that the PAS-positive glycoproteins are collected in the cysts (Call-Exner bodies) prior to release into the follicular antrum (Fig. 38) (Zamboni 1972, Bjërsing 1977, Guraya 1980a). This view is supported not only by the frequent observation of Call-Exner bodies with interrupted walls in the process of liberating their contents in the antrum, but also by the absence of such bodies in large follicles with fully distended antra (Zamboni 1972). All the recent studies carried out with the techniques of histochemistry, and transmission and scanning electron microscopy, have clearly shown that the PAS-positive droplets do not lie within the cytoplasm of granulosa cells (Motta 1965, Motta and Van Blerkom 1974, Zamboni 1972, Guraya 1980a) and therefore, the Call-Exner bodies are extracellular cysts to which granulosa cells are radially attached (Fig. 38). Finally their PAS-positive material is released into the follicular fluid (Fig. 38).

c) Plasma Membrane Specialization and Their Significance

During follicular growth, the granulosa cells forming the follicular epithelium undergo a series of complex changes in plasma membrane structure and function. As follicular growth is regulated by pituitary gonadotrophins (FSH, LH and prolactin), the membrane components responsible for both reception and translation of these hormonal stimuli perform an important function in the differentiative process leading to ovulation and luteinization of the follicle. The contact between granulosa cells becomes tenuous as they assume highly irregular outlines (Fig. 36) and develop various cytoplasmic processes which usually terminate by contacting, or invaginating, into the cytoplasm of another granulosa cell, or granulosa cell processes (Peluso et al. 1977a, Bjërsing 1978, Anderson et al. 1979, Cran et al. 1979, Albertini 1980). The cytoplasmic processes arising from cells of the corona radiata traverse the zona pellucida (Fig. 13), and some of these come into intimate contact with the oocyte microvilli (Zamboni 1974, 1976, 1980, Dvořák and Tesařík 1980), as already discussed in this chapter.

Junctional complexes. In the early antral follicles, the apposed cell membranes in the areas of contact associate by means of desmosomes (Fig. 39a, b) (Zamboni 1974, Amin et al. 1976, other references in Bjërsing 1978, Anderson 1979, Amsterdam and

---

Fig. 39a–g. Electron micrographs showing structure of junctional complexes in the rabbit granulosa. **a** Lanthanum-impregnated gap junction showing a 2–3 nm and microtubules *(arrow heads)* coursing parallel to the junctional membrane; **b** Transmission electron micrograph of a gap junction with associated microtubules *(MT)*; **c** En-face view of a lanthanum-impregnated gap junction in which polygonal arrays of subunits are outlined by the traces; **d** Freeze-fracture electron micrograph of a gap junction illustrating the particulate *P* face and pitted *E* face; **e** Thin section of portion of two granulosa cells from a type 3 follicle showing two macula adherens *(MA)*, junctions flanking a gap junction *(GJ);* **f** Plicated gap junctions filled with lanthanum tracer are shown from rabbit granulosa cells at 3 following injection of an ovulating dose of hCG; note the presence of 6–7 nm filaments *(MF)* within the junctional process; **g** Freeze-fracture replica of granulosa cell gap junctions composed of "loose" particle arrays interspersed between crystalline gap junction plaques *(arrowed)* (From Albertini 1980)

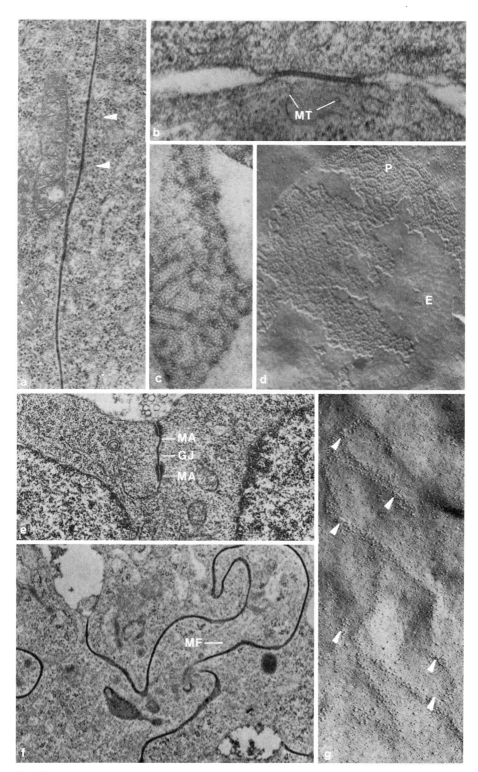

Fig. 39

Lindner 1979, Zamboni 1980). As the follicle develops, the number and size of the individual junctions increase and a large variety of shapes are seen (Albertini et al. 1975, Bjërsing 1978, Albertini 1980). In the follicles with very distended antra, containing oocytes nearing initiation of meiotic maturation, the prevalent means of cellular association is represented by two types of junctional devices morphologically comparable to (1) tight junctions (maculae adherentes) and gap junctions (or nexuses) which may be punctuated by intercomittant rudimentary desmosomes (Fig. 39e) (Espey and Stutts 1972, Merk 1971, Merk et al. 1972, 1973, Zamboni 1974, Bjërsing and Cajander 1974d, Albertini and Anderson 1974, Albertini et al. 1975, Amsterdam et al. 1975, Abel et al. 1975, McClellan et al. 1975, Anderson and Albertini 1976, Fukushima 1977, Coons and Espey 1977, Gilula 1977, Bjërsing 1978, Anderson 1971, 1979, Cran et al. 1979, Albertini 1980, Dvořák and Tesařík 1980, Toshimori and Oura 1982). The results of these various studies have indicated that the granulosa cells are electrically and (most likely) metabolically coupled by large gap junctions, whose size and frequency appear to be hormonally controlled (Merk et al. 1972). Towards the basal lamina the membranes of the adjacent cells of the granulosa appose very closely, forming which may appear to be tight junctions by transmission electron microscopy.

The follicular epithelium shows adhesive contacts which differ from true desmosome in that the intraperiod line in the extracellular space is absent (Albertini 1980). This junction, which is characterized by the association of filamentous dense material on the intracellular aspect of the plasma membrane, is called macula adherens (Fig. 39) (Albertini 1980). Although there occur changes in the incidence of maculae during growth of the follicle, a constant characteristic of this junction is its tandem placement with developing gap junctions on granulosa cells. Frequently, paired maculae flank small gap junctions in early stages of follicular growth.

Gap junctions or nexuses are highly differentiated portions of the plasma membrane between adjoining granulosa cells and thus appear as minute attachment sites that are frequently seen at the edges of cells throughout the granulosa. Both long zones of contact between cells (abutment nexuses) and spherical inclusions (annular nexuses) have been reported for the rabbit follicle (Fig. 39) (see Bjërsing 1978). Freeze-fracture replicas of gap junctions in early antral follicles show a number of clustered aggregates or plaques of 8–9 nm particles on the P-face which abuts the cytoplasm, and corresponding pits on the E-face or extracellular leaflet of the membrane (Fig. 39c, d) (Anderson 1979, Albertini 1980). As differentiation advances, large gap junctions may be seen occupying rather large areas of the membranes. The hexagonal pattern of gap junctions appears to result from the organization of the subunits comprising each particle (Paracchia 1973). The subunit particles and pits generally show a uniform packing periodicity of 9 nm, but on occasion extremely loosely packed arrays of particles are observed where the packing periodicity exceeds 10–11 nm (Fig. 39g). These loose particle arrays were originally interpreted as zones of junction formation, but more recent studies have shown that such patterns represent the structure of gap junctions in the coupled state, whereas crystalline, tight-packed aggregates represent uncoupled gap junctions (Paracchia 1977). According to Amsterdam and co-workers (Amsterdam et al. 1974a, b, 1976, Amsterdam and Lindner 1979), the hexagonal lattice structure can be easily achieved by submitting electron micrographs of freeze-cleavage replicas to optical differentiation analysis.

Merk et al. (1972) observed a correlation between the extent of thecal development and incidence of granulosa cell gap junctions whose number is greater if the theca interna cells are fully differentiated as steroid secretors, thus suggesting the role of steroid hormones of thecal origin in their formation. A similar suggestion has also been made by other workers (Bjërsing and Cajander 1974d, f). By using diethylstilbestrol, Merk et al. (1972) further confirmed this suggestion by showing that the frequency of occurrence of nexuses is oestrogen-dependent. Bjërsing and Cajander (1975d) observed that at 4 and 6 h after injection of hCG the granulosa cells in rabbit Graafian follicles are slightly dissociated around the whole follicle (see also Bjërsing 1978). The so-called abutment nexuses decrease and continue to do so up to ovulation, as also demonstrated for the granulosa cells of mouse ovaries treated with PMSG (Byskov 1979). Meanwhile, the number of anular nexuses has more than doubled (Bjërsing 1978).

Gap junctions (or nexuses) have been implicated as sites of molecular and ionic exchange and electrical and metabolic coupling between cells (Sheridan 1971, Gilula et al. 1972), which serve to coordinate cell function in many tissues. Both Co_2 and the Ca-transporting ionophore A 231987 (calcimycin) markedly reduce movement of labelled components, and the reported effects of Co_2 and Ca^{2+} support the suggestion that the gradients are a result of junctional transfer (Cran et al. 1981). In addition, morphological examination of Co_2-treated follicles shows a loss of gap junctions. The presence of gap junctions between granulosa cells and between cumulus cells and the oocyte (Zamboni 1980) has also suggested that this contact specialization mediates the intercellular exchange of regulatory factors during ovulation and luteinization (see references in Albertini 1980). Various electrophysiological and dye passage experiments support this suggestion and demonstrate that ions and small molecular weight substances are freely exchanged through granulosa cell gap junctions, a function known to be performed by gap junctions in many tissues (Gilula et al. 1978). The presence of tight junctions would seem to be incongruous with physiological experiments which have demonstrated that plasma proteins of very large size (in the neighbourhood of 1 million MW) have access to the follicular fluid (Shalgi et al. 1973). Under organ culture conditions, large proteins present in the follicular fluid may pass into the culture medium (Ménézo et al. 1978). Experiments using macromolecular markers in electron microscopy have confirmed that a blood-follicle barrier does not exist (Szöllösi et al. 1978). If truly present, tight junctions between granulosa cells would be either apparent or focal, thus allowing the fluid to have fairly free passage in and out of the follicle (Cran et al. 1976). Szöllösi et al. (1978) have illustrated this aspect in pig follicles, using horse-radish peroxidase as marker. The results obtained by using lanthanum nitrate also coincide fully with those obtained with peroxidase (Fig. 39).

The second important feature of granulosa cell gap junctions is the presence of intracellular vesicles of gap-junction membrane. These circular-sectioned profiles were earlier designated as annular gap junctions and are formed from deep invaginations of membrane-localized gap junctions which are pinched off during internalization (Espey and Stutts 1972, Albertini and Anderson 1975, Merk et al. 1973). Larsen and Tung (1978) have studied the origin and fate of cytoplasmic gap junctional vesicles in rabbit granulosa cells. Vesicles within the cytoplasm of granulosa cells in both thin section and freeze-fracture preparations consist of typical gap junctional membrane and are mostly composed of two membranes; in rare cases they may consist of a single membrane, identified as gap junctional due to the presence of typical 8.5 nm P-face particles and E-face pits. These vesicles appear to be formed by the endocytosis of junctional membrane from the surface as one invaginates into its neighbour at the junction (Fig. 39 f). Their association with microfilaments involves an active contractile process mediated by the intracellular actomyosin system of filament. The internalization of gap junctions during ovulation has been suggested to be related to the degredative breakdown of this membrane specialization (Albertini 1980). The presence of apparent acid phosphatase activity within the matrix of these gap junctional vesicles suggests that they may represent a stage in the specific degradation of gap junctional membrane (Larsen and Tung 1978).

Coons and Espey (1977) made a semi-quantitative study to determine changes in the abundance and size of surface nexuses and changes in the abundance of interiorized nexuses in growing and mature ovarian follicles during the ovulatory process. Mature follicles show larger granulosa cells than follicles in the early stages of antral formation (Guraya and Gupta 1979). Also, the granulosa cells of mature follicles have a slighly greater number of surface nexuses (without change in nexus length), and more annular gap junctions or interiorized nexuses as compared to immature follicles (Anderson 1979). These annular structures may contain any constituent(s) for the cytoplasm. In freeze-fracture replicas there are both 9 nm particles and pits. The biological significance of the internalization (phagocytosis) of certain gap junctions is still not known. However, Han (1979) has observed that time-dependent migration of [^{125}I]-hCG is endocytosed by granulosa cells and the endocytosed hormone might be sequestered into the lysosomes, multivesicular bodies, and/or transproted into the nucleus.

Thin-section electron microscope analysis of rat- and rabbit-cultured granulosa cells treated with concanavalin A (Con A) at 37 °C has demonstrated coordinated changes in the cytoplasmic disposition of microfilaments, thick filaments and microtubules during cap formation and internalization of lectin receptor complexes (Albertini and Anderson 1977). Con A receptor clustering is accompanied by an accumulation of subplasmalemmal microfilaments which assemble into a loosely woven ring as patches of receptor move centrally on the cell surface. Periodic densities develop in the microfilament ring, which is reduced in diameter as patches coalesce to form a single control gap. Microtubules and thick filaments emerge associated with the capped membrane. Capping is followed by endocytosis of the Con A receptor complexes. During the process, the microfilament ring is displaced basally into the cytoplasm, and endocytotic vesicles are transported to the paranuclear Golgi complex along microtubules and thick filaments. Ultimately, these vesicles accumulate near cell centre where they are embedded in a dense meshwork of thick filaments. Freeze-fracture analysis of Con A-capped granulosa cells has revealed no change in the arrangement of peripheral intramembrane particles, but large, smooth domains are conspicuous in the capped region of the plasma membrane.

The internalization of certain gap junctions may represent a mechanism for clearing the surface of "used" receptors, provided there is a turnover and reformation of the gap junctions. Recent studies suggest that the lysosomal system is involved in the decrease of gonadotropin receptors by a process which involves the inclusion of portions of the plasma membrane into vesicles (Dimino and Elfont 1980). These vesicles then combine with primary lysosomes, this resulting in the breakdown of receptor-hormone complexes. Cran et al. (1979) studied alterations in the junctional complexes of the membrana granulosa and cumulus in the sheep. In late antral follicles (3.0–5.9 mm diameter) gap junctions are frequent. At mid-estrus numerous annular nexuses are present in the granulosa but not in the cumulus.

Byskov (1979) has studied the alterations in the morphology of granulosa layers in the "healthy", "rescued" and atretic PMSG-treated large follicles of the mouse ovaries. The cells are often rounded but have developed numerous slender microvilli that protrude from their surfaces. As compared to untreated follicles, the cells in treated follicles are well separated from each other even close to the basement membrane. The number of desmosomes and gap junctions is apparently decreased as compared to untreated large follicles. Gap junctions are frequently seen between the cumulus cells. Many gap junctions seen between the microvilli within the zona pellucida appear to be internalized in the microvilli protruding from the cumulus cells. As the junctional specialization of the granulosa have been proposed to be involved in transport and exchange of ions and molecules and to function as electric and metabolic couplers, their decrease in large follicles of the PMSG-treated mouse (Byskov 1979) may be indicative of disturbance in the intrafollicular coordination systems. This may affect the differentiation of all follicular compartments. The gap junctions of granulosa are believed to accommodate many of the biochemical events associated with growth, maturation (meiosis), and ovulation. Amsterdam and co-workers (Amsterdam et al. 1975, 1976, 1979a) believe that gap junctions between granulosa cells, and between cumulus cells and the oocyte, may serve to propagate an LH-initiated signal towards the interior of the follicle, since only theca cells and mural granulosa cells, but not the periantral and cumulus granulosa cells, could be shown to possess LH receptors by autoradiographic analysis (Amsterdam et al. 1975, Oxberry and Greenwald 1982). Actually, there is a need for a good quantitation (morphometric analysis) of gap junctions for both the normal developing and atretic follicles. Basically, two methods of analysis have been used for the analysis of surface changes: (1) a point-counting method for determination of the relative and absolute amounts of gap junctional membrane present at three stages of follicular growth (early and mid-antral and pre-ovulatory follicles) (Wiebel and Bolender 1973); (2) planimetry for absolute analysis of surface, junctional, and cytoplasmic changes during follicular growth. Albertini (1980) has discussed the comparative aspects of the data obtained with these two methods, which are unable to resolve significant surface area alterations. However, it has become evident that a considerable amount of plasma membrane in the pre-ovulatory granulosa cell is occupied by gap junctions (approximately 30%) limiting the amount of surface membrane available for hormone reception.

Staehelin (1974) has reviewed the morphological information on intercellular junctions in various types of tissues derived from thin-sectioning, negative staining and freeze-cleave techniques, as well as from X-ray diffraction and biochemical investigations and correlated the structural parameters with known or proposed physiological functions. Gap junctions in the granulosa are believed to be involved in cell-to-cell attachment in cell-to-cell communication, in transport of hormones and nutrients within and through the follicular epithelium, as well as in hormone receptor relationships. The lack of a vascular system within the membrana granulosa appears to be replaced by the membrane specialization of the granulosa cell itself. Zamboni (1974, 1976, 1980) believed that the granulosa cell processes and junctional devices in the granulosa of a growing follicle ensure the isolation and stability of the follicular environment during critical phases of oocyte maturation. Merk et al. (1972) assumed that when granulosa cells become coupled by the nexus type of intercellular junction, they cease to exist as isolated units and become a communicating cell system, i.e., "functional syncytium", as also discussed by Staehelin (1974).

Besides the intercellular junctional devices, there is also present intercellular matrix, which gives positive histochemical reactions for complex carbohydrates with 1, 2-glycolytic acidic groups and D-mannoacyl and α-D-glucoacyl residues (Tadano and Yamada 1978). The complex carbohydrates contain hyaluronic acid, sulfate glycosaminoglycans and isomeric chondrotin sulfates. Tadano and Yamada (1979), using the electron microscope, have observed that in the intercellular matrix of granulosa cells in the adult mouse, dialyzed iron (DI) reactive structures containing acidic complex carbohydrates consist of a layer of a variable thickness coating their plasma membrane and reticular elements distributed in the spaces between the granulosa cells. The latter occurs in two types, one is clumped masses of irregular shapes and different sizes, whereas the other consists of filamentous figures radiating from the masses. The effects of digestion with streptomyces and testicular hyaluronidases upon the DI staining of the tissues indicate that DI reactive structures in the intercellular matrix contain at least three types of acidic complex carbohydrates.

Microvilli and blebs. The surface of granulosa cells is extremely irregular in outline due to the development of surface projections. The most common extensions are microvilli which may show a diameter of 80–120 nm and in longitudinal section may extend for 2–3 μm (Fig. 36). Besides the elongate surface projections, many blebs or bulbous extensions develop which, like the microvilli, also do not contain organelles. But abundant microfilaments are present in all these surface projections (Fig. 36). Changes in the extent and disposition of microvilli and blebs during folliculogenesis and atresia can be revealed by scanning electron microscopy. There is some evidence that, as the follicle grows, the number of microvilli seen in granulosa cells is increased (Motta and Van Blerkom 1974, Chang et al. 1977, Peluso et al. 1977a). Their functional significance in relation to absorption and secretion of substances within the granulosa must be determined in future studies.

Cilia. Besides the surface projections having microfilaments, granulosa cells also show primary cilia, generally one per cell that appear oriented towards the centre of the follicular antrum (Albertini 1980). In comparison to the microvilli, the primary cilium possesses internal structure, as it contains 9 + 0 array of doublet microtubules without dynein side-arms. The functional significance of the primary cilium in the granulosa needs to be determined more precisely but in other sensory tissues it serves a sensory function. It is interesting that the incidence of ciliation in the developing follicle is higher just before ovulation.

d) Cytological, Histochemical and Biochemical Features

Cytological characteristics. In the granulosa cells of developing follicles, electron microscopic studies have revealed the presence of a Golgi complex, granular ER in the form of tubules, small vesicles and cisternae, many free ribosomes, numerous clusters of polyribosomes, and rod-shaped or spherical mitochondria with simple lamellar cristae (Figs. 21, 36, and 40) (Guraya 1971a, Zamboni 1974, Bjërsing and Cajander 1974d, Abel et al. 1975, McClellan et al. 1975, Amin et al. 1976, Nowacki 1977, Mestwerdt et al. 1977a, b, 1979, Bjërsing 1978, Zamboni 1980, Dvořák and Tesařík 1980, Cran et al. 1979). These follicles are generally designated as primary, secondary and tertiary follicles (Mestwerdt et al. 1979). A striking morphometric characteristic of granulosa cells in this group of follicles in the human ovary is the small mean volume of cells and nuclei (Mestwerdt et al. 1979). The cytoplasmic structures, when examined in relation to cell volume, show abundant ER, the rough reticulum being twice that of the smooth reticulum. Another conspicuous feature is the small lipid volume in the cytoplasm of these cells. In secondary follicles of the sheep ovary, the rough and smooth ER proliferates forming an anastomosing network (Cran et al. 1979). In early antral follicles (0.4–2.0 nm diameter) the rough ER is composed of short cisternae; elongated mitochondria are present. The various cytoplasmic structures show irregular distribution. Smooth ER is generally in the form of vesicles. Cisternae of granular ER tend to be concentrated in some regions of the cytoplasm nearly devoid of other

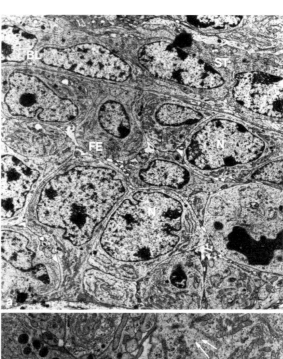
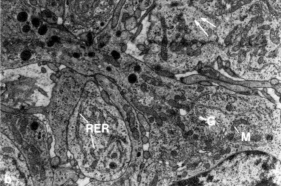

Fig. 40a, b. Electron micrographs of portions of trilaminar follicle from the guinea-pig ovary. **a** Showing follicular epithelium *(FE)*, basal lamina *(BL)* and surrounding stromal tissue *(ST)* or thecal cells. Follicle cell nuclei show electron-dense heterochromatin and nucleoli in the nuclei *(N)*, and various organelles in the cytoplasm (Guraya et al. 1974); **b** Higher power view of follicle cells showing Golgi complex *(G)*, mitochondria *(M)*, rough endoplasmic reticulum *(RER)*, numerous ribosomes *(arrows)*, follicle cell processes, etc.

components (Fig. 40). Numerous polysomes are present. The Golgi complex is well developed, consisting of long cisternae and small vesicles. Dvořák and Tesařík (1980) have reported the presence of some secondary lysosomes in granulosa cells of growing antral follicle of humans, which show few glycogen granules. Compound aggregates are absent. Nuclei are pleomorphic and contain large amounts of heterochromatin (Fig. 21) (Bjërsing 1978). Nucleoli are present. These cytological features of granulosa cells of growing follicles represent morphologic characteristics of protein synthetic active cells prior to the pre-ovulatory period (Mestwerdt et al. 1977b, 1979). Amin et al. (1976) have observed amorphous material of moderate electron density, similar to that of follicular fluid, in the cisternae of the granular ER, which have been suggested to represent secretory material. This material appears to reach the plasma membrane of granulosa cells by vesicles which fuse with it before discharging their content into the intercellular space. These ultrastructural observations also confirm the secretory function of granulosa cells during the growth of the follicle.

Recently, microfilaments (4–7 nm thick) have been demonstrated in granulosa cells of rabbit, dog, rhesus and human ovaries during the development of the follicle and during the pre- and post-ovulatory stages (Fig. 36) (Motta and DiDio 1974, Zamboni 1974, Abel et al. 1975, Mestwerdt et al. 1977a, 1979, Bjërsing 1978). The microfilaments are arranged in parallel bundles and may be located in a greater amount within large evaginations and microvillous-like projections of the granulosa cell cytoplasm (Fig. 36). These microfilaments are also present within the superficial areas of the granulosa cells. In all these cases, the microfilaments lie at locations where a contraction might move portions of the cells or partially change their shape (i.e., within the amoeboid and microvillous projections of the cytoplasm and within the superficial areas of the cells). Mestwerdt et al. (1979) have suggested that as these filaments are primarily seen in granulosa cells in the immediate vicinity of the zona pellucida in the human ovary, they are likely to be a factor in the formation of the zona pellucida. The microfilaments are probably comprised of f-actin, and appear to be more abundant in granulosa cells of ovulatory follicles (Motta and DiDio 1974, Cavallotti et al. 1975).

Albertini and Kravit (1981) have studied the isolation and biochemical characteristics of 10-nm filaments from cultured ovarian granulosa cells and colchicine-treated primary cultures of rat ovarian granulosa cells. Negative staining has revealed in electron micrographs an average filament diameter of 10.3 nm in the isolated fibre bundles, which, upon sodium dodecyl sulfate polyacrylamide gel electrophoresis, are observed to contain a major phosphorylatable polypeptide with a MW of 57,000, and several minor components including actin. Amenta and Cavallotti (1980) have demonstrated fluorescent antimyosin-like protein antibodies in the cytoplasm of oocyte and granulosa cells in the rabbit ovary. From the sites of their distribution as well as from their biochemical nature, Motta and DiDio (1974) have suggested that these microfilaments are capable of producing "in vivo" contractile movements of ganulosa cells and of partially causing change in their shape (see also Motta and Van Blerkom 1974). This suggestion is further supported by the recent observations on the effects of Ca^{2+} deprivation on cell shape in cultured ovarian granulosa cells (Batten and Anderson 1981). Ca^{2+} and Mg^{2+} deprivation results in cellular rounding which is reversible. Concomitant with the shape change, there is a dispersion of the structural proteins actin and α actinin. The arrays of large actin-containing bundles (stress fibres) are converted to a diffuse network, as observed by EM: α actinin, which is observed by immunocytochemistry to be in a periodic array along the actin bundles; the bundles are also disrupted and redistributed in the periphery of the cell upon rounding. These results indicate that Ca^{2+} and/or Mg^{2+} are necessary to maintain the integrity of stress fibres and/or restrict the movement of α actinin anchoring sites within the membrane. Batten and Anderson (1983) have studied distribution of actin in cultured ovarian granulosa cells. In addition to microfilaments, a number of microtubules (24 nm thick) have also been observed in the granulosa cells of rabbit ovary (Motta and DiDio 1974). They are believed to contribute to cell shape and elongation and also may be remnants of the mitotic spindle.

The abundant RNA, proteins and phospholipid bodies formed by the granulosa cells of developing follicles must be involved in their growth and differentiation processes (Guraya 1971 a). According to Mestwerdt et al. (1977 a), the metaplastic structures in the granulosa cell play a role in the normal development of the zona pellucida and the Call-Exner bodies, which have already been described. It is interesting to emphasize that the organelles specific to steroidogenic tissue, such as diffusely distributed lipoproteins, abundant agranular ER, mitochondria with tubular cristae and lipid granules, which are already described for hypertrophied theca interna cells (or thecal gland), do not show any appreciable development in the normal granulosa cells of growing follicles (Bjersing 1978). Mestwerdt et al. (1979) have also concluded that in the primary, secondary and tertiary follicles the granulosa cell layer rarely shows submicroscopic features characteristic of luteinization and steroid biosynthetic functions, respectively. This indicates that the granulosa cells of the developing follicle are not luteinized in vivo. Channing (1979 a) has suggested that there are inhibitors in the follicular fluid which are responsible for keeping the granulosa cells from luteinizing by regulating the cellular response to LH and FSH. These inhibitors will be discussed later in Sect. F of this chapter. This suggestion is supported by the fact that in cultures of granulosa cells from pre-ovulatory porcine follicles luteinization can be inhibited in the presence of fluid from small and medium follicles. The LH inhibitors can prevent in vitro luteinization of in vivo LH-primed mature granulosa cells. It has also been shown that oestrogens have no influence on the activity of the inhibitor, but that steroids may act on the secretion of inhibitor. Mestwerdt et al. (1977 a) have observed that the granulosa cells of tertiary follicle show the presence of single or grouped fat droplets in the cytoplasm, with metaplastic structures like filaments and/or microtubules being rare.

Enzymes. The granulosa cells of nonovulatory tertiary follicles in the human ovary show moderate activities of alkaline phosphatase and leucine aminopeptidase and LDH, whereas those of pre-ovulatory Graafian follicles show strong activity of LDH and alkaline phosphatase (Brietenecker et al. 1978). Enzyme activities (e.g. 3β-HSDH) indicative of steroidogenesis are generally not seen in the normal granulosa cells of developing antral follicles except in the basal portion (i.e. of granulosa cells) (Fig. 35 c) (Guraya 1971 a, 1973 a, 1974 b, Hoyer 1980, Parshad and Guraya 1983). Before the pre-ovulatory period, a weak activity for 3β-HSDH has been demonstrated in the granulosa cells by some workers (see Bjersing 1977 for references). The granulosa cells in large follicles of the Indian gerbil show some activity of 3β-HSDH when dihydroepiandrosterone was used as a substrate, but no visible activity appears with pregnenolone (Parshad and Guraya 1983); weak activites of 17α-HSDH are also developed in the granulosa of such follicles. In contrast, all workers have observed 3β-

HSDH activity in granulosa cells shortly after ovulation. The most important single factor determining a positive or negative activity appears to be the stage of the oestrous cycle at which the ovarian specimen is obtained (Bjërsing 1977). Using pre-ovulatory alterations of the ova as the criterion for the diagnosis of a Graafian follicle, Friedrich et al. (1974) have observed that granulosa cells show a weak-to-moderate 3β-HSDH activity, which in some cases equals the intensity seen in the thecal layer. Brietenecker et al. (1978) have observed that 3β-HSDH activity is lacking in primary and secondary follicles and in the granulosa cells of nonovulatory follicles, whereas the theca interna of such follicles shows moderate activity. Granulosa cells of human pre-ovulatory Graafian follicles shows 3β-HSDH activity (Brietenecker et al. 1978). Swartz and Schuetz (1980) have demonstrated the heterogeneity of intrafollicular somatic cells (granulosa cells) and ovulated cumulus masses as evidenced by 3β-HSDH activity. Granulosa cells lining the walls of vesicular follicles show different degrees of enzyme activity based on their distance from the basement membrane. Intrafollicular transformed cumulus masses and cumulus cells of ovulated masses within the oviduct do not react uniformly: some are positive for the enzyme and others are not. Parshad and Guraya (1983) have also observed histochemical differences in different layers of granulosa cells of developing follicles in the ovaries of the Indian gerbil. The inner layers of granulosa in pre-antral follicles with two to three layers of granulosa cells contain relatively more activity of NADH- and NADPH-tetrazolium reductases than their outer layers. During further growth of the follicle some follicles with incipient antrum formation show more NADPH-tetrazolium reductase activity in the middle layers of their granulosa, while others show both NADH- and NADPH-tetrazolium reductases in the outer layers, which also show more activities of succinate- and lactate-dehydrogenates (SDH and LDH) (Fig. 35a and b) and lipids than the inner layers. LDH tends to show more activity in the middle granulosa layers of some follicles. The layers of granulosa cells lying below the basal lamina, inner layers around antrum, and cumulus oophorus of pre-ovulatory Graafian follicles also show different responses in their histochemical features to the pentobarbitone induced atresia (Parshad and Guraya 1984). The results of all these studies have suggested that the cells comprising a specific cellular component of the ovary should be treated as individual entities and not as a homogeneous group in regard to their metabolic activities.

Granulosa cells of atretic follicles and in culture develop 3β-HSDH activity (Guraya 1973b), as will be discussed in Chap. VII. Brandau (1970) believes that the presence of 3β-HSDH activity in the granulosa of small and medium follicles may perhaps indicate that these cells act as target sites for the action of steroid hormones. They are unlikely to synthesize steroid in vivo because they lack the NADPH-producing enzymes as well as organelles typical of the steroid secreting cells. Miyamoto et al. (1980) have studied the distribution and activity of NADP-dependent G-6-PDH in the granulosa and theca cells of pre-antral and antral follicles in the nonpregnant and early pregnant goat ovary. In pre-antral follicles, granulosa cells show moderate enzyme activity, whereas in antral follicles they show strong activity. But the enzyme activity is very strong in theca interna, but weak in theca externa. This study has suggested that granulosa cells and theca interna of normal antral follicles and corpora lutea contribute to steroidogenesis. Theca interna and corpora lutea appear to make a major contribution in this regard. SDH activity is moderate in the granulosa cells of primary and pre-antral follicles but weak in those of antral follicles (Miyamoto et al. 1980).

Elfont et al. (1977) have observed that both the granulosa and theca interna cells of Graafian follicles in rats treated with PMSG-hCG show large quantities of reaction product (acid phosphatase) associated mainly with their Golgi complex, extensive agranular ER and some vacuoles, which indicates that gonadotrophic treatment stimulates the development of an extensive lysosomal system within steroidogenic cells of ovarian follicles. Since there are no morphologic indications of degeneration within these cells, these observations suggest that the lysosomal system may play a role in follicular steroidogenesis. β-glucuronidase, acid phosphatase and nonspecific esterase activities of granulosa cells are also increased during the growth of follicles in the guinea-pig ovary (Schmidtler 1980); the raction of nonspecific esterase is especially apparent in the cytoplasm of the cells surrounding the so-called Call-Exner bodies. The highest activity is seen in the content of these formations, whereas the liquor folliculi do not show considerable activity.

Silver (1978) has reported species variations in the distribution of cholinesterases in the ovaries of plains viscacha, cat, ferret, rabbit, rat, guinea-pig and roe deer. The granulosa cells react for acetylcholinesterases only in cat and rabbit. Some butyryl-cholinesterase is present in granulosa of all species except roe deer.

Steroid hormone synthesis. Despite great interest and attention to steroid (oestrogen) biosynthesis in the maturing follicle of the mammalian ovary, the cellular sites and mechanisms for follicle oestrogen production are still not clear. A theca cell origin has been suggested, as already discussed. Alternatively, a 2-cell-2 gonadotrophin theory has been proposed. In the latter, theca cells, in response to LH, are believed to secrete androgens which are aromatized to oestrogen in the granulosa cells in response to FSH. Here the significance of morphological, histochemical and biochemical characteristics will be discussed in relation to steroid biosynthesis in the follicle wall by taking into consideration the 2-cell-2 gonadotrophin theory, which has been primarily discussed from a biochemical point of view in several recent papers and reviews (see Armstrong 1980, Erickson 1982).

The results of various morphological (including ultrastructural), histochemical (or cytochemical) and biochemical studies, as integrated and discussed above, have indicated the biosynthetic activity in the granulosa cells of antral follicles, which may be associated with mitosis, formation of hormone receptors and maturation of their biosynthetic potential and steroidogenesis. The synthesis of proteins and glycoproteins by granulosa cells may be related to the formation of follicular fluid, in the supply of nutrient to the ovum, more recently in the formation of a factor inhibiting the effect of LH, etc. (see also Mestwerdt et al. 1979). These studies have provided relatively little evidence for the cytoplasmic characteristics related to the synthesis of steroid hormones in the granulosa cells of the developing antral follicle (see also Mestwerdt et al. 1979, Dvořák and Tesařík 1980). Various ultrastructural and histochemical findings still support the view that the theca interna is the main endocrine compartment of the mammalian follicle, which functions in response to LH, as already discussed in this chapter. But biochemical data are stronger than cytological and histochemical data for a role of granulosa cells in oestrogen synthesis.

LH receptor content in theca cells of developing Graafian follicles is progressively increased, indicating that theca cells become more responsive to LH (see also Carson et al. 1979). Magoffin and Erickson (1981) have demonstrated that LH induces undifferentiated theca cells to synthesize androgens in vitro and shown that prolactin

Summary: Roles of FSH, LH and Estradiol on Follicular
            Cell Function

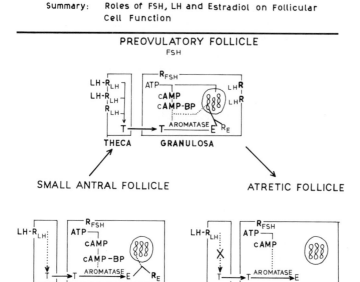

Fig. 41. Summary of changes in follicular cell function during the growth of small antral follicles to pre-ovulatory follicles. The stimulus for this growth appears to be small but sustained increases in LH which increase theca androgen *(T)* and thereby follicular oestradiol *(E)* production. Oestradiol enhances FSH/LH action by the induction of LH receptor *(R)*, cAMP binding sites (now characterized as the regulatory subunit of type II protein kinase) and presumably increased phosphorylation of key intracellular proteins. In the absence of a rise in LH, small antral follicles become atretic and never produce significant amounts of androgen or oestrogen; *BP* binding protein (Redrawn from Richards and Bogovich 1980)

prevents this stimulatory action of LH. The mechanism of this inhibitory action of prolactin needs to be determined. Autoradiographic evidence indicates that theca cells may contain prolactin receptors, raising the possibility that prolactin may have a direct action on theca cell function (Midgley 1973, Oxberry and Greenwald 1982). The increased content of LH receptors in the theca cell layer is suggested to serve to increase the amount of LH bound to theca cells, and thus may increase theca cell synthesis of steroid hormone (androgen) in the absence of increasing peripheral concentrations of LH (see Richards et al. 1979). Several in vivo and most recent in vitro studies have strongly suggested that the androgen secreted by the theca cells is aromatized by granulosa cells to oestradiol-17β in the presence of FSH (Bjërsing 1978, Anderson et al. 1979, Richards et al. 1979, Armstrong et al. 1979, Young Lai 1979, Fortune and Armstrong 1979, Armstrong 1980, Peters and McNatty 1980, Ermini and Carenza 1980, Farookhi 1981, Hirshfield 1982, Erickson 1982, Terranova et al. 1982); highly purified LH does not stimulate aromatization of androgen but it increases androgen synthesis by the theca cells, thereby increases the substrate content for the aromatase enzyme present in granulosa cells, and may also act directly on granulosa cells of developing pre-ovulatory follicles to increase the content of LH receptor (Fig. 41).

All these studies have indicated that FSH increases aromatase activity in granulosa cells (see Anderson et al. 1979, Armstrong 1980, Erickson 1982); very recently, Vernon et al. (1983) have made a study of aromatase activity in granulosa and theca cells of rhesus monkeys. Erickson (1982) has recently emphasized three important features of this action of FSH. First, granulosa cells are very sensitive to FSH in aromatase induction. Second, a lag phase of about 18 h is needed before FSH increases aromatase activity, after which activity increases rapidly. And third, pharmacological doses of highly purified LH and prolactin do not increase aromatase activity, indicating the regulation of aromatase activity is hormone specific for FSH. Once LH receptors appear on granulosa cells (i.e., in small antral and developing pre-ovulatory follicles) LH as well as FSH can increase the aromatase enzyme system (J. Richards, pers. commun). FSH action on aromatase is mimicked by cAMP and cAMP-inducing agents and is completely prevented by inhibitors of RNA and protein synthesis. Inhibitor of the enzyme aromatase also suppresses ovarian oestradiol secretion in rat (Hirshfield 1982). Collectively, these observations indicate that the role of FSH in regulating follicle oestrogen biosynthesis involves the maintenance of aromatase enzyme activity in the granulosa cells, and suggest that it proceeds by a process of cAMP formation followed by induction of specific RNA's and proteins. Results of some studies have indicated that calcium is involved in the regulation by FSH of granulosa cell steroidogenesis. Calcium is needed at least at two distinct levels, one controlling intracellular concentrations of cAMP, and a second located at a subsequent biochemical step(s) in the synthesis of progesterone (Carnegie and Tsang 1983a, b, Tsang and Carnegie 1983). Calmodulin, which serves as receptor for calcium, appears to be required in the stimulation by FSH of granulosa cell progesterone and oestradiol production in vitro (Tsang and Carnegie 1983).

According to Terranova and co-workers (Terranova 1981a, b, Terranova et al. 1982) granulosa cells of whole hamster follicles show a slight elevation in aromatase activity after 1–2 days of delay in ovulation, which may be due to a slight increase in serum concentration in FSH during delay (Terranova 1980). D'Amato et al. (1981) have shown that the steroid secretion profile in the perifused rabbit follicles depends upon the dose of gonadotrophin. Low and medium doses of LH with a constant high dose of FSH elevate significantly the secretion of progesterone and testosterone, only a high dose of LH with FSH increases oestrogen secretion. Progesterone secretion, unlike testosterone and oestrogen secretion, is not depressed following a challenge with a maximum dose of LH: FSH. Luteinization of perifused follicles appears to depend upon exposure to a high dose of gonadotrophin. Exposure of human pre-ovulatory follicles to hCG in vivo causes a shift in steroidogenesis from androgen to progesterone formation in isolated thecal cells and a marked increase in the progesterone production by the granulosa cells of pre-ovulatory follicles (Dennefors et al. 1983). Furthermore, the thecal cells, but not the granulosa cells, develop refractoriness to further stimulation with hCG in vitro. Terranova and Garza (1983) have also observed that LH surge enhances progesterone production and inhibits androgen production in theca of pre-ovulatory follicles of the hamster ovary and thus possibly regulates the onset of androgen/oestrogen synthesis during follicular development.

During the recent studies of Gore-Langton et al. (1981), LH-RH or an agonist is added simultaneously, to determine if the actions of FSH are impaired. Aromatase activity of the granulosa cells is reduced by exposure to LH-RH, which also impairs a similar action of cAMP. GnRH also inhibits FSH-stimulated pregnenolone biosynthesis, probably at the side-chain cleavage enzyme step, in addition to the reported

GnRH stimulation of progesterone metabolism and the GnRH inhibition of FSH-stimulated 3β-HSDH activity (Jones and Hsueh 1982a, b, 1983). Sheela Rani et al. (1983) have observed no interaction between GnRH-analogue and FSH in relation to progestin secretion by granulosa cells in vitro during the first 10 h, but at 24 h and later the presence of GnRHa clearly inhibits the steroidogenic response to LH and FSH. These results have suggested that the effects of GnRHa on granulosa cell steroidogenesis vary with the exposure time; the initial response is stimulatory and the later inhibitory. Furthermore, the response is also to some extent determined by the maturational stage of the granulosa cells.

Anderson et al. (1979) have observed that as the porcine follicle matures there is an increased ability of granulosa cells to convert androgens to oestrogen. This increase is greatest between granulosa cells obtained from medium versus small follicles. The ability to secrete progesterone also increases during follicular maturation but the increase in ability to secrete progesterone is greatest in granulosa cells from large versus medium follicles. Oestradiol synthesis by foetal monkey ovaries is correlated with antral follicle formation (Ellinwood et al. 1983). Evidence produced by several workers has suggested that androgen production may be the limiting factor in oestrogen biosynthesis by small antral follicles and that the increase in basal serum LH concentration seen at the end of pregnancy or during pro-oestrus is possibly responsible for the increased oestrogen by large antral follicles (Richards and Kersey 1979, Welschen and Dullaart 1976, Terranova 1981a, Uilenbroek 1981). Recent data of Veldhuis et al. (1982) have demonstrated important interactions of oestradiol and FSH in preparing granulosa cells for the high rates of steroidogenesis they ultimately express in the luteinized state (see also Veldhuis et al. 1983).

These various in vitro biochemical studies support the concept of a functional relationship between the theca interna and granulosa cells of the Graafian follicle (Fortune and Armstrong 1978, Schomberg 1979, Armstrong 1980, Erickson 1982) which was first introduced by Falck (1959) and later modified by Short (1962) and Bjërsing (1978, 1982). But the studies of Falck did not reveal which tissue actually produced the hormones. In other words, follicular oestradiol-17β is produced through the cooperation of both cell types of the growing follicle (Figs. 41 and 42). Each cell type contributes at least one factor to this synergism. Thecal cells are believed to provide an androgen precursor which is subsequently aromatized to oestrogen by granulosa cells under the stimulatory influence of FSH (see Schomberg 1979, Armstrong 1980, Erickson 1982 for references). This indicates the importance of the theca in the process of follicular oestrogen formation. Erickson (1982) has discussed three important aspects of the effect of LH on androgen biosynthesis. First, LH induces an enormous increase in androgen biosynthesis, whereas FSH and prolactin are without effect (Fig. 42). These observations reveal that LH plays a specific role in inducing the differentiation of ovarian androgen-producing cells. Second, the stimulatory effect of LH is dose-dependent, suggesting that the LH effect is physiological. And third, a rather long lag (about 2 days), is needed before LH induction of androgen production is observed, suggesting that the LH-induced differentiation of the theca cell may depend on molecular biosynthetic events.

The presence of LH receptors on thecal cells and FSH receptors on granulosa cells of small follicles (Fig. 41) (see Richards 1978, Richards et al. 1978, 1979, Carson et al. 1979, Ermini and Carenza 1980, Farookhi 1981) suggests that they have the capacity to

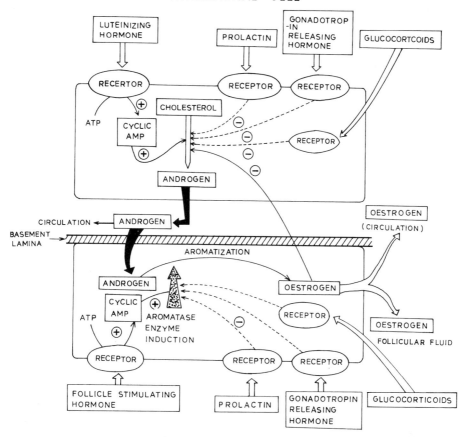

Fig. 42. Schematic illustration of sequence of action of various hormones which are known to influence follicular oestrogen production. The figure represents a hypothetical model based on the "two cell theory" for follicular oestrogen biosynthesis (Redrawn from Erickson 1982)

produce oestrogen before LH receptors are present on the granulosa cells. Edwards et al. (1980) have also observed that granulosa cells aspirated from the large pre-ovulatory follicles are active in the Δ^4 pathway and able to aromatize androgens to oestrogens, but do not undertake conversions involving the Δ^5 pathway. The evidence to date indicates that granulosa cells cannot synthesize androgens from cholesterol or progestins. It is possible that granulosa cells of developing follicles cannot synthesize C_{19} steroids, at least in part, because of their inability to synthesize steroids which precede testosterone. This possibility is supported by the observation that mitochondrial preparations of granulosa cells from porcine follicles show relatively little ability to convert 4-[^{14}C]-cholesterol to 4-[^{14}C]-pregnenolone and 4-[^{14}C]-progesterone (Dimino et al. 1979). This also correlates well with the relatively simple structure of their mitochondria and poor development of the smooth ER, which are believed to play an important role in the synthesis and storage of cholesterol and its subsequent transformation into pregnenolone and progesterone in steroid gland cells (Christensen 1975, Guraya 1974b, 1976a, b, 1978). This suggestion is further supported by the fact that the mitochondria of corpora lutea cells show tubular cristae (Guraya 1971a, Nicosia 1980b), and a much greater ability to convert 4-$[^{14}C]$-cholesterol to 4-$[^{14}C]$-progesterone (Dimino et al. 1979). Luteal mitochondria also show a greater ability to utilize fatty acids and isocitrate and citrate than do follicular mitochondria (Neymark and Dimino 1983). Recent studies have also suggested that ovarian cells may obtain much of cholesterol required for a steroidogenesis from LDL and HDL (Schreiber et al. 1983).

In regard to progesterone secretion, prostaglandin E_2 has been observed to cause an increase in progesterone production by human granulosa cells in vitro (see Bjërsing 1978). High intrafollicular prostaglandin E levels seen in the pre-ovulatory period are believed to cause a renewed increase in progesterone synthesis by the granulosa cells; the LH peak leads to desensitization of adenylyl cyclase (AC) only to LH and FSH, not to PGE (D'Amato et al. 1979, Hunzicker-Dunn et al. 1980), as will also be discussed in detail later.

Schomberg (1979) has discussed the steroidal modulation of steroid secretion. Studies in vitro and in vivo have indicated that steroid secretion by granulosa cells of the developing follicle is modulated in part by steroid hormones themselves. Androgens stimulate progestin synthesis in vitro by granulosa cells of all developmental stages independently of their role as an oestrogen precursor; generally, the action of oestradiol causes inhibition of progesterone secretion in vitro, but not necessarily in vivo. The nature of the relationship between steroid secretion by granulosa cells and the physiological processes of follicular development or atresia needs to be worked out in much greater detail. The cellular mechanism(s) of steroidal modulation of steroid secretion is unknown but apparently does not involve alterations in the number of granulosa cell LH receptors per se. We have yet to learn how the gonadotrophins and steroid hormones share the task of regulating ovarian morphologic and secretory responses (Armstrong 1980, Richards 1980, Erickson 1982).

In future studies, increasing attention should be paid to the interactions of steroid hormones in which rate-limiting enzymes in steroidogenesis may be directly affected. To cite one possible example: peak pre-ovulatory levels of oestrogen are followed by a drastic decrease as a result of the LH surge, as already discussed (see Terranova et al. 1982). It is not known if this is a direct effect of LH on specific enzymes or if the collapse in oestrogen is mediated by steroids produced as a result of LH activation. Recent results obtained by Terranova and Garza (1983) indicate that the LH surge enhances progesterone production and inhibits androgen production in theca of pre-ovulatory follicles and thus possibly regulates the onset of androgen/oestrogen synthesis during follicular development. Channing and Reichert (1983) have suggested that the exposure of the pre-ovulatory monkey follicle to the mid-cycle LH/FSH surge renders it unable to respond to FSH in terms of enhancement of oestrogen secretion. However, the cells can still respond in terms of increased progesterone secretion. Such a mechanism could in part account for a decrease of follicular oestrogen synthesis in the presence of increased follicular progesterone synthesis occurring immediately prior to ovulation. Welsh et al. (1983) have studied oestrogen augmentation of gonadotrophin-stimulated progestin biosynthesis in cultured rat granulosa cells. The results of this study suggest that intra-ovarian oestrogens may act locally to enhance the sensitivity of granulosa cells to FSH and LH, thereby increasing the biosynthesis of progestins and cAMP by the granulosa cells.

Guraya (1968e, f) has observed that theca interna cells of newly ruptured follicles are already filled with lipid droplets consisting mainly of triglycerides (neutral fats), and some cholesterol and/or its esters. The storage of cholesterol-containing lipids in the theca interna cells during ovulation has been attributed to the disruption of theca interna cells that takes place during ovulation (Guraya 1971a). Thus, the drop in oestrogen production after the pre-ovulatory peak may be due to some sudden alteration in the metabolism of cholesterol-containing lipids in the theca interna cells as they start to function in the storage of hormone precursor (Guraya 1971a). Apparently, LH causes decrease in 17α hydrolase and C_{17-20} lyase. What the theca and granulosa can or cannot do seems to be species-dependent. Therefore, more attention should be paid to species variations in regard to the functions of theca and granulosa. Variation in response among different species must be kept in mind while formulating general conclusions about the secretory functions of theca and granulosa of maturing follicle.

Hormone Receptors. Changes in tissue responsiveness of various components of the follicle to gonadotrophins and steroids are being extensively and intensively investigated (Hunzicker-Dunn et al. 1980, Richards 1980). Granulosa cells during follicular growth and maturation develop receptors for gonadotrophins (FSH, LH and prolactin) and steroid hormones (androgen and oestrogen) (Fig. 41) (Eshkol et al. 1976, Midgley and Richards 1976, Richards 1978, 1979, 1980, Richards et al. 1978, 1979, Carson et al. 1979, Darga and Reichert 1970, Peters and McNatty 1980, Ermini and Carenza 1980, Solano et al. 1981, Oxberry and Greenwald 1982, Guraya 1982c). The hormone receptors for gonadotrophins are located on the outer surface of the cell membrane of the follicular cells (Catt and Dufau 1977). Solano et al. (1981) have studied LH bioactivity and regulation of ovarian gonadotrophin receptors during the oestrous cycle of the rat. The presence and characteristics of LH receptors on human ovarian follicles and corpora lutea have also been reassessed during the menstrual cycle and pregnancy (Rajaniemi et al. 1980). Oxberry and Greenwald (1982) have made a detailed autoradiographic study of the binding of $[^{125}I]$-labelled FSH, hCG and prolactin to various compartments of the hamster ovary throughout the oestrous cycle. The binding of these gonadotrophins shows conspicuous changes during the growth, differentiation and ovulation of the follicle and corpora lutea. Variations in the number of available binding sites must represent changes in the number of receptors and/or changes in the number of receptors occupied by endogenous hormone. The binding of gonadotrophins is a saturable process, depending on time, temperature, pH and the concentration of ions in the assay medium, indicating that extracellular factors may affect hormone receptor interactions and consequently influence the cellular response to the hormone (see Hunzicker-Dunn et al. 1980). Epidermal growth factor is observed to modulate LH/hCG receptor binding in rat and porcine granulosa cell cultures (Symanski and Schomberg 1980, Mondschein et al. 1981). The receptor is species-specific, only human LH displacing hCG from it (Rajaniemi et al. 1980). It has a high affinity for hCG, and is sensitive to treatment with trypsin. Some receptors in the membranes may be masked by sialic acid, becoming accessible after treatment with neuraminidase.

The earliest inductive changes in granulosa cells result from oestrogenic stimulation where receptors for FSH in the membrane and oestradiol in the cytoplasm are acquired (Richards 1978, 1979, 1980, Richards et al. 1978, 1979, Peters and McNatty 1980, Ermini and Carenza 1980). A second distinct class of membrane receptors emerges for LH and a variety of cytoplasmic functions are induced (see Albertini 1980). Most of these alterations in granulosa cell characteristics can be attributed to either short- or long-term effects of FSH, although the actions of FSH are subject to modulation by intra-ovarian steroid levels as well as by a putative inhibitor of FSH binding (FSH-BI) (Darga and Reichert 1979). Darga and Reichert (1979) have observed a slight increase in $[^{125}I]$-h FSH binding as bovine follicle size increases. These observations are similar to those of Nimrod et al. (1976) who have reported that, in the rat, large follicles show a greater degree of binding than do medium follicles. But this is not true of intact cycling rat (Uilenbroek and Richards 1979). Nimrod et al. (1976) have not observed any difference in FSH binding to granulosa cells from medium or large follicles in the rabbit. However, Zeleznik et al. (1974b) have reported that in the porcine, granulosa cells from small follicles show the greatest binding capacity for rat FSH, and that this binding capacity decreases as the follicle size increases. Balachandran et al. (1983) have studied the binding of FSH to granulosa cells of follicles in PMSG-primed immature Swiss mice: 48 h after PMSG treatment FSH-binding was higher in the periphery than in the cumulus cells of the antral follicles. No localization in the atretic follicles could be seen by autoradiography 72 h after priming.

Carson et al. (1979) have provided evidence for alterations in specific binding of FSH and LH (hCG) to granulosa cells of ovine follicles as a function of follicular atresia and diameter. The detection of maximal specific binding of $[^{125}I]$-FSH to granulosa cells of smallest follicles studied suggests that granulosa cells of ovine follicles acquire receptors at an even earlier growth stage (1 mm in the pre-antral stage) than observed in this instance. Conversely, significant numbers of LH receptors are not seen on granulosa cells until follicles reach 4 mm diameter. FSH binding to granulosa of healthy follicles in the hamster ovary is also progressively increased with follicular size (Oxberry and Greenwald 1982). FSH stimulates the differentiation of the granulosa cells and promotes antrum formation (Richards 1980, Richards et al. 1978, 1979). Therefore, granulosa cells from immature follicles possess FSH receptor and thus respond to FSH but not to LH.

The major androgen found in the follicular fluid include androstenedione which originates from the theca, as already discussed. Follicles with more androgen than oestrogen are invariably those which show atretic changes. Nandedkar and Munshi (1981) have observed that dihydrotestosterone (1 mg) prolongs the dioestrous stage of the mouse cycle and there are 50% fewer large (normal) follicles than in controls, showing that it induces degeneration of ovarian follicles possibly by feedback on the pituitary. Parshad, R. K., and Guraya (1984a) have observed that administration of androstenedione to dioestrus rates results in increased population of ovarian antral follicles at pro-oestrus, decrease follicular atresia of this stage and limits the number of matured Graafian follicles leading to consequent decrease in the rate of ovulation during the cycle. It could not be determined whether the exogenous androstenedione has a direct effect on the follicle. However, the granulosa cells are endowed with the androgen receptor (Schreiber 1979, Ermini and Carenza 1980), the function of which is not clear. It is believed that an androgen receptor is not an obligatory requirement for follicular maturation, ovulation and corpus luteum formation. However, some recent studies have suggested a direct role for androgens in granulosa cell differentiation (Lucky et al. 1977). Schreiber (1979) has suggested that the cytosol and nuclear testosterone receptors may play a role in the regulation of granulosa cell proliferation during pre-antral follicular growth. Other recent studies have suggested androgen/FSH synergism in stimulation of progesterone synthesis in rat granulosa cells in vitro (Dorrington and Armstrong 1979, Nimrod and Lindner 1979, Anderson et al. 1979, Armstrong 1980). Schreiber et al. (1983) have reported that androgens synergistically augment the FSH stimulation of lipoprotein binding, lipoprotein degradation and cholesterol utilization for progestin production by cultured granulosa cells.

The results of various studies have indicated that growth of pre-ovulatory ovarian follicles involves hormonally induced proliferation and differentiation of both theca and granulosa cells, leading ultimately to an increased ability of follicles to produce oestradiol and to respond to the pituitary gonadotrophins (Richards 1980, Richards et al. 1978, 1979). The increased ability of follicles to synthesize and secrete oestradiol seems to depend on an increased ability of theca cells to produce androgen, as well as an enhanced ability of granulosa cells to aromatize androgens to oestradiol which, in turn, appears to be needed for FSH or FSH and LH to stimulate the development of functional receptors for LH in granulosa cells (Fig. 41). This suggests that the intrafollicular hormone oestradiol enhances the response of granulosa cells to the gonadotrophin. Therefore, production of oestradiol appears to determine which follicle will gain the mechanisms, including LH receptor, necessary for ovulation and luteinization (see Richards 1980, Richards et al. 1978, 1979, Baird 1983). This concept is supported by the fact that oestradiol administered to hypophysectomized immature rats stimulates granulosa cell proliferation (Louvet and Vaitukaitis 1976), causes the growth of large pre-antral follicles which are exquisitely sensitive to gonadotrophins, and increases the ability of FSH to act in vivo or in vitro to stimulate the appearance of granulosa cell LH receptor in developing antral follicles (Richards 1980, Richards et al. 1978, 1979). These in vivo and in vitro studies have also suggested that oestradiol functions to enhance FSH stimulation of cAMP in granulosa cells. Kolena and Channing (1971, 1972a) have also reported that the in vitro administration of either FSH or LH causes an acute stimulation of cAMP formation by these cells. Knecht et al. (1983a) have also indicated that the expression of functional FSH receptors during granulosa cell maturation is mediated by cAMP. Channing and Kammerman (1973) have observed that in association with follicular maturation there is an induction of LH receptors on the surface of granulosa cells. Both FSH and oestradiol are believed to be essential for the development of LH receptors in rat granulosa cells (Zeleznik et al. 1974a, Richards 1975, 1980, Nimrod et al. 1977). There is also a dose-dependent relationship between FSH stimulation of cAMP and FSH stimulation of LH receptor in the oestradiol-primed rats but not in hypophysectomized rats (Richards et al. 1979). Sanders and Midgley (1983) have also shown that induction of LH receptor by FSH is mediated through elevation of cAMP and that addition of exogenous 8-bromo-cGMP can mimic the effects of cAMP.

In recent studies on granulosa cells from immature rat ovaries (Goff and Armstrong 1977, Hamberger et al. 1978) and on granulosa cells from small follicles isolated from sheep ovaries (Weiss and Armstrong 1977), LH is observed to have a lack of effect in vitro on cAMP formation. In the rat, LH is not able to stimulate cAMP formation in granulosa cells until the pre-ovulatory stage. Exposure of intact pre-ovulatory follicles from this species to LH in vitro causes acute stimulation of cAMP formation (Nilsson et al. 1974) but, in addition, refractoriness to a new stimulation by the same hormone is induced for several hours (Lamprecht et al. 1973). A similar mechanism appears to be active in vivo with a decreased sensitivity to LH after the endogenous LH-FSH surge (Hunzicker-Dunn and Birnbaumer 1976a, Nilsson et al. 1977a).

Schwartz-Kripner and Channing (1979) have demonstrated that as the porcine follicle matures, there occurs a decrease in granulosa cell responsiveness to prostaglandin E_2 concurrent with an increase in LH responsiveness in terms of percent stimulation of cAMP accumulation. This alteration is most marked in the case of cells obtained from large follicles. This observation, taken in conjunction with previous findings of Kolena and Channing (1972b) of an additive effect of PGE_2 and LH in granulosa cells from medium-sized follicles, can be best explained by the presence of two separate receptors; one for LH and other for PGE_2. It is still to be determined whether there is a separate AC responding to each stimulatory agent.

It is interesting that responsiveness of granulosa cells to PGE_2 is somewhat similar to that of FSH; the responsivness of granulosa cells to FSH is greater in the case of granulosa cells harvested from small and medium-sized follicles of some species (pig and sheep) but not the rat, as already mentioned. A more than additive response in terms of cAMP production is seen, suggesting that the two agents may not act upon the same receptor. The relationship between FSH and PGE_2 in activation of adenylate cyclase needs to be determined in future studies (Ermini and Carenza 1980).

The maturation of the mechanism responsible for coupling the FSH receptor to adenylate cyclase has been suggested to require theca-derived androgen (Nimrod and Lindner 1979). Thanki and Channing (1979) have suggested that oestrogen (mainly oestradiol-17β) secreted by follicle (its theca and granulosa cells), at least in part, inhibits granulosa cells from secreting progesterone prematurely and allows them to divide until adequate numbers are present to form a corpus luteum. The granulosa cells of growing follicles have also been shown to possess oestrogen receptor (Richards 1978, Richards et al. 1978, 1979, Vaitukaitis and Albertson 1979, Peters and McNatty 1980, Ermini and Carenza 1980). Nuclear oestradiol receptor content has been shown to be preferentially localized in nuclei of isolated granulosa cells (Richards 1978, 1980, Richards et al. 1978, 1979). Furthermore, nuclear oestradiol receptor content is high during follicular development and low in association with atresia and early luteinization. These observations suggest that the effects of oestradiol on granulosa cell function are mediated via the translocation of the cytosol oestradiol–receptor complex to the nucleus. Binding of the oestradiol–receptor complex to nuclear "acceptor" sites may result in changed genomic expression, the synthesis of new messenger RNA (mRNA), and an alteration in any one of a number of the component parts of the FSH response system (Fig. 43) (Richards 1978, 1980, Richards et al. 1978, 1979). For example, oestradiol may function to change the number of FSH receptors in the granulosa cells. Richards et al. (1979), after discussing the possible sites at which oestradiol may act to increase the responsiveness of rat ovarian granulosa cells to FSH, have suggested that in the absence of oestradiol-mediate alterations in the number of FSH receptors per granulosa cell, oestradiol acts to enhance the ability of FSH to stimulate other intracellular effector systems, such as the production of cAMP and the

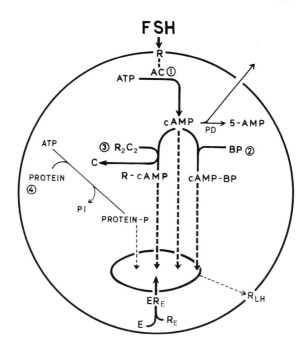

Fig. 43. Possible sites of oestradiol *(E)* action on components of the FSH-cAMP effector system in granulosa cells. Oestradiol acts on some other component(s) of the FSH response system to increase the ability of FSH to stimulate cAMP production *(1)*; to increase cAMP binding proteins (cAMP-BP) which are functionally distinct from *(2)* or identical to *(3)* the regulatory subunits of protein kinase (R_2C_2); or to induce synthesis of a specific protein *(4)* which is substrate for a cAMP dependent protein kinase; *R* receptor (Redrawn from Richards et al. 1979)

capacity of cells to bind cAMP (see also Richards 1980) (Fig. 43). This contrasts with the increase in ability of LH to stimulate cAMP that is associated with an increased number of LH receptors to porcine granulosa cells (Lee 1976). But it needs to be determined in future studies whether the effect of FSH to stimulate cAMP accumulation is mediated by altering the ability of FSH to stimulate the adenylate cyclase system or by decreasing phosphodiesterase activity (Hunzicker-Dunn et al. 1980). Jonassen et al. (1982) have, however, demonstrated that granulosa cells of intact immature rats show considerably more FSH- than hCG-stimulated adenylate cyclase activity. This suggests that FSH-responsive adenylate cyclase is a constitutive component of granulosa cells in pre-antral follicles which show a pronounced increase during the development of pre-ovulatory follicles, a change dependent on the synergistic actions of oestradiol and FSH. The causes of the differences in the response of adenylate cyclase to high concentrations of FSH at different stages of follicular development remain unclear. Veldhuis et al. (1983) have suggested that oestrogen and FSH might increase cellular cholesteryl ester stores, enhance the mobilization of existing cholesteryl ester for steroidogenesis, or augment cholesterol side-chain cleavage activity.

Just prior to ovulation granulosa cells undergo morphological and biochemical changes that modify their response to oestrogen (Bjersing 1978, Thanki and Channing 1979). Meanwhile, they begin to show an increased number of LH receptors (Richards 1978, 1980, Richards et al. 1978, 1979, Peters and McNatty 1980). Recently Lindsey and Channing (1979) have observed that as the porcine follicle matures there is a decrease in granulosa cell responsiveness to FSH concurrent with an increase in responsiveness to LH in terms of cAMP accumulation (see also Schwartz-Kripner and Channing 1979). This does not occur in the rat until after the LH surge (Uilenbroek and Richards 1979). Oxberry and Greenwald (1982) have reported that hCG binding is not detected in granulosa of pre-antral follicles but appears in basal mural granulosa of growing antral follicles on day 2 of the oestrous cycle of the hamster and then increases progressively to include all of the mural granulosa and becomes maximal at 12.00 h on day 4. This binding is slightly reduced at 16.00 h and 20.00 h on day 4.

Oestrogen together with FSH enhances mitotic activity and influences the number of LH and prolactin receptors in the granulosa cells (Fig. 41) (Midgley and Richards 1976, Richards 1978, 1979, 1980, Richards et al. 1978, 1979, Ermini and Carenza 1980). The presence of prolactin receptors in granulosa cells is well established (Rolland and Hammond 1975, Richards and Midgley 1976, Erickson 1982); mural granulosa cells of healthy antral follicles in the hamster ovary show high prolactin binding (Oxberry and Greenwald 1982). But their physiological significance still needs to be determined more precisely. Prolactin has been observed to act directly on the granulosa cells in culture to cause a marked inhibition of aromatase activity (Fig. 42) (Wang et al. 1980, Erickson 1982), leading to inhibition of FSH-stimulated oestrogen formation by the granulosa cells. The inhibitory action of prolactin is selective for aromatase, since prolactin does not alter FSH stimulation of progestin secretion (Wang et al. 1980). According to Oxberry and Greenwald (1982), the sharp reduction in prolactin binding to the theca and granulosa of large ovulatory follicle at the time of and at least 4 h after the LH and FSH surges suggests that either or both of these hormones may regulate prolactin binding, and/or that loss of prolactin binding may play a role in ovulation and/or luteinization. Solano et al. (1981) have demonstrated a significant increase in prolactin receptors at pro-oestrus in the rat, in addition to the increase of LH and FSH receptors at this time of the cycle; LH receptors are maximum. These studies have suggested a positive regulatory effect of serum prolactin on the ovarian lactogen receptors and this may be important in the control of corpus luteum function (Dunaif et al. 1982). The increases of serum progesterone are concomitant with increased prolactin receptor concentration indicating a permissive role of prolactin on steroidogenic function and secretion of the corpus luteum. Erickson (1982) has suggested that firstly, prolactin, in addition to its luteotrophic action in the corpus luteum, may also act on granulosa cells in developing follicles to inhibit FSH induction of aromatase enzymes (Fig. 42); and secondly, a classic negative feedback mechanism might function between the pituitary lactotrope and the oestrogen-producing granulosa cell. Studies using theca-cell culture system have also provided evidence of a direct inhibitory action of prolactin on the ovarian androgen-producing cell (Magoffin and Erickson 1981, Erickson 1982).

From the results of recent studies, following three important molecular aspects of follicular development during later stages, i.e. before the LH surge, have become apparent; (1) the number of specific LH (hCG) receptors is increased; (2) LH (hCG) stimulated AC activity is enhanced in contrast to FSH-stimulable AC, which seems to decrease, but this is not true of rat as FSH-responsive AC is increased in the pre-ovulatory follicle (Uilenbroek and Richards 1979, Jonassen et al. 1982); (3) both hormonally induced and spontaneous progesterone-secreting activity shows a marked increase. These three interrelated functions have suggested that the differentiation of the progesterone-secreting luteal cell is temporally regulated by the acquisition of receptors for LH (hCG) in conjunction with a coupled AC system. Results of various studies have also indicated that follicular maturation is closely accompanied by conspicuous increase in the responsiveness of AC to LH (see Hunzicker-Dunn et al. 1980, Jonassen et al. 1982). The increase in follicular responsiveness to LH functionally enhances those LH-induced effects which are mediated by cAMP without the need of any increase in peripheral levels of LH. The high degree of LH responsiveness in pre-ovulatory follicles is accompanied by the pre-ovulatory surge of gonadotrophins. Although a surge of gonadotrophins is required to induce ovulation, it needs to be determined whether the acquisition of a highly LH-responsive AC is also a prerequisite for the surge of LH to induce ovulation. However, when small antral follicles in rat ovaries are exposed to an ovulatory surge of hCG, they luteinize or develop atresia rather than undergo ovulation (Hunzicker-Dunn et al. 1980). This indicates that follicular readiness to ovulate may nead the presence of an AC which is highly responsive to LH. Furthermore, when a surge of gonadotrophins is induced by copulation, FSH forms only a small portion of the hormone complex (Dufy-Barbe et al. 1973, Goodman and Neill 1976).

It is well known that LH induces ovulation and luteinization of granulosa cells in large pre-ovulatory follicles, that luteinization is associated with a marked decrease in the number of receptors for both FSH and LH (Oxberry and Greenwald 1982), and that this decrease in gonadotrophin receptor content takes place by a mechanism in addition to the occupation of binding sites (see Richards et al. 1978, 1979). However, the pre-ovulatory follicles developing between Days 19–23 of pregnancy in the rat are characterized by increased LH receptor in granulosa and theca cells, increased FSH receptor in granulosa cells, and most notably by a marked increase in oestradiol accumulation (Richards et al. 1976, Richards and Kersey 1979). Solano et al. (1981) have observed that the LH surge is followed by a rapid decrease in ovarian LH and FSH receptors during the day of pro-oestrus in the rat, and loss of these receptors is not due to occupancy by endogenous hormone. A decrease of binding sites for LH and FSH has also been reported for the large ovulatory follicle of the hamster (Oxberry and Greenwald 1982). LH (hCG) might act to reduce its own receptor. Rajaniemi and Jääskeläinen (1979) have also suggested that the LH (hCG)-receptors in the rat granulosa cells are regulated negatively by hCG. This regulation appears to involve both occupied and free receptors. LH receptor induced in the granulosa cell by FSH in vitro has been observed to undergo down-regulation in response to LH (Schwall and Erickson 1983). Down-regulation is stimulated by very brief exposure to physiological concentrations of LH or hormones with LH-like activity. Solano et al. (1981) have suggested that negative regulation of gonadotrophin receptors and positive regulation of prolactin receptors are physiological components of the ovarian cycle.

Follicles ovulating in any given oestrous cycle of the adult rat are exposed to at least three to four prior surges of gonadotrophins (FSH and LH), whereas those ovulating in response to the postpartum gonadotrophic hormone surge actually develop under conditions of sustained tonic concentrations of gonadotrophins. Thus the functional significance of gonadotrophin surges in regulating follicular growth prior to ovulation is unclear. Follicles of early pregnancy in the rat possess the mechanisms (including LH receptor) necessary for ovulation and luteinization, but generally do not receive the LH surge (Richards and Kersey 1979). Rajaniemi et al. (1980) have concluded that human follicular development and luteinization involve a biphasic increase in the number of LH receptors in granulosa cells. The first increase is associated with maturation of these cells and the second with luteinization. Since LH can act to increase its own receptor in the presence of oestradiol and low amounts of FSH (Ireland and Richards 1978), it further supports the suggestion that LH plays a predominant role in the final stages of pre-ovulatory follicular growth to enhance oestradiol production via stimulation of theca cell androgen secretion (Fig. 42), as well as to facilitate an increase in its receptor by acting directly on granulosa cells (Fig. 41). However, the roles of LH and FSH in regulating the dynamics of theca-granulosa cell differentiation during early follicular development need to be determined more precisely. It has been recently observed that during culture in chemically defined medium containing FSH, granulosa cells acquire increased amounts of LH/hCG receptor and increased responsiveness to hCG in a manner similar to that in vivo (Sanders and Midgley 1982). Stewart et al. (1982) have suggested that the follicular fluid maturation stimulator acts in conjunction with FSH, LH and oestrogen, making the cells of specific follicles more responsive to FSH and LH. Follicular fluid is also observed to enhance oestrogen as well as progesterone secretion (Ledwitz-Rigby et al. 1981). A follicle, which is more responsive to gonadotrophins and converts more androgen will be more likely to mature than become atretic.

Richards (1978, 1979, 1980), after reviewing and integrating the results of various studies on the hormone receptors in the follicle, has suggested the presence of very complex molecular processes involved in the hormone regulation of hormone receptors in ovarian follicular development and differentiation (see also Midgley and Richards 1976, Richards et al. 1978, 1979, Guraya 1982c, Hsueh et al. 1983). Oestradiol acts on granulosa cells to increase receptor for oestradiol. FSH acts on oestradiol-primed granulosa cells to increase receptors for FSH and LH. LH acts on oestradiol-FSH-primed granulosa cells to bring about a decrease in receptors for oestradiol and LH, and concomitantly to effect an increase in receptor for prolactin (Fig. 41) (Richards et al. 1976, 1978, 1979, Richards and Williams 1976, Midgley and Richards 1976, Richards 1978, 1979, 1980, Ermini and Carenza 1980). This sequence of hormonally induced changes in hormone receptors has provided evidence for at least three types of receptor regulation (Richards 1978, 1979, 1980, Richards et al. 1978, 1979): (1) autoregulation, a process by which a hormone regulates (increases or decreases) the content of its own receptor (e.g. LH causes a decline in its own receptor: Rao 1979); (2) coordination regulation, a process by which a steroid (oestradiol) and a protein (FSH) interact to regulate the content of receptors for the same (FSH) or a different (LH) protein hormone, and (3) heteroregulation, a process by which one hormone affects the content of an entirely different hormone receptor. The effect of any given hormone appears to be related to cell type and stage of differentiation of that particular cell type. Hormone regulation of hormone receptors in granulosa cells appears to be an important aspect of follicular maturation (Richards 1978, 1979, 1980, Richards et al. 1978, 1979, Vaitukaitis and Albertson 1979, Guraya 1982c). This is supported by the fact that the effects of FSH can be greatly altered by the presence of oestradiol, suggesting that the initiation of follicular production of oestrogen and, perhaps equally important, the response of granulosa cells to oestrogen, determine whether or not a given follicle gains the mechanisms needed for responding to the stimulatory surge of FSH. FSH, more than oestradiol, is believed to act to increase the number of granulosa cell receptors for FSH. Richards et al. (1978) have suggested that both a steroid hormone (oestradiol) and a protein hormone (FSH), presumably acting via different intracellular mechanisms, each have limited capacities to promote granulosa cell multiplication. This indicates that stimulation by specific hormones at distinct developmental stages may be required for the apparent continuous nature of follicular growth until, in response to LH, the granulosa cells cease to proliferate and enter a state of nonproliferative differentiation. The effects of oestradiol on FSH stimulation of its own receptor precede those on FSH stimulation of the LH receptors. All these results have also shown that granulosa cells undergo progressive changes in response to oestradiol (Fig. 41). In its final stages of growth, the follicle then develops the mechanisms including the appearance of LH receptor that finally permit it to respond to the surge of LH to ovulate: to luteinize and to secrete progesterone (Richards 1978 1979, 1980, Richards et al. 1978, 1979). It still remains to be determined whether de novo synthesis of the LH receptor or activation of a pre-existing molecule is involved.

It is well recognized that granulosa cells in large Graafian follicles contain specific prolactin receptors (Fig. 41), as already discussed (Richards and Midgley 1976). However, the precise role of prolactin in regulating follicle development is unknown, as already stated. Prolactin, which is always present in the follicular fluid, is believed to influence the capacity of granulosa cells to secrete progesterone (McNatty et al. 1974). Wang et al. (1979) have reported that FSH stimulates the formation of prolactin receptors in rat granulosa cells, and have shown that the prolactin receptors are functionally coupled to progestin but not to oestrogen formation. In a related study, Erickson et al. (1979) have reported that FSH stimulates LH receptor formation in rat granulosa cells and demonstrated that the LH receptors (unlike those of prolactin) are functionally coupled to both oestrogen and progestin synthesis. High doses of prolactin are observed to inhibit the FSH and LH stimulation of oestrogen production in differentiated rat granulosa cells in vitro (Wang et al. 1981). The observation that prolactin markedly suppresses oestrogen secretion but does not do so significantly after the FSH- and LH-induced increases in progestin secretion (rather stimulates progesterone secretion) (Wang and Chan 1982) suggests that the prolactin-mediated inhibition of steroidogenic activity in the granulosa cells is specific for the aromatase system. Dorrington and Gore-Langton (1982) have also suggested that one contraceptive action of prolactin may reside in its ability to interfere with FSH effects on oestrogen biosynthesis (or FSH-induced aromatase activity) in rat granulosa cell cultures. Furthermore, prolactin inhibition is believed to occur at a site distal to gonadotropin-induced cyclic AMP synthesis. It is also believed that the action of prolactin on the granulosa cells may vary as a function of follicle development (Veldhuis and Hammond 1980). Wang et al. (1981) have suggested that the prolactin-mediated inhibition of oestrogen secretion from the granulosa cells could result in the termination of follicular growth and initiation of atresia in some follicles (see also Fig. 42).

LH-RH (or GnRH) and its analogues are now considered to exert actions independently of and at a distance from their well-known effects on the pituitary gland (Ermini and Carenza 1980). Various other studies have shown that GnRH and its agonistic analogues exert direct effects on ovarian granulosa cell function (for review see Knecht et al. 1983a). Many of these actions of GnRH are inhibitory (e.g. reduction of FSH-induced cyclic nucleotide accumulation and steroidogenesis in cultured rat granulosa cells), while other actions are stimulatory (e.g. induction of oocyte maturation and ovulation, stimulation of progestin cyclic nucleotide and prostaglandin accumulation and stimulation of phosphodiesterase activity in granulosa cells) (Knecht et al. 1983a, b, Erickson et al. 1983, Ranta et al. 1983, Davis and Clark 1983, Sheela Rani et al. 1983, Labrie et al. 1983, Hsueh and Jones 1982). The mechanisms by which GnRH or its agonists exert these effects in the granulosa are unknown (Fig. 42). GnRH and its analogues can inhibit the increase in LH receptors normally induced by FSH, and may also interfere with follicular development (Ermini and Carenza 1980). Specificity of the action of LH-RH agonists at the ovarian level is well supported by the reversal by LH-RH antagonists of the inhibitory effect of LH-RH agonists on both steroid formation and cyclic AMP acccumulation, as well as by the demonstration of similar LH-RH receptors in the anterior pituitary gland and ovary (Labrie et al. 1983, Hsueh and Jones 1982). Ovarian receptors could play a physiological role in the control of ovarian function by a locally produced LH-RH-like molecule. These studies have suggested that LH-RH (or GnRH) or a LH-RH-like peptide may be a potential bonafide "hormone" at the granulosa cells (Hsueh and Jones 1982). Rat granulosa cells show GnRH receptors which are regulated by both FSH and GnRH (Ranta et al. 1983). cAMP-phosphodiesterase is also regulated by FSH and GnRH (Knecht et al. 1983b). Davis and Clark (1983) have suggested that GnRH-induced changes in phospholipid metabolism represent an early event in the mechanism of action of GnRH. These changes may form important links between GnRH binding and altered Ca^{2+} translocation, enzyme activities and, ultimately, granulosa cell function. A factor known as gonadocrinin has been isolated from ovarian follicles, and shows properties similar to LH-RH (see Gore-Langton et al. 1981).

Epidermal growth factor (EGF) and fibroblast growth factor (FGF) both exert a dose-dependent inhibition of progesterone secretion by cultured porcine granulosa cells obtained from follicles of variable size (Channing et al. 1983b). Both factors stimulate cell growth as evaluated by cell numbers and cell DNA content. EGF also inhibits the expression in vitro of fundamental FSH-dependent processes essential for follicular differentiation, i.e. LH receptor induction and oestrogen synthesis (Schomberg et al. 1983). EGF appears to play an important endocrine role in controlling ovarian development by regulating the differentiation and steroidogenesis of granulosa cells (Jones et al. 1983). The cellular mechanisms by which the actions of EGF antagonize those of FSH need to be determined more precisely. However, Schomberg et al. (1983) have suggested that EGF action involves loci distal to the generation of cAMP. These results are consistent with the concept that modulation by EGF of LH/hCG receptor binding may represent a physiologically important control process which interacts with effectors, such as FSH, steroids, and insulin, to determine the overall course of follicle development. Arnaud et al. (1983) have studied the binding properties of EGF to rat ovarian receptors and the possible modulation of the level of these receptors by gonadotrophins and gonadal steroids in vitro. The results of all these studies support a physiological role of EGF in the regulation of reproductive processes.

Granulosa cells of several species of mammals show a requirement for insulin for optimal growth as well as for the expression of differentiated action (see Hammond et al. 1983b). May and Schomberg (1981) have also studied the effect of insulin on granulosa cell differentiation in vitro. Insulin is a requisite for the FSH-mediated induction of LH/hCG receptors in porcine granulosa cell monolayers maintained in serum-containing medium (see also May et al. 1980). Both FSH- and hCG-stimulated progesterone production is significantly enhanced by insulin after the first 2 days of culture. FSH-mediated LH/hCG receptor induction is dose-dependent with respect to insulin as is FSH-stimulated progesterone secretion. These results have indicated that gross morphological maturation of granulosa cells in vitro correlates with insulin-mediated biochemical differentiation, e.g. LH/hCG receptor induction and steroidogenic activity. The exact nature of the effects of insulin on the granulosa cells needs to be determined both in vivo and in vitro. The insulin concentrations used in these studies are higher than circulating insulin concentrations and thus are capable of activating receptors for insulin-like growth factors or somatomedins (Sm). The well-established insulin effects on ovarian function actually might be Sm effects. Both qualitative and quantitative studies have shown the consistent presence of high concentrations of Sm in follicular fluid (Hammond et al. 1983b). The possible effects of various hormones on the steroidogenic functions of theca and ganulosa of Graafian follicle are illustrated in Fig. 42.

Vaitukaitis and Albertson (1979) have observed that cAMP-dependent protein kinases appear to play an acute role in mediating the hormonal effects of hCG and possibly FSH (Fig. 43) (see also Richards 1978, 1979, 1980, Richards et al. 1978, 1979, Bjërsing 1978, Hunzicker-Dunn et al. 1980). Whether the protein kinase holoenzyme and the AC complex associated with specific FSH and hCG receptors are biochemically identical needs to be determined more precisely. However, Richards et al. (1979) have suggested that in granulosa cells having receptors for both FSH and LH, the receptors may be functionally coupled to the same AC system (Fig. 43) (see also Jonassen et al. 1982). Alternatively, the loss of hormone-stimulable cAMP accumulation may occur via intracellular mechanisms inactivating all AC systems equally. The hormone action is determined not only by the amount of hormone exposed to the cellular plasma membrane but also the responsiveness and the sensitivity of the AC system (Hunzicker-Dunn et al. 1979b, 1980). Although a number of hormone-specific alterations in protein kinase activities and multiplicities have been revealed (Richards 1978, 1980, Richards et al. 1978, 1979), the mechanism by which these kinases regulate the final hormonal response(s) is not known and thus needs further studies (Hunzicker-Dunn et al. 1980). Studies on the hormonal regulation of granulosa cell oestrogen and progestin biosynthesis by various regulatory agents have greatly facilitated our understanding of the control process involved in follicular maturation, and lutenization in vivo (Figs. 41 and 42) (Hsueh et al. 1983). The multiple hormonal control mechanism of granulosa cell differentiation also provides an interesting model for future studies on the diverse mode of hormone action.

Desensitization phenomenon. When intact pre-ovulatory follicles are exposed to LH they respond with an immediate activation of AC, resulting in increased production of cAMP and activation of the soluble PK (Fig. 27). This stimulation is accompanied by a desensitization of the cAMP system to further action by the same hormone (LH) (Lamprecht et al. 1973, Hunzicker-Dunn and Birnbaumer 1976a, 1980, Nilsson et al. 1977b). In other words, a rapid loss of sensitivity of AC to subsequent LH or hCG stimulation is induced (D'Amato et al. 1979). This phenomenon of AC desensitization was first noticed in ovarian tissues by Marsh et al. (1973) and Lamprecht et al. (1973), using an in vitro ATP-prelabelling technique to show hormone-stimulated cAMP accumulations. The explanation for this phenomenon appears to be the loss of available receptor sites (see Rajaniemi and Jääskeläinen 1979 for references). But no decreased AC occurs prior to a decrease in receptor (Jonassen and Richards 1980). Mills (1979) has observed that ovulated follicles from the rabbit ovary, which are insensitive to LH, do respond to cAMP with increases in the synthesis of steroid hormones, providing strong evidence that the cessation in response to LH at ovulation is due to a failure in the synthesis of cAMP. The LH-induced desensitization of follicular AC (Hunzicker-Dunn et al. 1979 a, b, 1980) appears to be the basis for the blockage of steroidogenesis in ovulated follicles of the rabbit (see also Richards 1978, Richards et al. 1978, 1979, for data on the rat ovary). Desensitization of luteinizing rat follicular cells is observed at 24 h but not 2 h after LH (or hCG) administration, whereas desensitization of rabbit follicular cells is observed within minutes. Both loss of receptors and desensitization form the possible mechanisms for regulating the responsiveness of granulosa cells to LH during early stages of luteinization. But it is necessary to distinguish the early mechanisms which decrease target cell responsiveness from the later more protracted mechanisms (receptor loss and desensitization) (Hunzicker-Dunn et al. 1980). The marked decrease in cAMP accumulation in vivo is believed to be associated with the activation of a phosphodiesterase, while the loss of receptors and desensitization may show changes in membrane structure and function which have taken place as a result of allosteric mechanisms, internalization of membrane components (including perhaps hormone receptor complexes) and/or cellular differentiation (Richards et al. 1978). Amsterdam et al. (1979a) have demonstrated by high resolution autoradiography that clusters of receptor-bound hCG are formed on the cell membrane of cultured rat granulosa cells, and that the bound hormone is subsequently internalized into lysosomes where degradation may occur (see also Lin et al. 1982, Rajendran et al. 1983). Lateral movement of receptor-bound hormone to form clusters, patches and caps following association with antibody to the hormone is demonstrated by using indirect immunofluorescence technique. The internalized hormone 9 (hCG) by pseudopregnant rat ovary has been characterized physically, immunologically, and biologically (Zimniski et al. 1982).

Desensitization of the cell to the hormone appears to precede hormone and receptor internalization. It is suggested that the primary mechanism of the desensitization phenomenon should be looked for at the level of the cell membrane, whereas the down-regulation of receptor density may be due to a later intracellular process. Restriction of receptor mobility within the membrane should be considered as one possible mechanism underlying the induction of desensitization. According to Rajaniemi and Jääskeläinen (1979), the discrepancy between the rate of disappearance of LH (hCG)-receptor and LH sensitivity of AC suggests that besides the loss of LH (hCG)-receptor, other mechanisms may also be involved in the desensitization of AC to LH stimulation. The extensive degradation of hCG to subunits and amino acids by the granulosa cells in vivo and in vitro appears to suggest that the receptor regulation involves the internalization and catabolism of the hormone or hormone receptor complex.

Recent studies have indicated that gonadotrophin receptors and AC occur in the cell membrane as separate entities that become associated to form an active complex during receptor occupancy by the homologous hormone (Dufau et al. 1979). The molecular mechanism by which desensitization occurs is not known even in terms of the involvement of a protein synthetic step, in which results in the literature are contradictory (see Lamprecht et al. 1977, Hunzicker-Dunn et al. 1979 a, 1980). Studies of possible requirements of AC desensitization have shown that it does not need PG synthesis, nor does it need RNA or protein synthetic events blocked by puromycin, cycloheximide, or actinomycin D (see Hunzicker-Dunn et al. 1980). Dibutryl cAMP, in the absence of LH, promotes a concentration-dependent decrease in LH-stimulated AC activity. However, the effect of dibutryl cAMP is time-dependent; incubations shorter than 2 h result in no decline in LH-stimulable AC activity. The various lines of indirect evidence have indicated that reversal or resensitization of the desensitized AC system to LH is mediated by a dephosphorylation reaction (Hunzicker-Dunn et al. 1979 a, 1980). The level of phosphorylation of membrane-associated components may, in part, regulate the activity of the AC system during the first phase of homologous desensitization. The second phase of desensitization, which takes place after the first hour following hCG-LH-receptor interaction in the rabbit follicle, is characterized by a loss of responsiveness to FSH as well as to LH and can be promoted by dibutryl cAMP (in the absence of LH). The causes of the differences in the response of AC to high concentrations of FSH at different stages of follicular development also remain unclear (Jonassen et al. 1982). The absence of desensitization in pre-antral follicles may be needed to permit a continuous nondisruptive pattern of follicular growth when small follicles are repeatedly exposed to gonadotrophin surges, whereas desensitization is required for the cessation of follicular growth and luteinization. Further studies are needed to reveal the physiological significance and the molecular mechanisms of desensitization of the folllicular AC system (Hunzicker-Dunn et al. 1980).

The AC system of pre-ovulatory follicles loses its capacity for being activated by a second challenge of LH when a large dose of gonadotrophin has been previously administered (Marsh et al. 1973, Lamprecht et al. 1973). Hunzicker-Dunn and Birnbaumer (1976 b) have reported that when oestrous follicles receive an ovulatory dose of hCG, desensitization is complete within 10–11 h following hCG and persists for 48 h. Within 72 h the corpus luteum, which is formed, develops an AC system that is sensitive to gonadotrophins. The rate of the decline in LH-stimulated AC activity in rat follicle is much slower than in rabbit follicle (Hunzicker-Dunn and Birnbaumer 1976 b). Concurrently, the mechanism by which follicle cells lose their responsiveness to LH may also differ in rats, as AC desensitization to LH in this species can be blocked with actinomycin D (Lamprecht et al. 1977).

The mechanism of desensitization in the physiological situation may involve actions directly on the AC enzyme and its coupling to gonadotrophin receptor (Hunzicker-Dunn and Birnbaumer 1976 a, b, Zor et al. 1976), and/or receptor occupancy and turnover (Conti et al. 1976). The theoretical desensitization phenomenon may result in a drop of follicular cAMP levels just before ovulation; a feature which may be related to the maturation process of the oocyte, as will be discussed in Chap. IV. Increase in the intrafollicular cAMP level following the gonadotrophin surge is necessary for the numerous effects of LH on the follicle, including the resumption of oocyte meiosis. Since cAMP in itself is known to inhibit oocyte maturation (Magnusson and Hillensjö 1977; see also Chap. IV), it is extremely important that the intrafollicular cAMP level then decreases or at least does not show higher levels that allow maturation of the oocyte (Hillensjö et al. 1978). Nilsson et al. (1977a) have not been able to show any decrease in follicular cAMP prior to ovulation in the rat. If desensitization and cAMP decreases are essential for luteinization, these events are expected to be demonstrable in follicular granulosa cells.

Hamberger et al. (1979) have observed that cAMP formation in rat granulosa cells is increased by the pre-ovulatory gonadotrophin surge analogous to the situation in the intact follicles but, in contract to follicular cAMP formation, it can be further raised by the addition of new hormone. This discrepancy must mean either that the desensitization mechanism of the follicle resides exclusively in the theca cells or that the experimental situation fails to demonstrate its presence in the granulosa cells. Based upon determinations in rabbit follicles which are popped prior to homogenization, retaining their theca layer but losing many of their granulosa cells, hCG-induced desensitization of the rabbit follicular AC system to LH is also associated (at least) with thecal cells (Hunzicker-Dunn et al. 1980). D'Amato et al. (1979) have observed that the refractoriness is not complete but instead it is differential with respect to cell type, as monitored by steroid secretion from pre-ovulatory follicles which are repeatedly challenged with an ovulatory dose of gonadotrophins. Desensitization in the granulosa cells may be part of a biphasic response with a different time course than in the follicles. Or the isolation procedure may cause an artificial response either by removing an inhibitory substance from the follicular fluid, or by breaking the contact between granulosa and theca cells. Also, another possibility is that the isolated cells and the intact follicles possess differences in receptor kinetics that interfere with the results, but this may not be likely after taking into consideration the data of Nimrod et al. (1977), who have found the same discrepancy. With these reservations, it appears possible that the pre-ovulatory LH-FSH peak is one of the factors that regulates the transformation of the primarily FSH-responsive granulosa cells into a primarily LH-responsive luteal cell. But it needs to be determined whether this effect is mediated by alterations in the coupling of AC to receptors, by alterations in receptor turnover or by other, hitherto unknown mechanisms.

Desensitization of the follicular system forms a physiological process. Follicle in the ovaries of rat and rabbit become functionally refractory to exogenous gonadotrophins when the AC is unresponsive to LH and the follicle in vivo is quiescent in steroidogenesis, i.e., in relation to production of oestrogen and testosterone, and produces only low or negligible levels of progesterone (see Hunzicker-Dunn et al. 1980). However, the significance of an AC system which is refractory to LH and FSH during luteinization needs to be determined more precisely.

D'Amato et al. (1979), using steroid secretion as an end-point during the gradual transition of a pre-ovulatory follicle to a corpus luteum, have investigated the desensitization phenomenon of pre-ovulatory follicle obtained from ovaries of oestrous rabbits. Initial gonadotrophin (LH-FSH) stimulation results in a statistically significant release of testosterone and oestradiol-17β, but not progersterone when compared to baseline values. Subsequent stimulation with LH: FSH results in significant but decreasing amounts of progesterone released from the follicles, whereas no significant response is seen for testosterone and oestradiol-17β for the duration of the experiment. A significant increase in progesterone secretion is concomitant with the initiation of luteinization of the granulosa cells, as seen histologically. This repeated response of progesterone to multiple injections of LH: FSH shows that both theca and granulosa cell layers are present. This is in contrast to the published reports of complete desensitization of the AC system in which popped follicles were used, resulting possibly in a significant loss of the granulosa cells (Hunzicker-Dunn and Birnbaumer 1976b). D'Amato et al. (1979) have suggested that the observed desensitization occurs primarily in the theca layer with a concurrent partial refractoriness of the granulosa cells, which at this time undergo morphological and biochemical luteinization.

E. Basal Lamina and Transfer of Substances into the Follicle

The mammalian ovarian follicle is avascular and thus every substance from oxygen to hormones, which reach it, must diffuse in the extracellular spaces from the vascular bed to the theca interna, and must pass across the basement membrane or basal lamina, and between the often densely packed granulosa cell layer. Therefore, it is interesting to know if at the border of the follicle a molecular filtration barrier exists similar to that existing in the testis of mammals (Guraya 1980b). The ovarian hormones secreted by the cells of the glandular layer of the theca interna can reach the follicle by way of the interstitial fluid and do not have to enter first the circulatory system (Szöllösi et al. 1978). The capillaries in the vascular layer, therefore, serve primarily for supplying the follicle with oxygen and chemical absorbed or produced at other sites of the body. Szöllösi et al. (1978) have observed that the vascular layer of the porcine ovary is noteworthy in that its capillaries are not fully continuous but large fenestrations are seen between endothelial cells, representing its lining. The spaces are large enough so that cellular elements may leave the vascular bed in addition to plasma. The follicle is thus surrounded by large blood lakes which abut directly against the basal lamina of the follicle (see also Daguet 1980). The molecular components transported by the plasma need thus to traverse only the thin, felt-like basal lamina (30 nm approximately) until they develop contact with the first follicle cells constituting granulosa. Therefore, the structure, chemistry and function of basal lamina will be discussed here.

The basal lamina of mature follicles contains in subjacent layers obvious, thick bundles of collagen fibrils. It is a bipartite structure composed of a homogeneous stratum upon which the peripheral layer of granulosa cells rests, and an outer region of collagen-fibres (Fig. 44) (Anderson et al. 1978). Blood vessels are not observed among granulosa cells, but a number of capillaries, frequently filled with red blood cells, are observed near the external surface of the basal lamina (Motta and van Blerkom 1974,

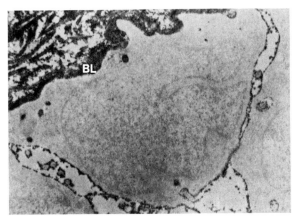

Fig. 44. The basal lamina *(BL)* and granulosa cells of a porcine ovarian follicle demonstrate the passage and distribution of horse-radish peroxidase throughout the follicle. The heavily staining layer is the basal lamina *(BL)* with a few adjacent collagen fibres (From Szöllösi et al. 1978)

Bjërsing and Cajander 1974f). The presence of a blood–follicle barrier, which regulates the transport of some substances from peripheral plasma to within the follicle, has been proposed (Zachariae 1958, Shalgi et al. 1973).

The transport of substances from serum into the intercellular spaces around granulosa cells and the antrum has been studied by many workers (see McNatty 1978). In order for serum-borne substances to enter into the follicle they must first pass out of the ovarian capillaries. A potentially high transfer rate of substances from the capillaries is possibly due to the specific arrangement of arterioles, capillaries and venules around the follicle. The fast appearance of labelled protein in the ovarian lymph after intravenous administration (3–4 min) supports the high capillary permeability. Ovarian capillaries are even permeable to high-molecular-weight compounds (McNatty 1978). Szöllösi et al. (1978) have observed that Graafian follicles incubated in toto with electron-dense markers, such as horse-radish peroxidase and lanthanum nitrate, are penetrated freely by the tracers through the granulosa cells and the cumulus oophorus, reaching the oocyte. This suggests that no blood–follicle barrier exists.

There does not exist any barrier to the transfer of low-molecular-weight substances from the blood capillaries to the follicular antrum. Labelled amino acids, water or elements such as $[^{131}I]$ or $[^{35}S]$ equilibrate rapidly with follicular fluid after administration into a peripheral site (see Edwards 1974, McNatty 1978). But some kind of barrier appears to be present to substances having high molecular weight, as proteins enter the follicle at different rates (von Kaulla et al. 1958, Mancini et al. 1962). The various substances including proteins, which derive from the serum, reach the antrum after passing through the thecal capillaries and intercellular spaces of theca and granulosa cells (Mancini et al. 1962, Cran et al. 1976).

The relative concentration of serum proteins in follicular fluids is almost in proportion to their molecular weights. The extra follicular substances, including serum proteins, appear to diffuse through extravascular regions within the ovary. Ultimately, they distribute in the spaces between theca cells, granulosa cells, and the antrum; some may even enter into the zona pellucida or the ooplasm (McNatty 1978). But some substances are unable to enter the follicle, as evidenced by the absence of large-molecular-weight proteins (71 million) in antral fluid (Shalgi et al. 1973, Andersen et al. 1976; Cran et al. 1976). From these observations Shalgi et al. (1973) has suggested that the follicular membrane (basement lamina) functions as a molecular sieve (Edwards 1974, Andersen et al. 1976). The presence of gap junctions between the cells of the follicular envelope further supports the suggestion that the basal lamina functions like a molecular sieve.

Carson et al. (1979) have observed that the relatively long incubation time (27–36 h) required to equilibrate $[^{125}I]$-labelled gonadotrophin between incubation medium and follicular fluid is most probably a result of passive diffusion of gonadotrophin across the theca externa and interna, the basal lamina and finally the granulosa. This is in marked contrast to the situation in vivo where gonadotrophins are transported to the follicle via the blood vessels which penetrate the theca interna to the basal lamina, thus eliminating the necessity to cross the whole thecal structure. No data is at present available in regard to the time needed for endogenous gonadotrophin to penetrate the follicle. However, the finding that levels of FSH in follicular fluid approximate but never exceed those in plasma suggests that the same diffusional process is operative in vivo (see Carson et al. 1979).

The transport of proteins and other large molecules through the follicular membrane is not only determined by their molecular weights, as LH and FSH have similar weights (33,000) yet show twice as much of their amounts in follicular fluid as in serum (Peters and McNatty 1980). Prolactin (MW 21,000) weighs less than either LH or FSH, but its presence in fluid varies greatly and may change from levels that are almost 60% lower to levels approximately 600% higher than those of LH (Peters and McNatty 1980). The size of gonadotrophin molecules should not be a determining factor for their passage through the intercellular spaces of granulosa as ferritin, with a Stoke's radius about twice that of LH and FSH, readily enters the follicle (Reichert 1972, Cran et al. 1976). From these observations it is evident that size is not only factor which determines the ability of a protein to enter the follicular membrane (McNatty 1978). The shape and charge of molecules may also be of great significance in this regard. It is possible that gonadotrophins do not possess the spherical configuration assumed when their stakes radii are calculated, and the charge on FSH is more than on LH (Geschwind 1963).

Ménézo et al. (1978) have recently studied in vitro exchange between the follicle and its culture medium. The continuous flow technique permitting better incorporation of amino acids was used. cAMP puric and pyrimidic bases can pass from the follicle into the culture medium. Glucose consumption and lactate accumulation do not vary significantly in relation to maturation stage or after the addition of hCG into cultures with PMSG. High-molecular-weight proteins are observed to transport from the follicular fluid into the culture medium, if the follicle remains healthy up to the end of culture. Atresia totally suppresses this permeability.

Our knowledge is still very meagre about the transport of substances from the follicle into the ovarian vein (e.g. steroid hormones) (Peters and McNatty 1980). The presence of steroid-binding protein is suggested to provide an explanation for the slow release of steroids from follicular fluid, as androgen-binding protein (ABP) with high affinity and low binding capacity has been demonstrated in follicular fluid (Vigerski and Loriaux 1976). The ABP binds preferentially dihydrotestosterone, testosterone and, to a lesser extent, oestradiol. No high-affinity binding proteins appear to be present for progesterone, androstenedione or oestradiol in spite of the fact that follicular fluid shows high concentrations of these steroids. It appears unlikely that the steroids are retained by low-affinity high binding capacity proteins, such as albumins. Giorgi et al. (1969) have observed that albumins in follicular fluid accelerate the outward transfer rate of steroids. Other factors appear to contribute to the slow clearance rate of steroids and other substances from the follicular antrum; for example the close proximity of the antrum to the source of synthesis of the steroid or protein coupled with the absence of a vascular network in the membrana granulosa may be significant (Peters and McNatty 1980).

The permeability of ovarian capillaries and follicles in prepubertal and sexually mature (pro-oestrus and oestrus) randomly bred Swiss Albino female mice has been studied by intravenous injection of either ferritin or horse-radish peroxidase (HRP) followed by examination with light and electron microscopes (Payer 1975). This study has revealed that capillaries in the interstitial and perifollicular regions are provided with a continuous endothelium that has constant permeability characteristics irrespective of sexual maturity or phase of the oestrous cycle. Horse-radish peroxidase leaves the capillaries primarily through interendothelial cell junctions and is present in all follicles within 30 s after administration of the tracer. Ferritin, on the other hand, is absent from endothelial cell junctions, and leaves the capillaries, at a slower rate than HRP, via cytoplasmic vesicular transport. Both tracers are found in the granulosa cells but rarely in the oocytes. The tracers reach the oocyte through the intercellular spaces between granulosa cells. These findings have demonstrated that the follicular apparatus of the mouse is permeable to ferritin and HRP, and that follicular regions such as the basal lamina of the folllicle and the zona pellucida do not stop or retard the passage of either tracer.

F. Follicular Fluid

Follicular fluid (liquor folliculi) is a liquid that accumulates in extracellular spaces within ovarian follicles (Fig. 11 g–i), as already described in Sect. D. Fluid-filled spaces first develop between granulosa cells during the early stages of follicular growth (Fig. 11). As the granulosa cells multiply there is a concomitant increase in the

Table 2. Characteristics of follicular fluid

Physical characteristics	Chemical characteristics
Straw-coloured	1. *Proteins:* Albumins, globulins, IgA, IgM, fibrinogen, LD-lipoproteins, heparin, bradykinin, galactosamine
Viscous	2. *Amino acids:* ASP, Thr, Glu, Glu-NH$_2$, Gly, Ala and Met
	3. *Enzymes:* Endopeptidase, proteinase, plasmin, aminopeptidase (cytosol), dipeptidase, alkaline phosphatase, adenosine triphosphatase, acid phosphatase, fructose bisphosphate aldolase, lactate dehydrogenase, aspartate aminotransferase, alanine aminotransferase, collagenase, hyaluronoglucosidase, pyrophosphatase
Viscosity varies with follicle growth	4. *Carbohydrates and glycoproteins:* Glucose, fructose, fucose, galactose, mannose, glucosamine, galactosamine, protein bound hexoses, hyaluronic acid, sulphated glycosaminoglycans, other isomeric chondroitin sulphates, heparin sulphates, neutral glycoproteins, plasminogen
Complex network of coarse fibrous mucopolysaccharides associated with fluid	5. *Gonadotrophins:* FSH, LH, prolactin as well as their subsunits
	6. *Steroids:* Cholesterol, androgens, progestagens, oestrogens
	7. *Prostaglandins*
pH 7.0	8. *Non-nitrogenous metabolites:* Lactic acid, sialic acid, citric acid
pCO$_2$ variable	9. *Elements and salts:* Sodium, potassium, magnesium, zinc, copper, calcium, phosphorus, sulphur, chlorine ions, inorganic phosphate
	10. *Inhibitors and stimulators:* Oocyte maturation inhibitor, (inhibits completion of oocyte meiosis), luteinization inhibitor (prevents or inhibits luteinization of granulosa cells), luteinization stimulator (stimulates luteinization of granulosa cells), FSH receptor binding inhibitor (depresses the binding of FSH to granulosa cells), inhibin or FSH-suppressing substance (depresses the secretion of FSH). Factors for capacitation and acrosome reaction of spermatozoa

production and accumulation of fluid: these small pockets of fluid eventually coalesce to form a follicular cavity (antrum) (Fig. 11). Addition of FSH together with oestrogen induces granulosa cell proliferation and the concurrent development of a fluid-filled antrum (McNatty 1978). With the growth of the follicle, the amount of follicular fluid increases to form a large antral cavity. In its development and size the antrum shows variations in different mammals (Mossman and Duke 1973, McNatty 1978). The follicular fluid provides the means by which cells of the avascular granulosa are exposed to a specific environment which differs from serum, and from that in adjacent follicles. Follicular steroidogenesis, oocyte maturation, ovulation and transport of the oocyte to the oviduct, as well as preparaton of the follicle for subsequent corpus luteum function, may depend on alterations in the physical and endocrine characteristics of the follicular fluid (McNatty 1978). Despite the widespread presence of follicular fluid in mammals, our knowledge of its physicochemical properties is confined to only a handful of species which include primates, some farm animals, the rat and the rabbit. This is due to difficulties involved in obtaining appreciable volumes of fluid for study. A general summary of physicochemical characteristics of follicular fluid is given in Table 2.

1. Types and Origin

Three kinds of follicular fluid have been distinguished. Depending on their source, they have been termed primary, secondary, and tertiary. The primary follicular fluid which is thought to originate from granulosa cells, is traversed by a reticulum and is probably fairly viscous (Gwatkin 1980). Just before ovulation, the antrum expands rapidly, with the accumulation of a less viscous secondary follicular fluid. The general consensus is that this secondary follicular fluid originates from peripheral plasma. The tertiary follicular fluid containing leukocytes and cell debris arises after ovulation and plugs the ruptured follicle (Lipner 1973, Gwatkin 1980). The morphological, histochemical and ultrastructural characteristics of granulosa cells (see Sect. D) reveal that they are metabolically active in protein synthesis (Guraya 1971a, 1973a, 1974b, Abel et al. 1975, McClellan et al. 1975, Nowacki 1977, Mestwerdt et al. 1977a, b).

2. Physical Characteristics

When follicular fluid is aspirated from excised ovaries, marked variations in colour, from almost colourless to intense yellow, may be observed. Generally, however, the fluid aspirated directly from follicles in situ is straw-coloured. The much wider variation of fluid from excised ovaries may be due to degenerative changes that occur in the follicular tissue because of anoxia (McNatty 1978). The pO_2 of follicular fluid seems to be more variable than that of blood. Fraser et al. (1973) have revealed that there is a marked decline in pO_2 of FF in human ovaries during the follicular phase of the menstrual cycle. Perhaps a reduced pO_2 is required for oocyte maturation to take place. Under in vitro conditions, a sharply defined pO_2 of 5% (38 mm Hg) is necessary for the maturation of hamster oocytes (Gwatkin 1980).

Follicular fluid is a rather viscous solution. The viscous nature of this fluid is due to mucopolysaccharides, or proteoglycan molecules, composed of sugars attached in chains of polypeptides (see Sect. F. 3). These mucupolysaccharides or proteoglycans have a capacity to form matrices in solvents. The study of ultra-thin ovarian sections stained with chromic acid-phosphotungstic acid has revealed that follicular fluid is a fine reticulated structure of mucopolysaccharides (Kang 1974). But recent studies on thick sections treated similarly have shown that the antral space contains, in part, a complex network of coarse fibrous mucopolysaccharides with narrow spaces containing soluble nonstructural components of the fluid (Kang et al. 1979). However, Apkarian and Curtis (1981), in a recent SEM cryofracture study of rat ovarian follicles, have observed that soluble proteins and glycoproteins of FF appear as a homogeneous granular precipitate. The mucopolysaccharide network is the major constructional framework of the antrum and provides the interstices for movement and retension of blood-borne macromolecules (e.g. tracers). The fibrous network is apparently the product of the granulosa cells and is continuous with the cell surface coat and the zona.

Presumably, the mucopolysaccharides ensure that the fluid is retained in a "jellylike" matrix within the intercellular spaces (McNatty 1978). The consistency (or viscosity) of the follicular fluid varies with the developmental stage of the follicle (Chang et al. 1976, McNatty 1978). In some species (e.g., cow, rat) viscosity of follicular fluid decreases as the follicle matures, whereas in primates (rhesus monkeys and women) it increases in the pre-ovulatory follicles (Peters and McNatty 1980). Pre-ovulatory bovine follicles have been shown to contain a thermolabile (at 100 °C), nondialyzable substance that can depolymerize the mucopolysaccharides in fluids of small follicles (McNatty 1978). This type of change in the viscosity can be induced in fluid of immature follicles by mixing it with that of rupturing follicles. In all Graafian follicles there is a viscosity gradient from the periphery to the interior. The cellular layers of the follicle are lined with a fluid consisting mainly of hyaluronic acid and therefore of a high viscosity. At ovulation, the egg mass is embedded in a follicular fluid which is much more viscous than the fluid escaping from the rupturing follicle (Edwards 1974).

The pH of follicular fluid is generally above 7.0 and is either similar to or lower than that of serum or plasma (Shalgi et al. 1972, Fraser et al. 1973, McNatty 1978). In women, the pH of follicular fluid correlates negatively with the partial pressure of carbon dioxide (pCO_2). Shalgi et al. (1972) have suggested that the pCO_2 is apparently the major factor regulating pH, although the presence of some nonvolatile fatty acids may also have some influence. Follicular pCO_2 values are either similar to or lower than those of peripheral venous blood (fluid, 17–54 mm Hg; blood 39–54 mm Hg), whereas the pCO_2 of human follicular fluid tends to be more variable (fluid, range 22–97 mm Hg; blood, range 31–47 mm Hg) (Shalgi et al. 1972). During the follicular phase of the menstrual cycle, many medium-to-large human follicles (10 mm diameter) show a pCO_2 in fluid which is similar to or lower than that in peripheral venous blood. In contrast to this situation, all follicles throughout the post–ovulatory period (or luteal phase) show higher pCO_2 than found in blood (Fraser et al. 1973).

3. Chemical Composition

The chemical composition and properties, and the possible physiological role of follicular fluid have been extensively studied (see references in McNatty 1978, Ledwitz-Rigby and Rigby 1979, Kang et al. 1979, Peters and McNatty 1980, Gwatkin 1980). Follicular fluid contains several proteins, amino acids, sugars, enzymes

(collagenase, hyaluronidase, transaminase, alkaline and acid phosphatases), mucopolysaccharides, gonadotrophins (LH/FSH/prolactin), steroids (oestrogens, androgens,
progestagens), salts, etc. (Table 2) (see Lipner 1973, Kang et al. 1979); and a cAMP
accumulation inhibitor has been reported in human ovaries (Kraiem and Lunenfeld
1979). Prostaglandins, which may be involved in ovulation, will be discussed in
Chap. V. The concentrations of proteins, steroids, carbohydrates, and mucopolysaccharides in the follicular fluid are not constant throughout follicular growth; some
increase whereas others decrease (see McNatty 1978, Ledwitz-Rigby and Rigby 1979).
The concentrations of the above substances also change with respect to the phases of
the oestrous or menstrual cycle (see Ledwitz-Rigby and Rigby 1979). Recently, several
studies have shown that certain follicular fluid constituents influence the functions of
granulosa cells, oocytes, follicular growth initiation, whole ovaries, the hypothalamic-
pituitary axis and Yoshida Ascites cells (see references in Ledwitz-Rigby and Rigby
1979). Although many of these sudies have been carried out with fluid from one species
and cells of another, it seems likely that follicular fluid may have several important
regulatory functions in the reproductive biology of the female. For example, Ledwitz-
Rigby and Rigby (1979) have hypothesized that follicular development may only
proceed to the pre-ovulatory state when the granulosa cells and fluid mature in a
coordinated manner.

   The proteins in follicular fluid are primarily low-molecular-weight proteins which
originate from the vascular system by transudation across the follicle basement lamina.
Large protein molecules, such as low-density lipoprotein, IgM, IgG, and fibrinogen,
are either absent or present in lower concentrations than in serum. Although there is
much evidence that the granulosa cells are engaged in biosynthetic activity, there is
some uncertainty as to the origins of proteins, enzymes and mucopolysaccharides and
other substances in follicular fluid (see Lipner 1973). Hexose-bound proteins occur in
the fluids of cows and mares, suggesting that mucopolysaccharides may derive from
the protein polysaccharides of plasma (McNatty 1978, Peters and McNatty 1980). On
the other hand, granulosa cells might be involved in the secretion of sulphated
mucopolysaccharides, as these substances could not be demonstrated in cystic follicles
which are devoid of granulosa cells. Moreover, Ax et al. (1979) have shown that
considerable amounts of mucopolysaccharides in follicular fluid originate from
granulosa cells. Apparently, FSH stimulates mucopolysaccharides synthesis, as is
evidenced by the linear incorporation of [^{35}S] into macromolecular substances in rat
ovaries with respect to the log-dose of FSH (Ax and Ryan 1979). These results suggest
that the studies of mucopolysaccharides constituents in follicular fluid may represent a
new approach in the study of follicular maturation, ovulation and the mechanisms of
gonadotrophic action. Apart from granulosa cells influencing the composition of
follicular fluid, differences that arise between the material and serum may arise from
the limited permeability of the membrana propria (the basement membrane) and
possibly from other unknown permeability barriers separating the thecal capillaries
from the follicular fluid (Gwatkin 1980). The permeability of the blood–follicle barrier
seems to be under endocrine control. The basement membrane is permeable to water,
certain drugs and albumin which may subsequently be taken up by the oocyte.
However, blood proteins of very high molecular weight are excluded from traversing
from the extravascular spaces into the follicular antrum. It is now well known that the
follicle wall does not prevent the passage of interstitial fluid into the antrum (see

Sect. E), although it does influence the rate of transfer of substances of different molecular weights and thereby changes the composition of antral fluid (see McNatty 1978). Thus, alterations in the fluid composition within the follicle are partly related to the permeability of the follicle wall. They are also partly related to the composition of interstitial fluid which, in turn, is related to alterations in the blood supply to the enlarging follicle. As the follicle enlarges, so does the extent of its vascular network, as already discussed in this chapter.

Increasing the local tissue concentrations of oestrogen within the enlarging follicles may increase the density, size, and/or permeability of capillaries associated with those structures (McNatty 1978). Shortly before ovulation, the high levels of LH may be a key regulatory agent in modulating the size and permeability of capillary vessels. During the pre-ovulatory period, increased vasodilation and permeability of capillaries in the pre-ovulatory follicle are believed to be specific steroid-mediated events induced by LH. Whatever the mechanism(s), it has become increasingly clear that increased blood flow and increased permeability of capillaries within the enlarging follicle would enhance the transfer of plasma substances into the surrounding extravascular spaces and also into the follicular cavity. Alterations in blood flow to the enlarging follicle would also influence the biosynthetic capacity of the follicle, which in turn influences the composition and concentrations of hormones within the follicular antrum.

a) Carbohydrates and Glycoproteins

Only a few carbohydrates have been measured in follicular fluid (see McNatty 1978, Peters and McNatty 1980, Gwatkin 1980). Glucose appears to be the major carbohydrate of follicular fluid. It forms about 80% of the total carbohydrate content in cows, whereas only trace amounts of fructose are present. The concentrations of follicular glucose vary throughout the ovarian cycle, with the highest amounts occurring in follicles at oestrus (Pascu et al. 1968). Lactic acid, sialic acid, citric acid, cholesterol, fucose and protein-bound hexoses have also been reported (Edwards 1974). Protein-bound hexoses in follicular fluid are lower than in plasma; the lowest levels occur in the largest follicles (McNatty 1978, Peters and McNatty 1980). The amounts of many non-nitrogenous metabolites (e.g. lactic acid, sialic acid, citric acid) and carbohydrates are reported to be lower in follicular fluid than in plasma or serum. But the low amounts described in various studies may not reflect the actual physiological situation, as the concentrations of many of these substances decline in follicular fluid of excised follicles after ovariectomy (see Edwards 1974). Actually, very little information about the correlative alterations in carbohydrate content with follicular growth is available.

Tadano and Yamada (1978) have demonstrated positive histochemical reactions for complex carbohydrate with 1, 2-glycolytic acidic groups and D-mannoacyl and α-D-glucoacyl residues in the follicular fluid of mouse follicles. The complex carbohydrates are hyaluronic acid, sulphated glycosaminoglycans, other isomeric chondroitin sulphates and neutral glycoproteins. Acid mucopolysaccharides in bovine follicular fluid are chondroitin sulphate-A and heparin sulphate, and their concentrations decrease with the progressive increase in the size of the follicle (Wada 1978). High concentrations of sulphated glycosaminoglycans are present in the follicular fluid (Ax and Ryan 1979, Yanagishita et al. 1979) where they are covalently bound to core proteins to form proteoglycans. Their synthesis may be regulated both positively and negatively by gonadotrophins (Gebauer et al. 1978, Mueller et al. 1978, Ax and Ryan 1979). The predominant sulphated glycosaminoglycan component of the proteoglycan of porcine follicular fluid is chondroitin sulphate-B (dermatan sulphate), although heparin-like molecules are also present in follicular fluid or synthesized by ovarian cells in culture (Gebauer et al. 1978, Yanagishita et al. 1979, Ax and Ryan 1979, Yanagishita and Hascall 1979). The proteoglycans from porcine ovarian follicular fluid have been isolated and characterized: they consist of about 20% protein, 50% dermatan sulphate and 20% oligosaccharides rich in sialic acid, galactose, mannose, glucosamine and galactosamine (Yanagishita and Hascall 1979). The amino acid composition of the proteoglycan is significantly different from that of cartilage proteoglycans, with a higher proportion of Asp, Thr, and Lys and lower amounts of Pro and Gly. The linkage of oligosaccharides to the protein core remains to be determined. The follicular fluid proteoglycan does not interact with hyaluronic acid. The proteoglycans are believed to be involved in cumulus expansion, as will be discussed in Chap. IV, and are correlated with antrum formation. The concentrations of sulphated glycosaminoglycans in follicular fluid appear to decrease during follicular maturation, being more concentrated in fluid from small pig follicles than from large ones (Ax and Ryan 1979, Schwartz 1982). LH terminates the synthesis of sulphated glycosaminoglycans by rat ovarian slices, an action that may be mediated by progesterone (Gebauer et al. 1978). However, Yanagishita et al. (1981) have observed that LH increases proteoglycan synthesis by rat granulosa cells in vitro, and that progesterone has no effect on this synthesis. These differences may be due to different experimental designs used in these two studies.

b) Proteins

The protein composition of follicular fluid has been studied in women, cows, pigs, sows, rabbits and rats (see McNatty 1978). The electrophoretic pattern of follicular fluid is essentially similar to that of plasma. There are at least two sources of protein in follicular fluid, namely the blood supply and the follicular cells (Andersen et al. 1976, Wada 1978). Most blood serum proteins are found in follicular fluid. The protein concentration of follicular fluid ranges from less than 50% to 100% of the protein concentration of serum (Manarang-Pangan and Menge 1971, Shalgi et al. 1973). The total quantity of proteins in human follicular fluid is approximately 58 mg/ml (Gwatkin 1980). The albumin content is as high as that of serum, or even higher. In spite of size differences in follicles, their proteins are similar (Andersen et al. 1976), but may vary with the stage of the ovarian cycle (Pascu et al. 1971). While the total amount of protein in different follicles is similar to that in serum, the relative concentrations of individual proteins often differ widely (McNatty 1978, Peters and McNatty 1980, Gwatkin 1980). This is supported by the fact that albumins and β-globulins are invariably at higher levels in fluid than in serum, while IgA, IgG, fibrinogen, α- and γ-globulin are usually present in lower amounts. Macroglobulins and substances with molecular weight in excess of 1 million are undetectable (see references in Peters and McNatty 1980). In human follicular fluid the concentration of individual proteins decreases with increasing molecular weight (Gwatkin 1980). IgG (MW 1.5×10^5) is found at 74% of the serum concentration, but IgM (MW 9.0×10^5) and B_1-lipoproteins (MW 1.2×10^6) are either undetectable or present in trace amounts. Similar results have also been described for the cow (Andersen et al. 1976).

Studies on the concentrations of individual follicular proteins have not always produced consistent results for each species (e.g. albumins; McNatty 1978). In many cases, the differences are difficult to evaluate because studies are often made on pooled fluids from follicles of unknown morphology, using different methods. Generally, the levels of individual proteins continue to alter throughout follicular growth and show marked differences between healthy and atretic and cystic follicles (McNatty 1978). With the increase in size of the healthy follicle, the follicular concentration of larger-molecular-weight proteins is increased (Andersen et al. 1976) possibly due to increased permeability and clearance of substances from capillaries surrounding the follicle. Cystic and atretic follicles, despite the stage of the ovarian cycle, also show higher concentrations of large-molecular-weight proteins in comparison to nonatretic follicles or plasma (McNatty 1978, Peters and Mc Natty 1980). Some of these proteins may be derived from degenerating cells within the follicle or from leukocytes or plasma cells which invade atretic follicles (Marion et al. 1968).

Human follicular fluid contains a heparin-like anticoagulant that may be important in controlling clotting at ovulation (Gwatkin 1980). Heparin is part of a family of complex polysaccharides. A heparin-like anticoagulant has been demonstrated in the follicular fluid of various mammalian species (McNatty 1978). Further evidence for "heparin-like" substances in follicular fluid has become available when the anticoagulant activity of fluid is neutralized with either protamine sulphate or toludine blue. Although follicular fluid has only low levels of fibronogen relative to those in plasma (Shalgi et al. 1973), this is sufficient to form a clot when an excess of thrombin is added to follicular fluid from mares (see McNatty 1978). Attempts have been made to identify proteins specific to follicular cells by immunizing rabbits with pooled follicular fluid. Subsequently, the follicular fluid antibodies that were not found to cross-react with anti-bovine antiserum were those directed against fibrinogen and it's split product D + E only. ABO isoagglutinins occur in human follicular fluid, and antisperm antibodies have been reported in bovine follicular fluid after immunization against sperm (Gwatkin 1980). A bradykinin-like substance demonstrated in rabbit, cow and human follicular fluid, decreases the tone of isolated rabbit and bovine fallopian tubes (Gwatkin 1980). It has been suggested that this substance could enter the fallopian tubes at ovulation and affect ovum transport. As already discussed, IgG may be present in follicular fluid and be liberated into the oviduct ampulla at ovulation (Gwatkin 1980). Significant IgG levels have been reported in the hamster egg–cumulus complex (Johnson 1973). Antisperm antibody has been observed in the follicular fluid of an infertile woman (Edwards 1974) and in the follicular fluid of cows immunized against sperm (Menge 1970). Gwatkin (1980) has suggested that follicular fluid antibodies, directed against spermatozoa and against such egg antigens as the zona pellucida, may be responsible for some cases of infertility and could be exploited for the control of fertility.

Shalgi et al. (1977) have compared the follicular bursal and ampullar fluids and plasma of rats, in regard to concentration of protein, electrophoretic pattern of proteins in disc electrophoresis and pattern of proteins after isoelectric focussing. The order of decreasing protein concentration is: follicular fluid → plasma → bursal fluid

→ ampullar fluid. Proteins of follicles and bursa are similar or identical to plasma proteins, but oviducal proteins show a different pattern.

Amino Acids. Velazquez et al. (1977) have observed that human follicles have amino acid concentrations that are generally higher than those in blood plasma: only the concentrations in cystic follicles were found to be lower than in plasma. Of the various amino acids studied Asp, Thr, Glu, Glu-NH$_2$, Gly, Ala and Met were at similar concentrations in follicular fluid and plasma. Collectively however, the basic amino acids were generally almost twice as concentrated in follicular fluid, compared to plasma.

Dave and Graves (1978) analyzed follicular fluid and serum of normally cycling and oestrous-synchronized cows for total and free amino acids. Sixteen total and free amino acids were observed. In follicular fluid the concentration of both total and free amino acids was lower, with the exception of cystine, than in blood serum of the same cow. Glutamic acid was the major component in every fluid. In percentage, the content of basic amino acids was higher in the follicular fluid than in the blood serum, indicating a selective transfer of the amino acids from the preferentially blood into the follicular fluid. Alternatively, these substances may be secreted by follicle cells. Synchronization with progesterone followed by PMSG injections increases freely available amino acids in the follicular fluid more than the blood serum. Their presence has been attributed to local synthesis of the protein by the granulosa cells.

c) Enzymes

The follicular fluid contains many enzymes including proteases and peptidases which show changes in their concentrations during the growth of the follicle (Lipner 1973, Edwards 1974, Espey 1974, McNatty 1978, Gwatkin 1980). Most of the enzymes occur in other tissues. In women, the activity of enzymes (proteases and peptidases) in fluid from healthy follicles is higher than in fluid from cystic follicles or in serum (Caucig et al. 1972); the highest activity is seen in small antral follicles, indicating that enzyme activity may not necessarily be related to the steroidogenic capacity of a follicle. Many of the enzymes perform intracellular functions (i.e. the aminotransferases, phosphatases, esterases and dehydrogenases), and possibly derive by leakage from degenerating cells.

Lysosomal enzymes, such as hyaluronidase (Zachariae and Jensen 1958), endopeptidases (Jung 1965) and collagenase (Rondell 1970), are present in follicular fluid and may play a role in follicular rupture (Jung 1965). An increase in colloid osmotic pressure due to hyaluronidase depolymerization of mucopolysaccharides may explain the pre-ovulatory changes in follicles (Zachariae and Jensen 1958). Other enzymes may also contribute to follicular rupture which will be discussed in Chap. V. The injection of collagenases into a follicle can induce rupture (Espey and Lipner 1965), and increased collagenase activity has been observed in the dome of the follicle at the time of ovulation (Espey and Rondell 1968). However, in vitro, the isolated ovarian collagenase did not increase the distensibility of strips from pig follicles (Espey and Stacey 1970). Perhaps several enzymes acting synergistically may be involved in the distension and rupture of the follicle.

Plasmin is a substance which is thought to the involved in follicular rupture. Plasmin and plasminogen have been measured in bovine follicular fluid and it has been suggested that plasmin may decrease the tensile strength of the follicle wall, degrade the basement lamina and activate procollagenases to allow ovulation (Beers et al. 1975). Plasminogen is a glycoprotein which is present in follicular fluid at concentrations similar to those in the serum and appears to derive from the latter. The source of plasminogen activator is granulosa cells and it is released in response to LH of follicular fluid.

d) Elements

The concentrations of sodium and potassium in follicular fluid are of great physiological significance, as they are mainly responsible for the regulation of the osmotic pressure of the fluid. In women, their concentrations are observed to be similar to those in blood (McNatty 1978); the osmotic pressure of fluid from large antral follicles is the same as that of plasma. Sodium concentrations in the follicular fluid of the cow only slightly exceed those in the serum, whereas potassium concentrations are consistently higher in fluid than in serum. The cause of species variations in their concentrations is not known. In follicular fluid the concentrations of other ions Mg, Zn, Cu, Cl, and inorganic phosphate are similar to those in serum in most species (see McNatty 1978, Peters and McNatty 1980, Gwatkin 1980).

Chang et al. (1977) have observed that concentrations of various elements such as Na, K, Cl, Mg, Ca, P and S are similar to those of plasma in the pig. Knudsen et al. (1978) have studied the concentrations of hydrogen ions, oxygen, carbon dioxide and bicarbonate in porcine follicular fluid. Wada (1978) has observed that Na concentration of bovine follicular fluid is lower than that of blood serum, and the K concentration of follicular fluid decreases with increasing follicular size. K/Na ratio falls progressively with increasing follicular size. Knudsen et al. (1979) have observed that Na and osmolal concentrations of follicular fluid in cyclic pigs do not vary significantly, as they are similar to those in the plasma. The K concentration is greater in small (days 12–13) and medium-sized (day 16) follicles than in plasma or large (day 18–oestrus) follicles of cyclic cows. Post-mortem changes appear to cause higher K and osmolality and lower Na values in the follicular fluid. Burgoyne et al. (1979), using electron probe microanalysis, have determined concentrations of Na, Cl, K, Ca, Mg, S and P in samples of follicular fluid, ovarian vein serum and peripheral venous serum obtained from virgin rabbits at 2 h intervals up to 10 h after hCG injection. Throughout this 10 h period the elemental composition of follicular fluid is essentially the same as that of the blood serum. There occurs significant drop in follicular Ca^{2+} relative to blood during the 10 h period, which may suggest Ca^{2+} involvement in the regulation of oocyte maturation, as will be discussed in Chap. IV. Significant differences were also noted for K and P between follicles within rabbits, but the significance of this finding is unknown.

e) Inhibitors and Stimulators

Many studies have suggested that the follicular fluid contains nonsteroidal substances which can modify several different processes (Channing and Batta 1981, Channing et al. 1982, Schwartz 1982a, 1983, Rigby et al. 1983, Ward et al. 1983a, b). These include oocyte maturation (Channing and Tsafriri 1977, 1978, Amsterdam et al. 1979b, Channing 1979b, Channing and Batta 1981, Channing et al. 1982), functional luteinization of granulosa cells (Channing and Tsafriri 1975, Channing 1979b, Channing and Batta 1981, Channing et al. 1982), cellular enzyme activity (Moore et al.

1975) and FSH secretion (de Jong and Sharpe 1976, Channing 1979 b, Channing and Batta 1981, Channing et al. 1981, 1982, Grady et al. 1982). The putative intra-ovarian regulators, which have been studied in follicular fluid from bovine, porcine, and/or human ovaries, include an oocyte maturation inhibitor (OMI), a luteinization inhibitor (LI), a luteinization stimulator (LS), an FSH receptor-binding inhibitor (FSHRBI), inhibin (an FSH-suppressing substance also called "folliculostatin") (Channing 1979 b, Channing et al. 1982, Schwartz 1982 a, 1983, Channing and Segal 1982, Ward et al. 1983 a, b). In addition, gonadocrinins and gonadostatins are also thought to be present in follicular fluid (Esch et al. 1983). These terms are generic names for factors that stimulate or inhibit secretion of both FSH and LH from the pituitary.

The discovery of these novel intra-ovarian peptidic substances has caused great excitement and controversy, as evidenced by the publication of several review articles and books (see Channing and Batta 1981, Channing et al. 1982, Schwartz 1982 a, b, 1983, Grady et al. 1982, Channing and Segal 1982, Greenwald and Terranova 1983, Franchimont et al. 1983, de Jong et al. 1982, 1983, Esch et al. 1983). Various biological activities mentioned above fall into two major groups. One group consists of stimulatory actions, and other group of inhibitory biological actions.

Follicular fluid from porcine or bovine or human follicles contains an inhibitor(s) which inhibits maturation of oocytes in culture. The fluid from small antral follicles is found to be more effective than from large antral follicles (Tsafriri and Channing 1975 a, Tsafriri et al. 1976 a, Gwatkin and Andersen 1976, Silverman et al. 1982, Channing et al. 1983). This inhibitory factor(s) has been termed the Oocyte Maturation Inhibitor (OMI) which is also called "Meiotic Arresting Substance" (Sato et al. 1982). Concentrations of OMI decrease with increasing follicle diameter ($P < 0.05$) in the porcine ovary, independently of the stage of the oestrous cycle (van de Wiel et al. 1983). Its concentrations in the largest follicles become low before the onset of oestrus, and are essentially unaltered 24 h later. Granulosa cells, but not fibroblasts or theca cells, inhibit oocyte maturation in vitro (Tsafriri et al. 1976 b, Gwatkin and Andersen 1976, Tsafriri and Bar-Ami 1982), suggesting that OMI is produced by the granulosa and/or cumulus cells (Channing et al. 1983 a, Sato et al. 1982). Sato et al. (1982) have isolated the meiotic arresting substance OMI from porcine granulosa cells. The general consensus is that granulosa cells secrete OMI, and their potential to secrete OMI in culture decreases as the follicle matures (Channing et al. 1983 a). Addition of FSH and prolactin to cultured granulosa cells stimulates OMI secretion, whereas addition of testosterone or dihydrotestosterone brings about a decrease in OMI secretion (Channing et al. 1983 a). Purification of OMI has been difficult, possibly because of the uncertainties of a bioassay system which depends on the assessment of oocyte germinal vesicle breakdown. Cameron et al. (1983) have found that South African clawed toad *(Xenopus laevis)* oocytes provide a simple, readily available, year-round bioassay material for testing follicular OMI. The OMI is a protein with a molecular weight between 1,000 and 10,000, and has been identified in bovine, porcine, and human follicular fluid (Tsafriri et al. 1976 a, Channing 1979 b, Channing and Batta 1981, Channing et al. 1982). OMI separated from porcine granulosa cells is peptide of which the molecular weight is estimated to be 1,450 to 3,000 (Sato et al. 1982). The prevention of oocyte maturation by the follicular fluid can be overcome by addition of LH to the culture medium (Channing and Tsafriri 1977, 1978, Gwatkin and Andersen

1976), as will be discussed in Chap. IV. Since the inhibition is reversed by the addition of LH to the medium, a role for this peptide in the regulation of oocyte maturation has been proposed (Tsafriri 1978a). But unequivocal evidence that such an inhibitor has a physiological role in vivo has not been obtained (see detailed discussion in Chap. IV).

Luteinization inhibitor (LI) is present in greater amounts in the fluid of small follicles as compared to large follicles, and acts to keep granulosa cells from luteinizing in terms of morphology and progesterone secretion (Channing 1979b, Channing and Batta 1981, Channing et al. 1982). This suggestion is supported by the fact that fluid from small follicles inhibits the luteinization of granulosa cells from large follicles (Channing and Tsafriri 1975). LI is believed to act to inhibit maturation of LH receptors and acute LH stimulation of cAMP accumulation (Channing 1979b). This substance (LI) may be related to a factor with similar effects, possibly produced by the egg (El-Fouly et al. 1970, Nekola and Nalbandov 1971). However, the presence of this egg factor (or luteostatic factor) has not been established (Lindner et al. 1974). The results of other studies also do not support the presence of a luteostatic factor produced by the oocyte (see Hillensjö et al. 1981). McNatty (1978) has suggested that until a substance which inhibits steroidogenesis is isolated from follicular fluid, the concept of luteinization inhibition must be treated with some reservation. Shemesh (1979) has suggested the presence of a prostaglandin synthetase inhibitor as well as a luteinization inhibitor in the bovine follicular fluid from mid-cycle follicles, but that these substances disappear in the pre-ovulatory follicle.

Amsterdam et al. (1979b) have observed that the follicular fluid can inhibit specifically the LH-sensitive adenylate cyclase, suggesting that it may play a role in regulating hormone action in the ovary. The observation that ovarian inhibitor activity decreases concomitantly with follicular maturation supports the contention that elimination of inhibitor(s) may be part of the preconditioning of the follicle for subsequent sensitization to LH. Since LH overcomes the inhibitory effects of follicular fluid, it is also possible that pre-ovulatory follicles contain high levels of LH which overcome these inhibitory effects. It needs to be determined more precisely whether the follicular inhibitor acts directly on adenylate cyclase which suppresses cAMP and thus PGF and progesterone, or whether two different inhibitors are involved. The activity of prostaglandin synthetase inhibitor decreases with maturation and the failure of the follicle to change this inhibitor may result in a cystic or atretic follicle. This concept is supported by the recent findings of Shemesh (1979), who has observed that only ovulating follicles lack the inhibitory activities, whereas activity still exists in the other follicles on the day of ovulation. Ledwitz-Rigby and Rigby (1979) have observed that addition of follicular fluid to culture media has striking effects on granulosa cell morphology, progesterone secretion and LH stimulation of cAMP accumulation (see also Ledwitz-Rigby et al. 1982). Ledwitz-Rigby (1983) has demonstrated a stimulatory effect of follicular fluid on oestrogen secretion by immature porcine granulosa cells. At least two factors influencing luteinization are present in porcine follicular fluid. One enhances progesterone secretion by granulosa cells from small, immature antral (1–2 mm diameter) follicles. The other inhibits progesterone secretion by granulosa cells from medium (3–5 mm) and large pre-ovulatory (6–12 mm) follicles. The inhibitor (LI) occurs predominantly in fluid from small follicles, whereas luteinization stimulator (LS) is most active in fluid from large follicles, but also exists in the fluid from small follicles. Luteinization inhibitor (LI) causes primarily an alteration of cyclic nucleotide metabolism. Ledwitz-Rigby et al. (1982) have hypothesized that differences in the follicular fluid content of "maturation stimulating molecules" may determine which follicles are able to produce adequate oestrogen and respond to the gonadotrophins maximally, and which are not. Maximal response to gonadotrophins and the actions of the oestrogen produced within the follicles may prevent atresia. Osteen and Channing (1983) have observed that PFF modulates the induction of LH/hCG receptors and progesterone secretion in cultures of immature granulosa cells exposed to FSH, insulin, thyroxine and cortisol. The follicular fluid environment becomes more permissive of LH/hCG receptor induction and progesterone as follicle maturation progresses (see Sect. D).

The follicular fluid components which regulate luteinization of cells, require much more study. Apparently, LI not only inhibits luteinization, but also prevents the accumulation of LH receptors (Channing et al. 1982, Schwartz 1982a). Thus the follicles, which develop a responsiveness to LH and subsequently ovulate, require a higher ratio of LH receptors to LI. By contrast, follicles, which are to undergo atresia, do not acquire LS in sufficient quantity. What regulates the development of these factors is not known at present, nor is their mechanism of action understood. LI and LS have not been isolated. They are believed not to be steroids, as charcoal extraction does not remove activity. They both appear to have a molecular weight greater than 10,000 (Channing et al. 1982, Schwartz 1982a). The purification of these components of the follicular fluid will be needed before the extent of intragonadal regulation of granulosa cell dynamics can be fully understood.

Follicular fluid from bovine follicles inhibits endogenous DNA-dependent RNA polymerase activity in Hoshida-ascites cells (Moore et al. 1975), but the inhibitory influence has not been attributed to proteolytic enzymes, although decreased activity due to the presence of phosphatase or ribonuclease activity in follicular fluid was not ruled out. The fluid component that causes this activity is heat-labile, and it is postulated that it is a protein. The possible relationship, if any, of this activity to the other follicular/oocyte inhibitors is unknown.

Inhibin, a hormone first suggested by McCullagh (1932), came under intensive study in the last decade. It has also been named inhibin-m or inhibin-f, folliculostatin and gonadostatin (see Williams and Lipner, 1982). Bovine and porcine follicular fluid has been shown to contain inhibin-like activity, i.e. it contains a substance (MW > 10,000) that specifically suppresses FSH secretion in castrated animals (de Jong and Sharpe 1976, Lorenzen and Schwartz 1979, Channing 1979b, Williams et al. 1979, Shander et al. 1979, Miller et al. 1979, Channing et al. 1979b; Channing and Batta 1981, Hermans et al. 1981, Williams and Lipner 1982, Schwartz 1982b, 1983, Anderson and Hoover 1982, De Greef et al. 1983, Franchimont et al. 1983, de Jong et al. 1982, 1983). Exogenous FSH can overcome this PFFI blockade (Schwartz 1983a, b). The presence of FSH-inhibiting or suppressing follicular factors has also been shown in follicular fluid from sheep, porcine, equine and human sources (Daume et al. 1978, Lorenzen et al. 1978, Lorenzen and Schwartz 1979, Channing and Batta 1981, Hermans et al. 1981, Channing et al. 1982, Grady et al. 1982, van de Wiel et al. 1983, Anderson and Hoover 1982, Tsonis et al. 1983a, b) and it suppresses FSH in rats, mice, horses, hamsters and monkeys (Channing and Batta 1981, Channing et al. 1982, Schwartz 1982a, b, 1983, Grady et al. 1982). About 2 h are required for it to suppress FSH significantly in vivo and the effect of a single intravenous injection disappears after about 10 h. It cannot inhibit ovulation if administered before the ovulatory surges of LH or FSH, but it can inhibit growth of smaller follicles which are dependent upon elevated levels of FSH. Follicular fluid obtained from small and medium porcine follicles shows a higher concentration of inhibin than fluid from large follicles. Van de Wiel et al. (1983) have also observed that concentration of inhibin shows a tendency to decrease with increasing follicle diameter on days 10, 15 and 18 but not on day 5 of the cycle in the pig. Its concentrations in the largest follicles are low before the onset of oestrus and are essentially unaltered 24 h later. A positive correlation exists between OMI and inhibin concentrations, whereas correlation between inhibin and progesterone concentrations is negative. The inhibin content of follicular fluid from individual ovine follicles varies with follicular fluid volume, but not with degree of atresia, as observed by morphological criteria (Tsonis et al. 1983a, b). It shows marked variation between follicles, possibly reflecting a combination of the number and activity of granulosa cells within the follicle and the exact rate of inhibin from the follicle. Follicular fluid obtained from human ovaries during the follicular phase contains more inhibin than follicles sampled during the luteal phase. Chappel et al. (1979) have demonstrated the presence and activity, both in vivo and in vitro, of an inhibin-like substance present within the hamster ovary, which is believed to act synergistically with oestradiol (at the level of the anterior pituitary gland) to regulate the secretion of FSH. Fujii et al. (1983) have observed that follicular fluid inhibin activity is relatively constant throughout the oestrous cycle of the rat except for a well-developed surge at pro-oestrus. Ovarian steroid appears to be an inadequate feedback explanation for the regulation of serum FSH levels (Schwartz 1982b, 1983). But the ovarian follicular fluid is remarkably effective in suppressing FSH specifically in a number of in vivo models as well as in vitro on a pituitary cell culture. Equine follicle has also been shown to have an inhibin-like activity (Channing et al. 1981). The fluid of atretic follicles contains less inhibin-like activity compared to fluid from viable follicles. The decrease in inhibin-like activity in the atretic follicles could reflect granulosa cell degeneration (Tsonis et al. 1983a, b) since granulosa cells have been suggested to be the source of follicular inhibin-like activity in a number of species (Channing et al. 1980, 1982, Channing and Batta 1981, Hermans et al. 1981, Grady et al. 1982, Schwartz 1982b, 1983, Anderson and Hoover 1982, de Jong et al. 1982, 1983, Henderson and Franchimont 1983, Franchimont et al. 1983). In vivo studies with the cyclic rat have suggested that inhibin secretion by the ovary can be affected by the gonadotrophins, and that the inhibin secretory response to gonadotrophin stimuli may be a function of the characteristics of the follicular population on the ovary (Anderson and Hoover 1982). In vitro studies on the secretion of inhibin by porcine granulosa cells have suggested that the capacity of granulosa cells to secrete inhibin in suspension culture increases with the stage of maturity of the follicle from which the granulosa cells are taken. This pattern is the reverse of that observed for inhibin activity in porcine follicular fluid (Anderson and Hoover 1982). Hormonal control of inhibin secretion needs to be determined in future investigations (Anderson and Hoover 1982). However, in vitro experiments of Franchimont et al. (1983) show that inhibin secretion by granulosa cells is stimulated by aromatizable, nonaromatizable and synthetic androgens. This stimulatory effect appears to be a direct one and mediated through androgen receptor activation. The oestrogens do not influence inhibin secretion, progesterone decreases inhibin secretion by granulosa cells. The direct role of gonadotrophins is unclear. Prolactin does not appear to modify inhibin secretion by granulosa cells. Lee (1983) has proposed that PMSG treated immature female rats can be used as a model system for studying control of inhibin secretion.

The probable site of inhibitory action of inhibin is at the level of the anterior pituitary gland as it can act to inhibit cultured anterior pituitary cells in their ability to secrete basal as well as LH-RH-stimulated release of FSH in vitro (Channing 1979 b, Shander et al. 1979, Schwartz 1982 b, 1983). It can also inhibit LH-RH stimulation of FSH secretion in the rat in vivo. Wise et al. (1979) have shown that porcine follicular fluid (PFF) changes the responsiveness of the pituitary gland to LH-RH. It inhibits FSH secretion by 70%–100%, depending upon the method of LH-RH infusion. PFF has only a minor effect on LH-RH-induced LH secretion (Ying et al. 1982, Sairam et al. 1982). Charlesworth et al. (1983) have observed that the effect of PFF on FSH is highly selective and shows a clear dose-dependence in suppressing FSH in the face of high endogenous or exogenous GnRH levels in vivo, suggesting that secretion of FSH in vivo is probably less dependent than LH on GnRH. De Greef et al. (1983) have observed that the inhibin-like activity in bovine follicular fluid does not suppress the plasma levels of FSH by affecting its plasmic clearance or by influencing the hypothalamic release of LH-RH, but that it has a direct effect on the adenohypophysis in inhibiting the release of FSH. PFF also contains another factor which can decrease the levels of LH-RH in the hypophysial stalk plasma. PFF affects FSH, but not LH (Hermans et al. 1982). All these observations support the concept that nonsteroidal substances of gonadal origin play an important role in the regulation of pituitary gonadotrophin secretion (see also Hermans et al. 1981, Schwartz 1982 b, 1983). Franchimont et al. (1983) have observed that a reduction of FSH and, to a lesser degree, of LH secretion by the pituitary cells is associated with a dose-dependent increase of cyclic GMP. The variations of cyclic GMP are simultaneous with those of FSH and LH secretion, suggesting that cyclic GMP is a possible candidate as second messenger mediating the inhibitory effect of inhibin. cAMP variations are not related to the dose of inhibin and to the amplitude of gonadotrophin. Administration of porcine follicular fluid early in the menstrual cycle of the rhesus monkey causes a decrease in the number of granulosa cells appearing in the follicle at mid-cycle (see Channing et al. 1979 b, Channing at Batta 1981). Kling et al. (1983) have suggested in vitro inhibition of gonadotrophin-stimulated aromatase activity by a PFF preparation. This observation needs to be confirmed in future studies.

Chemical purification of ovarian inhibin is needed for a more rigorous assessment of its physiological role (Sairam et al. 1982, de Jong et al. 1982, 1983, Franchimont et al. 1983, Williams and Lipner 1982). Ovarian inhibin will also provide a useful means to selectively inhibit FSH secretion without causing inhibition of LH so that the physiological role of FSH may be elucidated. The inhibin-F from the follicular fluid of pig has been found to be trypsin-senstitive, suggesting that it is a peptide (protein) (Lorenzen et al. 1978, Lorenzen and Schwartz 1979, Grady et al. 1982). It is associated with a molecule (or complex) with MW over 70,000. Lorenzen and Schwartz (1979) have observed that porcine follicular fluid (or its active agent, inhibin) can suppress serum FSH in intact or sham-operated females (rats) of all ages (e.g. during the neonatal period, after weaning, and during puberty and adulthood) (see also Hermans et al. 1981, Grady et al. 1982, Schwartz 1982 b, 1983). Inhibin in bovine follicular fluid is associated with a glycoprotein with an apparent molecular weight of 65,000 (de Jong et al. 1983). A comparison of the results from the in vitro bioassay for bovine and ovine inhibin preparations with data from the immunoassay has yielded a good correlation between bio- and immunopotencies (de Jong et al. 1983).

All the known characteristics of inhibin indicate that it may be identical to testicular inhibin (Franchimont et al. 1983) but that follicular fluid may be a more potent source of the material. The high potency and specificity of inhibin in suppressing FSH but not pituitary LH secretion has focussed much attention on its posssible hormonal status. Unlike the other intrafollicular factors discussed (OMI, LS, and LI), inhibin must act via the bloodstream to suppress pituitary function, i.e. it must be a hormone. It is secreted by the granulosa cells in culture (Anderson and Hoover 1982); it may be found in ovarian vein blood (Schwartz 1982a, b, 1983, Anderson and Hoover 1982, Kimura et al. 1983). As steroids or low MW substances "leak" only slowly from the follicular antrum (t $\frac{1}{2}$ > 9 h), the mechanism by which a granulosa cell-derived inhibin is secreted into the bloodstreams presents certain problems (Anderson and Hoover 1982). And unless a more sensitive and rapid assay for its measurement is developed, the verification of its hormonal status will remain controversial for some time (Hermans et al. 1981). Recently, Kimura et al. (1983) have suggested that the selective surge of FSH occurs as a consequence of the decrease in inhibin secretion from the ovary, which is perhaps due to the ovulation doses of hCG altering the functional activity of the granulosa cells in the large Graafian follicles. The inhibin hypothesis cannot rest on firm foundation until the substance(s) has also been isolated in pure form to homogeneity and its structure eventually identified (Sairam et al. 1982). The molecular weight of inhibin, as determined by using extracts or fluids, although informative, should be viewed with considerable caution. Future purification work may resolve many of these problems.

Whether inhibin also has an effect on the follicle itself is unknown. Fluid from large bovine follicles retards growth initiation of follicles in immature mouse ovaries (Peters et al. 1973), but it has still to be determined whether this is due to the "inhibin-like" activity of the fluid.

Ying and co-workers (Ying and Guillemin 1979, Ying et al. 1981) reported the presence of biological activity similar to that of LRF in porcine follicular fluids or in acid extracts of rat ovaries pretreated with PMSG. This LRF-like activity is designated as gonadocrinin (Esch et al. 1983).

Darga and Reichert (1979) have shown that a low-molecular-weight inhibitor of FSH binding to bovine granulosa cell receptors is present in follicular fluid. This factor is called FSH receptor binding inhibitor (FSHRBI) (Reichert et al. 1982). These studies have indicated another mechanism by which hormone–receptor interactions might be regulated. The material is assessed by its ability to inhibit binding of radio-labelled FSH to granulosa cells in vitro. The concentration of FSHRBI increases with increasing size of the bovine follicles from which follicular fluid is collected. FSHRBI appears to be a peptide with a molecular weight of less than 5000. There seems to be no question that such factors exist, but it has as yet not been possible to demonstrate in vivo effects, possibly due to our incomplete understanding of their biological and pharmacological properties, as well as due to lack of sufficient quantities of purified material for such studies. Further investigations are needed to purify and characterize the follicular fluid FSHRBI in order to assess its nature and possible physiological significance more precisely (Reichert et al. 1982). Miller et al. (1979) have observed that bovine follicular fluid has an inhibitory effect on ovulation in the bovine, which requires both protein and steroid factors. This may have been the result of a direct effect at the ovarian level or, more likely, an effect mediated via the hypothalamus and/or the pituitary. Follicular fluid also contains some stimulators of granulosa cell replication (Hammond et al. 1982, 1983). Hammond et al. (1982, 1983) have hypothesized that growth-promoting stimuli lead to a uniform increase in polyamine synthesis which influence granulosa cell replication. Whilst an LH receptor binding inhibitor (LHRBI) has been identified in corpora lutea from rat, human and porcine ovaries (Schwartz 1982a, Channing et al. 1982, Ward et al. 1983a, b, Kumari et al. 1982), it has not been found in follicular fluid.

Follicular fluid from several mammalian species has been shown to induce the capacitation and acrosome reaction of sperm in vitro: these changes are essential before the spermatozoon can fertilize the egg (Gwatkin 1980). Whether follicular fluid is actually involved in the induction of these processes in vivo has not been established. However, it is possible that follicular fluid, entrapped in the cumulus oophorus or secreted by the cumulus itself, could perform such a function.

Both dialyzable and nondialyzable factors have been shown to be responsible for the in vitro induction of the capacitation and acrosome reaction of hamster sperm by bovine follicular fluid (Yanagimachi 1969). The dialyzable factor has since been distinguished as albumin (Lui et al. 1977); which may either have a direct physical effect on the sperm membrane (Blank et al. 1976) or carry enzymes and steroids (Peters 1975) that modify its structure. These results are in agreement with the observation of Hoppe and Whitten (1974), indicating that albumin is needed for the fertilization of mouse eggs in vitro. The nature of the dialyzable factor is not yet known although it is known to stimulate sperm motility (Yanagimachi 1969) and to be partly inorganic (Gwatkin 1980). It has also been noted that treatment of human sperm with human follicular fluid in combination with cAMP induces metabolic (Hicks et al. 1972) and surface changes (Rosado et al. 1973) that are similar to those involved in capacitation and the acrosome reaction. Such treatment also capacitates rabbit sperm (Rosado et al. 1973).

f) Hormones

Human follicular fluid contains gonadotrophins (FSH, LH and prolactin as well as their subunits), androgens, oestrogens and progestins (Edwards 1974, McNatty 1978, McNatty and Baird 1978, Greenwald 1980). Some of these hormones have also been reported for other mammalian species (McNatty 1978, Fujii et al. 1983). In the human ovary, FSH and LH, but not prolactin, always occur at concentrations lower than those found in blood. Gonadotrophins control the development of the follicles, as already discussed in this chapter. While LH is usually considered to be the "ovulating hormone," FSH undoubtedly influences the composition and volume of follicular fluid as well as the follicular morphology. Follicular fluid is rich in steroids, and steroid concentration in follicular fluid greatly exceeds that in blood. The levels of steriod hormones also vary according to the stage of follicle growth. The ratio of follicular to plasma concentrations of all protein hormones (including subunits) is not constant; it varies with the size of follicle and stage of ovarian cycle and/or stage of follicular maturation. Except for humans and sheep, little is known about gonadotrophin concentrations in follicular fluid from other species. FSH levels increase in large follicles despite decreasing levels of FSH in plasma. Not surprisingly, more oestradiol is present in follicles with fluid that also contains FSH, since, as already discussed in this chapter, FSH plus oestradiol enhances granulosa cell mitosis and is thus associated with follicular growth. LH, progesterone and oestradiol show clear peaks at the late follicular phase before ovulation; the concentration of prostaglandin F, known to be involved in ovulation, increases rapidly (McNatty 1978, Gwatkin 1980, Greenwald 1980). As the follicle enlarges throughout the follicular phase, there is an overall decline in the concentrations of prolactin in follicular fluid. When follicles of different sizes and at different stages of the menstrual cycle in women are compared, the highest levels of oestradiol are seen in the very presumptive pre-ovulatory follicles during the late follicular phase. In women, just before ovulation the presumptive pre-ovulatory follicle contains peak concentrations of FSH, LH, progesterone, and probably PGF_2 and declining levels of androgen, prolactin, and oestrogen.

The antral follicle secretes a large amount of steroid hormones which also accumulate in the follicular fluid (McNatty et al. 1975, McNatty 1979, Greenwald 1980). The latter contains some steroid hormones in amounts which are 40,000 to 1 million times higher than those in blood. The major steroids present in the follicular fluid include progestins, androgens and oestrogens which vary in concentrations during follicular growth, suggesting the changing rates of their secretion by the surrounding tissue (Short 1964, Edwards 1974, McNatty 1978, Peters and McNatty 1980, Greenwald 1980). The hormonal milieu within each follicle is characteristic for that follicle (McNatty 1978). Most antral follicles of the human ovary show more androgen than oestrogen (androgenic follicles), only a few follicles show more oestrogen than androgen (oestrogenic follicles), and it is these follicles that continue to mature. McNatty and Baird (1978) have analyzed the concentrations of oestradiol and androstenedione in human follicular fluid as a function of stage of the menstrual cycle, the number of granulosa cells in the follicle, and the presence or absence of FSH in the fluid. In follicles with detectable amounts of FSH, and more than 50,000 cells per follicle, the amounts of oestradiol exceed the amount of androstenedione at all phases of the menstrual cycle. In follicles having similar numbers of granulosa cells, but no detectable FSH, the amount of androstenedione exceeds that of oestradiol at all stages of the cycle. In follicles showing less than 50,000 cells androstenedione was the dominant steroid regardless of whether FSH was present or not. These latter follicles are defined as cystic and can equally be called atretic. In a later study McNatty et al. (1979) observed that the amount of oestradiol in follicular fluid was related to the optimum number of granulosa cells in the follicle. If the number of granulosa cells is reduced, oestradiol amount is lowered and oocyte viability is reduced. Furthermore, human follicles which are atretic, i.e. which have less than a requisite number of granulosa cells for that size of the follicle, have a high androgen/oestrogen ratio in the follicular fluid (McNatty 1980). In the mare follicle neither a drop of oestrogens nor a rise in androgens is associated with atresia, and decreased steroidogenesis occurs only after degeneration is advanced (Kenney et al. 1979). Recent studies on the occurrence of testosterone in mare follicular fluid, in a concentration which is two orders of magnitude higher than that in peripheral plasma, suggest that the follicle may contribute to the production of circulating testosterone (Silberzahn et al. 1983). In the sheep ovary, several follicles become "oestrogenic" and more antral structures reach their final stages of growth than is the case in women.

Other steroids or steroid precursors, which have been studied in follicular fluid, include cholesterol and pregnenolone. In general, the concentrations of steroids in follicular fluid change throughout follicular growth. During the growth phase, androgens and oestrogens are the steroids seen in follicular fluid in the highest concentrations. Just prior to ovulation, the pre-ovulatory follicle may also show very high concentrations of progestins (Brietenecker et al. 1978, Greenwald 1980). During the luteal phase, the follicles of women, ewes and mares may show large amounts of progestins and androgens (e.g., norandrostenedione and/or androstenedione) (McNatty 1978).

Binding-proteins for some of hormones have also been observed in follicular fluid (Giorgi 1969, Gwatkin 1980). The results of various studies, as discussed here, show that follicles maintain their own hormonal environment, and the rate of diffusion of steroids from the follicular fluid into the blood is extremely low, as is evidenced by the extraordinarily high levels of oestrogen which may sometimes be present. The binding of steroids by proteins in follicular fluid has been offered as an explanation as to why high concentrations of steroids are retained in the follicular fluid (McNatty 1978). An androgen-binding protein has been reported in the follicular fluid of patients with the polycystic ovary syndrome (Vigersky and Loriaux 1976) and a cortisol-binding protein has been seen in porcine follicular fluid (Mahajan and Little 1978). In cows and women without polycystic ovaries, the androgen-binding globulin is present in fluid in concentrations similar to, or lower than, those in plasma. The basis for differences in the concentration of androgen-binding globulin in cystic and noncystic follicles is not known (McNatty 1978). Proteins that bind oestrogen have also been detected in the follicular fluids of cows and women (McNatty 1978). Movement of steroids into follicular fluid from nearby follicular structures is also restricted, as can be evidenced by the low progesterone levels found in follicles during luteal phase, in the presence of an ipsilateral corpus luteum.

The physiological significance of large amounts of various steroid hormones stored in the follicular fluid needs to be determined more precisely. Intuitively, one feels that high local levels of steroids are involved in various negative and positive feedback interactions at the ovarian level. However, it is possible that the elevated concentrations of oestrogen in follicular fluid merely represent a storage phenomenon and that this hormone is believed to be of greater significance after its post-ovulatory release and subsequent actions on the oviduct. At present, one has to accept that there really is no direct in vivo proof that follicular fluid steroids have the profound effects on the ovary that have been postulated.

Chapter IV
Ovum Maturation

Ovum maturation in mammals is accompanied by the resumption of meiosis. This maturation in mammalian oocytes is initiated during prenatal life or shortly after birth. The oocyte reaches the diplotene stage of the prophase just before or immediately after birth. At this stage, by a mechanism yet not fully understood, the meiotic process is arrested ("first meiotic arrest") with the nucleus (germinal vesicle) in prophase I of meiosis (Guraya 1977a). Meiotic maturation refers specifically to the process of nuclear progression from the diplotene (dictyate) stage of the first meiotic prophase to metaphase II (Fig. 45). Upon appropriate stimulation from circulating pituitary gonadotrophins (FSH and LH) or, in some species, after removal from Graafian follicles and culture in vitro, dictyate-stage oocytes resume meiosis, resulting in the "final" maturation of the mammalian oocyte (Tsafriri 1978a, 1983, Lindner et al. 1980, Tsafriri et al. 1982a, b). In most mammalian species studied, meiosis is resumed within the mature, pre-ovulatory follicle(s). FSH and LH cause the expansion of the antrum and maturational changes in the theca and granulosa cells, as discussed in Chap. III, although the oocyte nucleus remians morphologically unaffected until the onset of the so-called LH surge. FSH and LH – either directly or indirectly through steroid intermediates (especially oestrogen) – also regulate the proportions of follicles which undergo ovulation or atresia (see Chaps. III and VII) (see also Harman et al. 1975, Hay and Cran 1978, Erickson 1982). But the details of hormonal interaction with the oocyte, which leads to meiosis, remain to be clarified (Tsafriri 1983, Tsafriri et al. 1982a, b). However, at ovulation, a number of oocytes, depending on the species, are released at the metaphase of the second meiotic division (Fig. 45); in a few species of mammals dictyate oocytes are normally ovulated and maturation takes place in the oviduct (Tsafriri 1978a).

Ovum maturation, which involves the formation of polar bodies by meiosis (Fig. 45), has been extensively studied in vivo and in vitro in recent years with modern techniques of electron microscopy, histochemistry and biochemistry (Tsafriri et al. 1976b, Tsafriri 1978a, b, 1979, 1983, Leibfried and First 1979a, Schuetz 1979, Hillensjö et al. 1979, Readhead et al. 1979, Wassarman et al. 1976, 1979a, Van Blerkom 1980, Tsafriri et al. 1982a, b). The results of all these studies are controversial, as very divergent views have been expressed about the structural and chemical changes and control mechanisms involved in ovum maturation under various in vivo and in vitro conditions (Tsafriri et al. 1976b, Tsafriri 1978a, 1983, Channing 1979a, Baker 1979, Lindner et al. 1980, Tsafriri et al. 1982a, b). Species variations in this regard have also been reported. Resumed meiosis in the mammalian oocyte involves a sequence of well-defined nuclear (nuclear maturation) and cytoplasmic (cytoplasmic maturation) changes, including dissolution of the oocyte nuclear

membrane, mixing of nucleoplasmic and cytoplasmic components, the first reduc-
tional division at metaphase I with the abstriction of the first polar body and finally
"arrest" of meiosis at metaphase II ("second meiotic arrest") (Fig. 45). Although the
cytoplasmic maturation of the oocyte is an essential step on the path to successful early
mammalian development, the exact nature of this process is still not well understood
(Lindner et al. 1980, Tsafriri et al. 1982a, b, 1983). The second meiotic metaphase is
generally completed after fertilization (Fig. 45), or after the provision of an activating

OOCYTE MEIOSIS

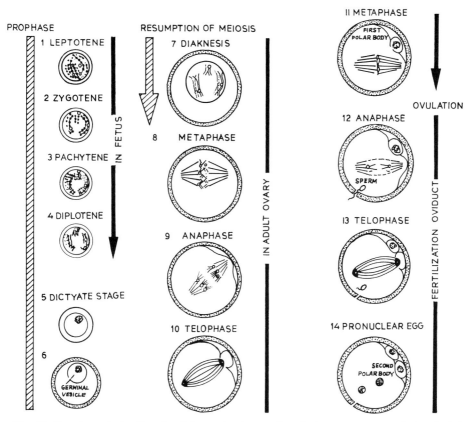

Fig. 45. Oocyte meiosis. For simplicity, only three pairs of chromosomes are depicted. *1–4*
Prophase stages of the first meiotic division which occur in most mammals during foetal life. At
zygotene *(2)* the homologous maternal and paternal chromosomes begin to pair, and at
pachytene *(3)* they are paired along their entire length, thus forming bivalents. During pachytene,
each homologue cleaves longitudinally to form two sister chromatids, so that each bivalent forms
a tetrad. During this stage, interchange of genetic material between maternal and paternal
chromatids occurs by crossing over. At diplotene *(4)* the chromosomes begin to separate,
remaining united at points of interchange, the chiasmata. The meiotic process is arrested at this
stage ("first meiotic arrest"), and the oocyte enters the dictyate stage. When meiosis is resumed,
the first maturation division is complete *(7–11)*. Ovulation occurs usually at the metaphase II
stage, *(11)*, and the second meiotic division *(12–14)* takes place in the oviduct only following
sperm penetration (From Tsafriri 1978a)

stimulus-parthenogenesis. The morphological (including ultrastructural), cytochemical and biochemical characteristics of dictyate oocytes have already been discussed in
Chap. III (see also Tsafriri 1978a). Here only results of recent morphological
(including ultrastructural), cytochemical studies on changes which occur during the
formation of first polar body (or ovum maturation in vivo and in vitro) will be
discussed. The role of various factors in the regulation of oocyte maturation will also
be discussed.

A. Nuclear Changes and Expulsion of Polar Bodies

The morphological changes, which occur during oocyte maturation, include nuclear
maturation and expulsion of the first polar body, morphological aspects of maturational process in the cytoplasm and modification of the oocyte follicle cell
relationship. Numerous studies have described the morphological changes in maturing
mammalian oocytes (see Tsafriri 1978a). Here I will emphasize the relatively recent
findings.

   The primary oocyte remains in diplotene throughout follicular growth (Fig. 45). The
nucleus contains a full (diploid) complement of chromosomes; the oocyte at this stage
is comparable to a somatic cell in G_2. The period of meiotic arrest of the oocyte persists
(which is constant for each species) up to ovulation, when the oocyte (primary oocyte)
resumes meiosis [or germinal vesicle breakdown (GVB) occurs in response to surge of
gonadotrophic hormones] (Fig. 45). Since GVB is the first prominent change seen, it is
widely used as a criterion for the "resumption of meiosis" or "resumption of
maturation." Meiotic maturation occurs only in a small fraction of the oocytes in the
ovary, as the majority of them degenerate at various stages of their growth and
differentiation, as will be discussed in Chap. VII. Several steps in the process of meiotic
maturation can be recognized, which include condensation of chromosomes into
distinct bivalents, disappearance of nucleolus, dissolution of the nuclear (germinal
vesicle) membrane, alignment of chromosomes on the metaphase I spindle, separation
of homologous chromosomes and emission of the first polar body, and arrest of
nuclear progression at metaphase II (Fig. 45) (Tsafriri 1978a, Wassarman et al. 1976,
1979b). For the mouse, these changes occur about 13 h before ovulation, whereas in
man and pig the corresponding times are 36 and 42 h, respectively (see Tsafriri 1978,
Baker 1979, Wassarman et al. 1979b). The other species variations in this regard are
described by Tsafriri (1983).

   A few hours before ovulation the nucleus leaves its central position, moving to the
periphery of the cell. With the initiation of meiosis both in vivo and in vitro, the
nucleolus is transformed into a a highly compact and electron-dense body, losing its
reticular structure. It is progressively reduced in size and ultimately disappears from
view before the separation of the first polar body (Fig. 45). Meanwhile, convolution of
the nuclear envelope occurs, followed by its disruption and, finally, fragmentation into
cisternae which are indistinguishable from the elements of endoplasmic reticulum
(Zamboni and Mastroianni 1966a, Baca and Zamboni 1967, Zamboni 1970, 1972,
Hertig and Barton 1973, Tsafriri 1978a, Baker 1979, Dvořák and Tesařík 1980). These
alterations of nucleolus and nuclear envelope are also closely accompanied by the

maximum condensation of chromatin into chromosomes (diakinesis), which are released in the ooplasm after the nuclear envelope is fragmented (Fig. 45) (see also Zybina et al. 1980). The chromosomes are composed of thin filaments in close apposition to one another but not uniformly oriented. All chromosomes are surrounded by filamentous material, which is intimately associated or continuous with some of the chromosomes (Zamboni and Mastroianni 1966 a, Vasquez-Nin and Echeverria 1976, Tsafriri 1978 a, Baker 1979).

Motlik and Fulka (1976) have provided the details of morphological characteristics during GVB in pig oocytes in vivo and in vitro. For their better characterization, the whole process has been divided into four well-defined stages which are based on the chromatin changes, and on nucleolus and nuclear membrane disappearance. In the intact germinal vesicle (GVI) the nuclear membrane and nucleolus can be clearly distinguished and chromatin forms a ring or horseshoe around the nucleolus. In the second stage (GV II) a few orceine-positive structures (chromocentres) develop on the nuclear membrane. During the GV III, slightly stained chromatin clumps, placed especially around the nucleolus, and the beginning of strand formation, are typical. In the GV IV, the nuclear membrane becomes less distinct and the nucleolus disappears completely. Chromatin appears in the form of an irregular network or as individual bivalents. According to these criteria germinal vesicle breakdown in vivo is completed in most pig oocytes between 20–24 h after hCG injection. In culture, a comparable stage of development is reached between 16–20 h. McGaughey and Van Blerkom (1977) also observed that in in vitro experiments GVB occurs much earlier. Daguet (1980) observed that the oocyte GV of pre-ovulatory follicles in sow, having a diameter of 2 mm (during the first 2–3 days of the follicular phase), shows uniformly dispersed chromatin with no condensation in vivo. The GV of oocytes from pre-ovulatory follicles of 2 mm diameter (from days 2 and 3 of the follicular phase to day 5) presents condensed chromatin in the shape of a crown or horseshoe surrounding the nucleous and irregular chromatin clusters in the nucleoplasm; these clusters are generally seen lying against the nuclear membrane. These observations are, more or less, in agreement with those of Motlik and Fulka (1976). Sato et al. (1979) studied the formation of nucleus and cleavage of pig follicular oocytes cultured in vitro. Formation of nucleus occurs at 57–60 h in oocytes with corona cells and at 27 h in oocytes without corona cells. One or two nuclei are predominantly seen in the oocytes. Extrusion of the second polar body is not seen, but oocytes having two polar bodies are observed. Cleavage occurs at 72 h of culture in modified Krebs Ringer bicarbonate solution. Motlik et al. (1978 a) studied GVB in bovine follicular oocytes cultivated in vitro. The intact GV of control oocytes does not correspond to the findings in porcine oocytes (GVI). Instead, it is characterized by a nuclear membrane and a few chromocentres in finely granular nucleoplasm (GV II). After 2 and 3 h of cultivation GV III predominates (50% and 64% respectively). Faintly staining chromatin, after a longer interval in the form of filaments, also appears around the chromocentres. After 4 h 35.8% of the oocytes are still at stage GV III, but the GV of 46.6% of the oocytes already contains filamentous bivalents (GV IV). After 5 h GV IV (59.7%) is typical, with a less distinct nuclear membrane and condensing bivalents. GVB is completed in 30.4% of the oocytes after 5 h of cultivation and in 92.50% after 6 h. The result of all these studies demonstrates that the nuclear components undergo conspicuous changes with the initiation of meiosis in oocytes cultured in vitro.

Since the nucleus occupies an eccentric position in the oocyte as meiosis begins (Fig. 45), when the nuclear envelope breaks up before the alignment of chromosomes in metaphase, these chromosomes are released into an area devoid of ooplasmic organelles. The linear arrangement of chromosomes at the equatorial plate of the meiotic spindle leads to the formation of metaphase I (Fig. 45) (Hertig and Barton 1973, Tsafriri 1978 a). In metaphase I the chromosomes lie parallel and the spindle perpendicular to the cell surface, centrioles are lacking and the microtubules of the spindle apparatus converge at foci at the periphery of the spindle poles (Fig. 45) (Szöllösi 1972, Tsafriri 1978 a). The spindle is composed of microtubules which always show kinetochore at their point of attachment to the chromosomes. In the mammalian oocyte meiosis occurs in the absence of centrioles (Zamboni 1972, Hertig and Barton 1973, Tsafriri 1978 a, Baker 1979). Spindle microtubules at the poles may be associated either with clusters of compactly arranged vesicles (mouse: Zamboni 1972) or with electron-dense fibrillar structures (see Tsafriri 1978 a) or with tubular profiles resembling endoplasmic reticulum in their morphology (rabbit: Zamboni and Mastroianni 1966 b). However, in human oocytes the polar ends of the spindle do not appear to form association with any definite structure (Zamboni et al. 1972). In one exceptional case, multiple centrioles are seen in a rhesus monkey oocyte matured in vitro (Zamboni 1975).

The subsequent meiotic stages evolve in an essentially similar fashion in the various mammalian species investigated (Fig. 45) (see reviews by Zamboni 1972, Tsafriri 1978 a, Baker 1979). A separation of homologue sets of chromosomes without a centromeric division takes place (reduction division). One set of homologues remains in the cell which is now called the secondary oocyte, whereas the other set moves into a small bleb of cytoplasm which is extruded and forms the first polar body (Figs. 45 and 46 a, b) (Hertig and Barton 1973, Dvořák and Tesařík 1980). During telophase, the oolemma becomes indented around the spindle until the mid-body is formed. Vacuoles enclosed by membranes resembling the oolemma appear to lie in association with the mid-body. It has been suggested that these vacuoles coalesce to form the plasmalemma of the oocyte and first polar body, which is originally connected with the oocyte through the mid-body whose remnants can be seen in nearby regions of the perivitelline space (Zamboni 1972). The extruded first polar body is morphologically similar to the oocyte, as it contains scattered chromosomes, microtubules derived from the first metaphase spindle, cortical granules, a few mitochondria and smooth vesicles (Fig. 46 c). It can easily be distinguished by the presence of cortical granules. Occasionally, degenerative changes can be seen the polar body (Zamboni et al. 1972). The secondary oocyte and the first polar body each carry half (haploid) a complement of chromosomes. During reduction division, the amount of DNA is reduced to half, so that the secondary oocyte in this regard is comparable to a somatic cell in G_1. The chromosomes in the secondary oocyte are arranged in a metaphase-like configuration (Fig. 46 b) in which they remain until after fertilization. These electron microscope studies have further confirmed that pre-ovulatory resumption of meiosis by the dictyate oocyte includes germinal vesicle breakdown (GVB), diappearance of the nucleolus, and the progress to the metaphase of the second meiotic division (Fig. 45).

At ovulation a secondary oocyte arrested at the metaphase of the second meiotic division ("second meiotic arrest") is released from the follicle and enters the oviduct where fertilization takes place. Meiosis is completed after a spermatozoon has

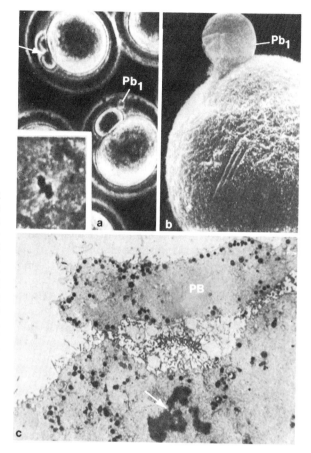

Fig. 46a and b. Phase contrast micrographs of mouse oocyte during the abstriction of the first polar body *(Pb₁)*. Occasionally, the first polar body divides *(arrow)*, but one of the polar bodies thus produced either degenerates rapidly or is resorbed by the oocyte. The appearance of the first meiotic metaphase spindle under the phase contrast microscope is shown in the *inset* (From van Blerkom 1980); **c** Electron micrograph of human oocyte with the second metaphase chromosome *(arrowed)*; PB the first polar body with numerous cortical granules (From Dvořák and Tesařík 1980)

penetrated the zona pellucida and entered the secondary oocyte: each chromosome splits into the sister chromatids by a division of the centromeres (Fig. 45). The division results in the mature ovum and a second polar body. The chromosome number is not further reduced by this division (Fig. 45); the nucleus of the mature ovum is haploid. The DNA content is, however, again reduced and now contains the same amount as found in mature spermatozoon (Guraya 1985).

The second polar body can be distinguished from the first one by the absence of microvilli and cortical granules. This is due to the fact that the second polar body is formed after penetration of the sperm into the ovum, i.e. after the liberation of the cortical granules and disapparance of microvilli (see Zamboni 1972, Hertig and Barton 1973). Under certain circumstances completion of the second meiotic division in the rat can be induced by application of a cold shock in vitro or in vivo and by some forms of anaesthesia (Thibault 1949, Austin and Braden 1954). Explantation of tubal oocytes, which are in the metaphase of the second meiotic division, leads to the formation of the second polar body within 1 h (Zeilmaker and Verhamme 1974). This process does not depend on the activity of the cytochrome oxidase system, as KCN and uncoupling agents do not inhibit it. Zeilmaker (1978) has also reported some data on the formation of a second polar body 24–32 h after incubation in vitro of ovarian oocytes (see also Zeilmaker and Verhamme 1976).

The behaviour of nuclear complement in the two polar bodies also differs greatly. In the first polar body it remains in the form of isolated chromosomes which are not surrounded by the nuclear envelope, whereas in the second a nucleus is constituted with a double nuclear envelope. The nuclear complement of the ovum also behaves exactly in a similar fashion at the corresponding stages of meiosis. The oocyte chromosomes remain close to the polar body for only a short time after the first meiotic division has been completed (Zamboni and Mastroianni 1966a, Baca and Zamboni 1967). After ovulation they occupy a position which is usually opposite to that of the first polar body (Stefanini et al. 1969). Zamboni (1972) believes that the location of the oocyte chromosomes with respect to the polar body could be an important criterion for evaluating whether an oocyte has attained maturation in vitro or is already mature before it has been placed in culture. Zamboni et al. (1972) investigated the fine morphology of human oocyte maturation in vitro. The sequence of meiotic changes is basically the same as described in vivo (Baca and Zamboni 1967). Komar and Roszkowski (1979) studied GVB and the resumption of the first meiotic division in 36 oocytes (57.0%) out of a total of 63 oocytes obtained from 11 patients in various phases of the menstrual cycle. Twenty-nine of these (46.0%) reached the metaphase stage of the second meiotic division, while the remaining seven (11.1%) were blocked at the metaphase stage of the first meiotic division. Suzuki et al. (1981) observed that of human follicular oocytes with germinal vesicles, 79% resume meiosis within 48 h in modified Ham's F-10 medium for oocyte and granulosa culture, and 59% of them reach the second metaphase stage of meiosis. A cytogenetic analysis of cultured follicular oocytes of cow has revealed a wide morphological heterogeneity in nuclear chromatin structure (Ernst et al. 1980). In 22–26 h 86% of oocytes undergo meiosis, while 36% and 38% reach metaphase I and II, respectively. In 38–44 h 76% of oocytes reach metaphase II. During maturation some abnormalities are seen in chromosomes, such as despiralization, polyploidy, strong condensation and sticking. Of oocytes that reach metaphase II, 60% show no obvious signs of chromosome degeneration.

B. Cytoplasmic Changes

Cytoplasmic changes also occur during the meiotic maturation of oocyte. It involves rapid alterations in the distribution of organelles, such as lysosomes in the rat (Ezzell and Szego 1979) and mitochondria in the mouse (van Blerkom and Runner 1983) and cow (Kruip et al. 1983). Zamboni and Thompson (1976) have described the alterations that occur in the mitochondria and endoplasmic reticulum of maturing oocytes in the rhesus *Macaque*. The mature oocytes show mostly granular endoplasmic reticulum, which is in the form of narrow cisternae closely associated with mitochondria. The latter show a pale matrix and only a few plate-like cristae arranged in shelf-like fashion. Their matrix becomes dense and the cristae are increased in number and meanwhile become tubular in oocytes which have resumed or completed meiotic maturation. The mitochondria usually form rosettes around elements of endoplasmic reticulum, which are now of the agranular variety and in the form of dilated saccules. These complexes frequently include also lipid droplets which consist mainly of phospholipids and some triglycerides (Guraya 1967d, 1970b). According to Zamboni and Thomson (1976), the presence of mitochondria, granular endoplasmic reticulum complexes in immature oocytes, suggests that, prior to resumption of meiosis, oocytes are involved in protein synthesis, and the presence of mitochondria-agranular endoplasmic reticulum-lipid complexes in maturing or mature oocytes may be indicative of synthesis and/or metabolism of steroid hormones during meiosis. These steroid hormones may be important for oocyte maturation and/or follicular luteinization. But the present author is of the view that the mitochondria-agranular endoplasmic reticulum-lipid complexes may be indicative of utilization of lipid droplets for some other metabolic activities either during meiosis or embryogenesis, rather than steroidogenesis.

In mouse and rat the period immediately preceding the GVB is accompanied by a marked translocation of organelles (van Blerkom 1980). In these two species the first grossly observable indication of resumed meiosis is the clustering of organelles around the oocyte nucleus. Cytochemical studies of the perinuclear area in rat oocytes have revealed the presence of a substantial amount of lysosomes, suggesting lysosomal involvement in the dissolution of the nuclear membrane (Ezzell and Szego 1979). But electron microscope examination of this area in the mouse and the rat has shown the predominance of mitochondria rather than of lysosomes (van Blerkom 1980). The actual dissolution of the nuclear membrane in the oocytes of both species appears to involve the formation of numerous vesicles that may be recycled by the cell during the establishment of the male and female pronuclei (van Blerkom and Motta 1979).

Other major cytoplasmic changes include the development of perinuclear micro-tubule organizing centres during GVB (Calarco et al. 1972) and the accelerated production of cortical granules during the later stages of resumed meiosis (Zamboni 1970, Zamboni et al. 1972, Santhananthan and Trounson 1982). In the mouse, cortical granules (Nicosia et al. 1977), microvilli (Eager et al. 1976, Nicosia et al. 1978) and surface glycoproteins (Johnson et al. 1975) vacate areas of cortical cytoplasm and overlying plasma membrane from which the first and second polar bodies will emerge. This change in cortical and plasma membrane properties may alter the characteristics of the plasma membrane (fluidity) in a highly regional fashion, thus insuring polar body abstriction at a particular location, or it may restrict the site(s) of fertilization (Nicosia et al. 1978), or do both.

Szöllösi et al. (1978) suggested that the control of cortical maturation of the oocyte depends on or is at times correlated with the interruption of the gap junctions. This suggestion is supported by the fact that, in closely timed series, it has been demonstrated in rabbit oocytes that cortical granules populate the egg cortex only .between 6–6½ h after mating. This process occurs timewise or coincides with the interruption of gap junctions between the corona cells and oocyte, as it is to be discussed later on. Under appropriate culture conditions, the same process is reproduced in bovine follicles (Ménézo 1976, Ménézo et al. 1976). If in cultured follicles the peri-oocyte reaction (elongation of corona cells, withdrawal of foot processes and elimination of gap junctions) does not occur, cortical granules are present in large clusters several microns from the cortex. Within these clusters, the diameter of the granules and their density may greatly vary. This holds true regularly for control follicles as well as for those maintained for 48 h in culture with PMSG alone. When, however, hCG is added to the medium during the final 24 h of culture the peri-oocyte cell reaction and the interruption of gap junctions take place. Under these conditions centrifugal displacement and alignment of the cortical granules are observed. The membrane of the cortical granules almost develops contact with the inner leaflet of the oocyte cell membrane fully parallel to their localization during in vivo maturation. The myosin-like protein demonstrated immunohistochemically in the mammalian oocyte (Amenta and Cavallotti 1980) may be involved in the displacement of cortical granules to the egg periphery, and in their subsequent eruption during the cortical reaction (see Guraya 1982a). The absence of cortical granules from the oocyte cortex recovered from oestrous rabbits, and the fact that they shift to the egg periphery during 48 h of culture in a medium containing FSH-LH, have been already reported (Nicosia and Mikhail 1975). The relationship with the simultaneously occurring reaction of the peri-oocyte cells was, however, not observed. Suzuki et al. (1981) observed an increase in cortical granules and tubular aggregates and a decrease in the Golgi apparatus during the maturation of human oocytes in vitro. Scanning electron microscopy of the oocyte surface has revealed a decrease in number and length of microvilli. Ultrastructurally, mitochondria with laminar cristae, enlarged nuclei, dispersed Golgi apparatus, smooth endoplasmic reticulum, and lipid droplets are seen in the ooplasm; granular cytoplasmic protrusions are observed on the oocyte surface. Suzuki and Yanagimachi (1983) have made freeze-fracture observations on the oolemma and cortical granules in the ovulated hamster oocytes with their cumulus cells so as to determine whether they possess special features for fusion with the sperm.

Recent electron microscope studies of mouse oocytes during resumed meiosis have also demonstrated polarity in the distribution of cortical granules in the ooplasm (van Blerkom 1980), and the establishment of such a polarity may be related to the breakdown of the germinal vesicle that accompanies resumed meiosis. All these observations suggest that cortical maturation is an important aspect of oocyte physiology, as the cortical granules and the cortical reaction at the time of sperm penetration play most probably a role in the defence of polyspermy (Guraya 1982a). In case they are not able to take up their peripheral position, polyspermy could result. The gap junctions between corona cells and oocyte appear to inhibit this aspect of cytoplasmic maturation of the mammalian oocyte. It has also been suggested that as the oocyte approaches ovulation, and possibly during the first meiotic metaphase, a factor, or set of factors, capable of decondensing spermatozoal chromatin, is either activated or synthesized de novo (see van Blerkom 1980), as will also be discussed later. This ability of the cytoplasm appears to develop progressively and is indicative of the maturational process.

C. Alterations of the Oocyte Follicle Cell Relationship (or Oocyte-Cumulus Communication)

The cumulus cell–oocyte complex develops in the immature state prior to the pre-ovulatory gonadotrophin surge, as already discussed in Chap. III. During this period the cumulus cells remain lightly packed around the oocyte, which is maintained at the germinal vesicle stage. The cumulus cells proliferate, presumably as a result of the action of FSH and oestrogens (Richards 1980), and they possibly also participate in the maintenance of oocyte meiotic arrest (Rice and McGaughey 1981, Tsafriri 1983,

Tsafriri et al. 1982 a, b), as will be discussed here. After the pre-ovulatory gonadotrophin surge, the cumulus cell–oocyte complex undergoes significant maturational changes. The oocyte undergoes germinal vesicle breakdown and the completion of the first meiotic division, as already discussed in Sect. A. Although oocyte maturation is triggered by the pre-ovulatory gonadotrophin surge, little is known about the details of the hormonal interaction with the oocyte which leads to meiosis (Tsafriri et al. 1982 a, b, Tsafriri 1983). However, several lines of evidence have suggested that the cumulus granulosa cells surrounding the oocyte may be the primary target cells for the meiosis-inducing hormone, as synchronously with oocyte maturation the cumulus oophorus undergoes morphological transformations. Some biochemical changes occurring in the cumulus cells after gonadotrophin stimulation have also been reported recently, but much more work is needed in this regard in order to define the role of cumulus oophorus in oocyte maturation.

Completion of the first meiotic division just precedes, or occurs simultaneously with modifications or alterations of the morphological and physiological relationships between the oocyte and the cells of the cumulus oophorus, and corona radiata (Norman and Greenwald 1972, Zamboni 1970, 1972, 1976, Hertig and Barton 1973, Hillensjö et al. 1975, 1976, Dekel et al. 1976, Magnusson et al. 1977, Ahrén et al. 1978, Tsafriri 1978 a, Szöllösi et al. 1978, Schuetz 1979, Dvořák and Tesařík 1980, Beers and Dekel 1981, Numazawa and Kawashima 1982, Leibfried and First 1982). LH or PMSG initiates changes in the cumulus oophorus cells (Dekel and Kraicer 1974, Hillensjö et al. 1976, Dekel et al. 1977. Ahrén et al. 1978) and meanwhile their oxygen consumption is decreased both in vivo and in vitro before any morphological changes occur. These changes have been studied both in vivo and in vitro. The possible role of the coupling between the cumulus cells and the oocyte through gap junctions in the regulation of the resumption of meiosis has also been discussed in some recent revievs (Moor and Cran 1980, Tsafriri 1983).

1. Changes In Vivo

In mature oocytes some very few follicle cell processes are seen in the zona pellucida; instead, an increased number of empty channels are seen. Actually, just before ovulation, microvilli begin to disappear until only a few blunt processes remain and the remainder of the oolemma assumes an irregular or sometimes undulating profile (Hertig and Barton 1973, Zamboni 1976, Tsafriri 1978 a, Schuetz 1979, Dvořák and Tesařík 1980). Meanwhile, the multitude of complex cytoplasmic processes of corona radiata cells, which have been described to traverse the zona pellucida to reach the oocyte surface or even to penetrate for various distances into the ooplasm of growing oocyte, as described in Chap. III, decrease in number and then disappear. As a result of these morphological changes, the cumulus cells form an investment of a more loose nature than in early stages of follicle growth. This change is believed to be brought about by retraction, as well as by degeneration of granulosa cell processes, as judged by the frequent presence of globular bodies of degenerated cytoplasm in the perivitelline space of mature oocytes (Zamboni 1970, 1972, 1976, Hertig and Barton 1973). Recent studies of Zamboni (1976, 1980) have further extended and confirmed these observations. The denudation of the zona pellucida occurs by retraction of the granulosa cell processes, as supported by the increased number of empty channels that can be seen in the zonae pellucidae of mature oocytes and the rounding up of the cells of the cumulus oophorus, and/or by their degeneration followed by phagocytosis, as demonstrated by the frequent occurrence of portions of granulosa cell cytoplasm within phagocytic vacuoles of maturing or mature oocytes in large antral and pre-ovulatory follicles. The loss or retraction of cellular processes traversing the zona pellucida, and of joining structures, also results in loosening the compact cumulus oophorus. Granulosa cell processes packed with microfilaments form the basis of contraction, as already discussed in Chap. III.

In rabbits, in which LH release follows copulatory stimulus for 1 h, it has been shown that approximately $5–5\frac{1}{2}$ h after copulation the gap junctions between corona cells and oocyte are interrupted (Szöllösi 1975a). The gap junctions are eliminated also in cultured bovine follicles when a preliminary culture of 24 h duration with PMSG is continued for an additional 24 h after the medium is supplemented with hCG. The exact time when the junctional complexes are interrupted has not yet been determined under culture conditions. In the case of follicles cultured for 48 h with PMSG alone, gap junctions between corona and oocyte are retained. Gilula et al. (1978) studied cell-to-cell communication in cumulus–oocyte complexes from rat ovarian follicles before and after ovulation. Numerous gap junctions are formed on the oocyte surface by cumulus cell processes that traverse the zona pellucida and contact the oolemma. The entire cumulus mass is also connected by gap junctions via cumulus-cumulus interactions; the specialized junctions may be important in maintaining meiotic arrest (Lindner et al. 1977). In the hours preceding ovulation, the frequency of gap junctional contacts between cumulus cells and the oocyte is decreased and the cumulus is disorganized. Electrophysiological measurements have indicated that ionic coupling between the cumulus and oocyte is progressively decreased as the time of ovulation approaches (Tsafriri et al. 1982a, b, Tsafriri 1983). These findings further support the suggestion that cell communication is a characteristic feature of the cumulus–oocyte complex, and this communication is apparently terminated near the time of ovulation (Moor and Cran 1980). It has been postulated that the interiorization of gap junctions (Motta and van Blerkom 1974) described between the oocyte and granulosa cells in mammalian follicles approaching ovulation is also a consequence of physiological uncoupling "in vitro" (Dvořák and Tesařík 1980). Scanning electron microscope studies by Phillips et al. (1978) have also shown that cumulus cells of the pre-ovulatory cumulus– oocyte complex are associated with one another and with the oocyte by long processes. At ovulation these cellular processes are withdrawn and space between cumulus cells is filled with hyaluronidase-sensitive material. Ahrén et al. (1978) observed that the structure of the cumulus oophorus isolated before the LH surge is compact, consisting of densely packed granulosa cells, and there is no lysing effect of hyaluronidase. Following the LH surge the cumulus is gradually altered into a dispersed structure of cells embedded in a hyaluronidase-sensitive matrix. The alterations in cumulus morphology and sensitivity to hyaluronidase, referred to by these workers as cumulus maturation, start somewhat later than oocyte maturation. These observations are also in agreement with a more detailed chronological study of the maturational changes in the cumulus of the adult pro-oestrous rat (Dekel and Kraicer 1974, Dekel et al. 1979).

Mumford et al. (1981) studied proteinase activities of the golden hamster eggs and cells of the cumulus oophorus. Eggs show aminopeptidase and elastase-like activities, but no detectable trypsin-like activity. Aminopeptidase, endopeptidase, trypsin-like and elastase-like activities are present in cumulus cells. These enzymes may be playing some role in oocyte–cumulus changes. The temporal pattern of the termination of communication between the cumulus and oocyte may indicate that communication provides a mechanism for regulating the maturation of the oocyte during follicular development before ovulation (Lindner et al. 1980, Tsafriri et al. 1982a, b, Tsafriri 1983). All these findings have suggested that the release of the oocyte from follicular suppression of meiosis is a result of the uncoupling of cumulus–oocyte communic- ation, possibly induced by the ovulatory hormone. The influence of LH on the resumption of meiosis may be secondary to its action on the sequestration of the oocyte. However, several observations are not in agreement with this suggestion. The disruption of oocyte–cumulus cell communication follows rather than precedes GVB (Tsafriri 1983). Tsafriri (1983) has stated that junctional association between the oocyte and the cumulus is essential for maintaining meiotic arrest; it is doubtful whether breakdown of junctional association between the oocyte and cumulus cells is the physiological trigger of the resumption of meiosis.

The exact timing of the ultrastructural changes in oocyte follicle cell association in relation to the meiotic changes in the oocyte in vivo still needs to be determined more precisely. However, study of the morphological alterations in the cumulus-granulosa cells corresponding to oocyte maturation has indicated that following criteria can be used to determine cumulus maturation: (1) degree of cellular dispersion, (2) presence of blebbing cells and (3) effect of hyaluronidase on cell detachment from the cumulus complex (Hillensjö et al. 1976, Dekel and Kraicer 1977, Ahrén et al. 1978).

The oocyte–follicle cell uncoupling is the morphological correlate of the LH- stimulated break of the inhibitory influence of granulosa cells on oocyte maturation (Lindner et al. 1980, Tsafriri et al. 1982a, b; Tsafriri 1983). Hillensjö et al. (1976) suggested that gonadotrophins trigger both oocyte and cumulus maturation, but the morphological response is expressed faster in the oocyte. The site of LH action triggering various alterations in vivo is yet to be established more precisely (Lindner et al. 1980, Tsafriri et al. 1982a, b, Tsafriri 1983, Schultz et al. 1983). In a series of experiments, Ahrén et al. (1978) observed that acute alterations in the respiratory activity occur in the cumulus cells and oocyte in relation to the pre-ovulatory morphological maturation in the rat. The pre-ovulatory cumulus decreases its respiration at about the same time as the oocyte increases its respiration. The metabolic shift in the cumulus cells produced by LH is seen before any alteration occurs in the morphology of the cumulus complex. An early step in the LH action on the cumulus appears to be the generation of cAMP. The morphological alterations caused by LH can be mimicked by db cAMP as well as by phoshodiesterase inhibitors (Dekel and Kraicer 1977, Hillensjö 1977, Tsafriri et al. 1982a, b, Tsafriri 1983). It is also interesting that alterations in oxygen consumption by the oocyte and the cumulus, respectively, develop much more rapidly following artificial release of the cells into culture – in about 1 h – as compared to the lag period seen when LH is given in vivo – about 4 h (Ahrén et al. 1978). This difference in time suggests that the action of LH in vivo has to counteract an inhibitory influence, which is quickly removed by the artificial isolation, but which needs a longer time for inactivation in vivo. The presence of a follicular oocyte maturation inhibitor (OMI) has been suggested for several species (Tsafriri 1978a, b, 1983, Lindner et al. 1980, Channing et al. 1982, Schwartz 1982,

Tsafriri et al. 1982a, b) (see also Chap. III). The timing of the early biochemical response to the gonadotrophic stimulus may differ from the sequential changes seen at the morphological level. Further studies are needed in order to determine more precisely the mechanisms involved in the gonadotrophic stimulation of both resumption of meiosis and alterations in cumulus respiration. It also needs to be revealed whether the decrease in cumulus cell respiration is necessary for oocyte maturation. The decreased respiration may reflect a Crabtree effect related to increased gycolysis, making more pyruvate available for the oocyte (Magnusson et al. 1977). It is also possible that the decreased respiration is due to a specific inhibition by LH of mitochondrial activity in the cumulus cells. The metabolic adjustment in the cumulus cells will then make available a higher amount of oxygen to the oocyte, which may be of importance for the resumption of oocyte maturation, as will be discussed later on.

Direct contact of the oocyte with the follicular fluid is established during the maturation stage as the cells of the corona radiata and cumulus oophorus tend to separate and create intercellular spaces into which the fluid can penetrate (Balboni 1976). Shortly before ovulation a loosely connected group of cells consisting of the oocyte and the accompanying cumulus cells floats in the follicular fluid. The oocytes recovered from the ovarian follicles at this stage, surrounded by corona radiata cells, are invested by a large number of cumulus cells embedded in the viscous matrix of the follicular fluid in the native state and are called pre-ovulatory cells (Edwards and Steptoe 1975). The fluid is markedly increased in viscosity in some species. The various alterations of the oocyte–follicle cell relationship previously described are a prelude to the process of egg denudation, which is completed in the lumen of the fallopian tube after fertilization (Zamboni 1970, 1972, Hertig and Barton 1973).

Recent immunofluorescent and transmission and scanning electron microscope studies have revealed the presence of a mosaicism in distribution of microvilli and concanavalin-A receptor sites in the plasma membrane of mouse oocytes arrested at metaphase II of meiosis (Eager et al. 1976). Van Blerkom (1980) has discussed the functional significance of this putative macromolecular polarity in the mouse oocytes. Coincident with nuclear progression through meiosis, a maturational process also appears to occur in the plasma membrane of oocyte, which, after its completion, gives to it the ability to bind and fuse with the plasma membrane of a spermatozoon. But the degree to which the plasma membrane of oocyte changes in preparation for fertilization probably varies significantly among species. The molecular-macromolecular aspects of such alterations in the plasma membrane of oocyte during ovum maturation need to be investigated in detail. However, von Weymarn et al. (1980a) have observed that after resumption of meiosis in vitro, a specific differentiation developes on the vitelline membrane (oolemma). The same differentiation in surface organization is also seen in the in vivo precociously matured oocytes. Phillips et al. (1978) have observed that after in vitro sperm–egg fusion in hamster the small region of the oolemma overlying the sperm nucleus is free of microvilli. This area increases in size and at 1 h after sperm–egg fusion bulges to form a large incorporation cone.

2. Changes In Vitro

Changes in oocyte–follicle cell relationship similar to those in vivo have also been observed to occur in oocytes maintained in culture (Zamboni et al. 1972). The role of the attached cumulus cells in the maturation of liberated oocytes cultured in vitro continued to be controversial (Lindner et al. 1980, Moor et al. 1981, Eppig 1982, Tsafriri et al. 1982a, b, Tsafriri 1983). Their beneficial effect on the maturation of oocytes in culture has been reported for several species of mammals (see also Tsafriri 1978a, Tsafriri 1983). This beneficial effect has been attributed to the peculiarly specific energy source requirement of denuded oocytes. When pyruvate is added in the medium, the denuded oocytes mature readily in vitro and are indistinguishable from cumulus-enclosed oocytes.

LH has been shown to acutely decrease oxygen uptake by pre-ovulatory cumulus cells (Dekel et al. 1976). In a further study of this effect, it has been observed that cumuli isolated after 1.P. injection of 10 µg LH show a significant decrease occurring 4 h after the injection (Ahrén et al. 1978). The inhibitory effect of LH on cumulus respiration shows a lag phase just as the morphological alterations do after the exposure to hormone. Also, when LH is added to cumuli in vitro for various periods of time, there occurs a decrease in respiration, showing that LH acts directly upon these cells (Ahrén et al. 1978). This decrease, however, develops more quickly than in vivo, and a new steady-state respiration, amounting to approximately 50% of the control respiration, is observed by 1 h of culture. By this time there are yet no morphological signs of LH action in the cumulus (Ahrén et al. 1978).

FSH but not prolactin also decreases cumulus cell respiration in vitro (see Ahrén et al. 1978). This role of FSH is further supported by its binding to the cumulus of hamster follicle (Oxberry and Greenwald 1982). FSH specifically stimulates synthesis of hyaluronic acid by mouse cumulus in vitro (Eppig 1979). It is believed that the cumulus expands as the result of the gonadotrophin-induced deposition of glycosaminoglycan, possibly hyaluronic acid, between the cumulus cells. This is supported by the fact that the expansion of mouse cumuli oophori in vitro can be stiumulated by highly purified FSH, but not by highly purified LH or hCG (Eppig 1979). The action of FSH is essential for follicular growth prior to the pre-ovulatory gonadotrophin surge (see Chap. III). Yet FSH, which promotes follicular growth, does not induce cumulus expansion in vivo, even though the cumulus cells of early antral follicles are capable of this response in vitro (Eppig 1980). This suggests that some follicular component inhibits the precocious expansion of the cumuli oophori until after the pre-ovulatory gonadotrophin surge (Eppig 1980). Sulphated glycosaminoglycans present in high concentrations in the follicular fluid as the carbohydrate moiety of proteoglycans (see Chap. III) suppress FSH-induced cumulus expansion in vitro (Eppig 1981a). The most active glycosaminoglycans are heparin, heparin sulphates and chondroitin suphate B (dermatan sulphate) (Eppig 1981a). Sulphated glycosaminoglycans are believed to inhibit precocious mucification of the cumulus oophorus that could be potentially elicited by FSH present in antral follicles. Their action is believed to be distal to FSH-stimulated elevation of cAMP (Eppig 1981b). Eppig et al. (1982) have, however, shown that the sulphated glycosaminoglycans (heparin, and chondroitin sulphate B) used at concentrations that completely inhibit cumulus expansion do not prevent FSH-stimulated elevation of intracellular cAMP. Highly purified FSH stimulates the incorporation of [³H]-choline into isolated cumulus cell–oocyte complexes, and this incorporation is the result of cumulus cells (see also Eppig and Ward-Bailey 1982). FSH-stimulated [³H]-choline incorporation appears to be mediated by cAMP as it is also stimulated by db cAMP, cholera toxin, and 3-isobutyl-1-methyl-xanthine.

Unlike the inhibitory effect of sulphated glycosaminoglycans on FSH-stimulated cumulus expansion, neither heparin nor chondroitin sulphate B inhibits [³H]-choline incorporation into cumulus cell–oocyte complexes. In addition, heparin does not inhibit FSH-stimulated anabolism of the [³H]-choline into phosphatidyl choline. In various aspects, FSH-stimulated choline incorporation resembles FSH-stimulated cumulus expansion (Eppig and Ward-Baily 1982). These results have suggested that the action of sulphated glycosaminoglycans on FSH-stimulated functions of cumulus cells is differential, and that its inhibitory action on cumulus expansion is exerted distally to FSH-stimulated elevation of intracellular cAMP. The physiological and molecular aspects for the differential action of sulphated glycosaminoglycans on cumulus expansion and choline incorporation need to be determined more precisely. The exact role of these molecules in oocyte–cumulus changes and ovulation remains to be elucidated, but they do appear to be secretion products of granulosa cells per se. It appears that after the LH surge concentrations of sulphated glycosaminoglycans decline to levels insufficient to block cumulus expansion. The basis for their differential action on cumulus cells may be related to the sequence of follicular maturational process. It needs to be determined whether these ovarian proteoglycans affect cumulus cell function in the same way as commercially prepared sulphated glycosaminoglycans from other sources. Future studies must be directed to reveal the developmental programme for glycosaminoglycan and proteoglycan synthesis by granulosa cells in situ, as well as their concentrations and properties in follicular fluid during various phases of follicular growth. Then an attempt must be made to correlate these observations with the hormonal milieu and the sequence of maturational events in the oocyte–cumulus cell complex.

Where it is not possible to show the presence of specific receptors of LH on the oocyte, specific LH/hCG receptors are reported on cumulus cells. This finding is further supported by the fact that gonadotrophin stimulation of cumulus progesterone secretion has been reported for the rabbit (Nicosia and Mikhail 1975), pig (Hillensjö and Channing 1980) and rat (Hillensjö et al. 1980, 1981). Hillensjö et al. (1981) have shown that 20α-dihydroxy-progesterone secretion is stimulated by FSH and LH, but that de novo synthesis of oestradiol is low and nonstimulable by gonadotrophins. But cumulus cells can aromatize exogenous testosterone. Qualitatively, the pattern of steroid synthesis and its control by LH and FSH appear to be similar in the cumulus cells and membrana granulosa cells (Ahrén et al. 1979). Racowsky and McGaughey (1983) have observed that supplementation of all media with testosterone and FSH significantly stimulates the synthesis of oestradiol by the porcine cumulus cells, compared with that of control groups. The synthesis of progesterone, however, is significantly stimulated by testosterone and FSH only in the BSA and serum-supplemented media. It needs to be determined whether steroidogenesis in the cumulus cells has any specific role in the development of oocyte. Oocyte nuclear maturation (meiosis) appears to be independent of steroidogenesis (Lieberman et al. 1976), as will be discussed in Sect. D. However, oocyte cytoplasmic maturation may be regulated at least in part by steroid hormone (Moor and Trounson 1977, Tsafriri et al. 1982a, b,

Tsafriri 1983). Tsafriri (1983) has suggested that the meiosis-inducing action of LH is somehow mediated by cumulus cells.

Very divergent views have been expressed about the function of cumulus cells during in vitro maturation of mammalian oocytes. Meinecke and Meinecke-Tillmann (1978) have made experimental studies of the function of cumulus cells during in vitro maturation of porcine oocytes. About two-thirds of the oocytes from tertiary or Graafian follicles show a granulosa cell reaction in growth media containing follicular fluid taken from Graafian follicles. Denudation of the eggs decreases the maturation rate drastically. Maturation of denuded eggs takes place only in media containing follicular fluid obtained from Graafian follicles. About one-third of the in vitro matured oocytes develop to morula or the blastocyst after fertilization in vitro. Binov and Wolf (1979) studied the spontaneous maturation of intact, cumulus-free and zona-free mouse oocytes in complex and simple media. The rates and frequencies of maturation from the germinal vesicle to the metaphase II stage are similar for all oocytes, suggesting that these investments are probably not critical for maturation. Results from recent experiments have shown that the oocyte and cumulus cells remain highly coupled to one another even though the oocyte has resumed meiosis (Moor et al. 1981, Eppig 1982).

Fukui and Sakuma (1980) studied maturation of bovine oocytes cultured in vitro in relation to ovarian activity, follicular size and the presence or absence of cumulus cells. They have observed a highly significant difference only for oocyte maturation rates between oocytes with cumulus cells and those without cumulus cells (P < 0.001). The results of their experimental studies have indicated that the presence of cumulus cells is apparently the most important factor studied for in vitro maturation of bovine follicular oocytes. Minato and Toyoda (1982), using various media, have studied the induction of cumulus expansion and maturation division of porcine oocytes in vitro.

Foote and Thibault (1969), using in vitro techniques, showed that the nuclear maturation of oocyte depends upon the physiological or mechanical isolation of oocyte from granulosa cells, suggesting the inhibitory effect of granulosa cells on meiotic maturation. However, spontaneous maturation occurs in co-cultures of mouse granulosa cells and oocytes (Tsafriri 1978a); nevertheless, inhibitory action of granulosa cells on oocyte maturation has been shown in co-cultures of porcine granulosa cells with oocytes. This effect is dose-dependent and cells from small, medium and large follicles are active in this respect. The inhibitory effect of porcine granulosa cells is not reversible by LH, FSH, PGE_2, or dibutyryl cAMP (db cAMP) (Tsafriri 1978a). It has been suggested that there is an intrafollicular inhibitor whose effects can be mediated by LH, PGE_2, and cAMP within the follicle (Tsafriri 1978a b, 1983, Lindner et al. 1980, Schultz et al. 1983). Since the oocyte and surrounding cumulus cells are coupled via junctions, as already discussed in Chap. III, which may permit "biochemical communication" between the two cell types (Wassarman et al. 1979a, Salustri and Martinozzi 1980), the granulosa cells may maintain intracellular cAMP levels in the oocyte which prevent the resumption of meiosis (Lindner et al. 1980, Beers and Dekel 1981, Tsafriri et al. 1982a, b, Tsafriri 1983). Disruption or change of these channels of communication at the time of ovulation (or at the time of isolation of oocytes) may lead to a lowering of cAMP levels in the oocyte and the onset of meiotic maturation, as will be discussed in detail in Sect. D. LH appears to be essential for oocyte maturation in rodents; although FSH is also effective. Tsafriri and co-workers (Lindner et al. 1980, Tsafriri et al. 1982a, b, Tsafriri 1983) have suggested that the effect of LH on the resumption of meiosis is mediated by cAMP, the meiotic process is inhibited by continuous exposure of the oocyte to high concentration of cAMP, and it is to be determined more precisely whether LH induces maturation by a surge of cAMP in the germ cell itself, or indirectly through a rise in cAMP in some other follicular compartment. Beers and Dekel (1981) have recently postulated that, in the follicle, the oocyte may be maintained in meiotic arrest by a steady supply of cAMP transferred to the oocyte from the cumulus by the process of intercellular communication. Then under the influence of gonadotrophins, perhaps by a cAMP-dependent mechanism which needs very high concentrations of the nucleotide, communication is interrupted, terminating the supply of cAMP to the oocyte. Results obtained by Salustri and Siracusa (1983) suggest that metabolic cooperation with the cumulus oophorus and meiotic resumption are both regulated by FSH through variations of intracellular levels of cAMP. However, Schultz et al. (1983) have suggested that the putative oocyte meiosis inhibitor (OMI) is activated in response to elevated levels of cumulus cell cAMP. The inhibitor (OMI) may then enter the oocyte via gap junctions to maintain meiotic arrest. The putative OMI factor may be involved in regulating cAMP levels in the oocyte.

Further investigations are required to determine more precisely the role of granulosa (or cumulus) cells in the regulation of maturation. The nature of products of granulosa cells, which act as inductors or inhibitors of oocyte differentiation during the course of induced or spontaneous meiotic maturation, needs to be determined more precisely (Lindner et al. 1980, Tsafriri et al. 1982a, b, Tsafriri 1983).

D. Effects of Hormones, Drugs, and Various Other Factors on Oocyte Maturation

In recent years, the effects of hormones and various other chemical substances on ovum maturation in vivo and in vitro have been extensively studied to determine the regulation of oocyte maturation (or meiosis). Divergent views have been expressed not only for the oocytes of different species but also for the oocytes of the same species.

1. Gonadotrophins

The development of follicular antrum is regulated by gonadotrophins, as already discussed in Chap. III (see also Richards 1980). The correlation between antrum formation and acquisition of meiotic competence in rodents may suggest that the latter is also dependent upon gonadotrophins. Hypophysectomy of immature rodents, prior to the acquisition of meiotic competence supports this hypothesis (see Tsafriri et al. 1982a, b, Tsafriri 1983); FSH (or PMSG) increases the number and the percent of competent oocytes, suggesting that hypophyseal gonadotrophins are involved in the development of the capacity to complete the first meiotic division.

The resumption of meiosis in vivo occurs only if the oocyte resides in a mature Graafian follicle, as already described, and it is dependent on the levels of gonadotrophic hormones in the serum and follicular fluid during pro-oestrus (so-called LH-surge) (Tsafriri 1978a, b, Peters and McNatty 1980). Administration of sodium pentobarbitone (Freeman et al. 1970, Ayalon et al. 1972) or antiserum to the β-subsunit of LH (Tsafriri et al. 1976d), or hypophysectomy (Vermeiden and Zeilmaker

1974) of rats prior to the pre-ovulatory gonadotrophin surge prevents not only ovulation but also oocyte maturation (see also Tsafriri et al. 1982 a, b, Tsafriri 1983). If an entire follicle is removed from the ovary before the LH surge has taken place, the oocyte remains arrested in the dictyate stage throughout the culture period and also degenerates unless the LH is placed within the explant medium. By contrast, when the follicles are explanted after the endogenous surge of gonadotrophins, the oocytes mature in hormone-free medium (Tsafriri 1983). If an oocyte (with cumulus) is removed from a follicle and explanted, it undergoes meiosis without addition of hormones to the culture medium. This further suggests that within the protected environment of the follicle, completion of meiosis is prevented until the LH pre-ovulatory surge removes an inhibitor or renders it inactive.

Meiosis is triggered in rat pre-ovulatory follicles by LH, hCG, FSH (Tsafriri et al. 1982 a, b, Tsafriri 1983). All these hormones induce a rise in the follicular cAMP. Foote et al. (1978) have observed that treatment of 40 pre-pubertal heifers with PMSG, FSH, as well as with each of these gonadotrophins plus LH, results in follicular development and oocyte maturation. Resumption of oocyte meiosis is greater in the PMSG-treated heifers than those given FSH. Treatment with LH increases the number of oocytes undergoing meiosis in both PMSG- and FSH-treated heifers. The percentage resuming meiosis increases with increased follicle size in the PMSG groups, but not in the FSH groups.

There can be little doubt that the onset of pre-ovulatory maturation of the oocyte and its follicle in vivo is related to the "LH surge" although the precise sequence of events that follows this stimulus remains to be established. It should be mentioned that gonadotrophins are only required when the oocyte resides within its follicle: when the oocytes are removed into a chemically defined medium they resume meiosis spontaneously to metaphase II. The incidence of this change is dependent on the status of the follicle from which the egg is recovered (see Tsafriri 1978 a, b, Lindner et al. 1980, Tsafriri et al. 1982 a, b). This suggests that the effect of the LH surge on the maturation is not a direct one, but an effect on the environment of the oocyte. Since the meiosis-inducing action of LH appears to be exerted through the mediation of granulosa cells, it appears that meiotic competence of the oocyte is acquired somewhat earlier than full responsiveness to LH is reached by the surrounding granulosa cells. The relationship between follicular and oocyte maturation is reviewed by Thibault (1977). It will be interesting to mention that the oocytes of hamster Graafian follicle show FSH receptor (Oxberry and Greenwald 1982), whereas those of rat follicles contain prolactin receptor (Dunaif et al. 1982). Both the receptors are believed to be involved in ovum maturation. But the mechanism of their involvement needs to be determined.

Experiments on cultured follicles of pro-oestrus rats have revealed that the follicle-enclosed oocyte remains in the dictyate stage throughout a 12–24 h culture period in the absence of hormone, whereas addition of gonadotrophin to the culture medium induces meiosis (Tsafriri et al. 1972, Lindner et al. 1974). Also, when pre-ovulatory follicles of the PMSG model are incubated in a chemically defined medium (modified bicarbonate buffer with glucose and albumin) for 2–10 h the follicle-enclosed oocytes remain immature and the cumulus preserves its compact nature, while, upon addition of gonadotrophins to the medium, oocyte and cumulus maturation occur similarly as seen in vivo (Hillensjö 1976, Hillensjö et al. 1976). Under the same conditions LH markedly stimulates glucose uptake and lactate formation by the follicles. db cAMP mimicks the effect of LH on cumulus maturation (Hillensjö 1977) and follicular lactate formation (Ahrén et al. 1978), but as long as the nucleotide is present it prevents oocyte maturation. Pre-incubation of the follicles in db cAMP followed by transfer to plain control medium, however, enables oocyte maturation (Hillensjö et al. 1978).

Erickson and Sorenson (1974) observed that mouse oocytes removed at the diplotene (germinal vesicle) stage from large antral follicles reach metaphase II, while those from small antral follicles remain at the diplotene stage, and are apparently "blocked" at metaphase I, indicating some influence of previous exposure to endogenous gonadotrophins. Similar results are also obtained when whole mouse ovaries, suitably pre-treated with FSH or PMSG in vivo, are maintained in organ culture in medium containing LH or hCG (Baker and Neal 1972). In the absence of added gonadotrophins very few oocytes, irrespective of the size of the follicles in which they reside, undergo meiotic maturation unless they have been pre-treated in vivo with superovulating doses of PMSG (Baker and Neal 1972, Lindner et al. 1974). Recent studies have also indicated that the induction of meiosis is mediated by LH (Tsafriri 1978a, 1983, Baker 1979, Biggers and Powers 1979, Lindner et al. 1980, Tsafriri et al. 1982a, b). It is now well established that oocyte maturation beyond the diplotene stage can be induced by treating whole ovaries, or isolated follicles in organ culture, with LH or hCG (see Tsafriri 1978a, 1983, Baker 1979, Lindner et al. 1980, Tsafriri et al. 1982a, b). These studies have also demonstrated a relationship between follicle development and the capability of the oocyte to complete the first meiotic division following a hormonal stimulus (see also Sorensen and Wassarman 1976). Whereas oocytes in large follicles reached the second metaphase stage on resumption of meiosis, oocytes within small or medium-sized follicles were arrested in culture at the first metaphase stage (see Baker 1979, Lindner et al. 1980, Tsafriri 1983, Tsafriri et al. 1982a, b). All these studies have revealed the role of the stage of follicular development in both the acquisition of maturation competence by the oocyte and the sensitivity of the oocyte to the maturation-inhibiting action of the follicle (see also Baker and Hunter 1978).

Further evidence for the maturation of somatic components of the follicle affecting the oocyte has been obtained from an experiment in which mature pigs were treated with hCG either on the day preceding the LH surge (day 20 of the cycle), or when the follicle is much smaller and immature (day 17) (Hunter et al. 1976). Administration of hCG on day 20 caused ovulation of normal oocytes at metaphase II which, when submitted to spermatozoa, developed the normal block to polyspermy and showed full development of a single male and female pronucleus. In contrast to this situation, injection of pigs on day 17 of the 21-day oestrous cycle resulted in ovulation of immature eggs at the germinal vesicle stage. These oocytes did not develop block to polyspermy, sperm heads in the ooplasm did not swell to form male pronuclei, and the cumulus cells remained immature and spherical in shape (Hunter et al. 1976, Baker and Hunter 1978). The immaturity of the follicles on day 17 of the cycle was not only supported by the size of the follicle and shape of cumulus cells but also by the concentrations of oestrogens in the follicular fluid, which were only one-third of those in control animals or those treated with hCG on day 20. These results have clearly indicated that there are several cellular and endocrine events that occur simultaneously in the follicle during pre-ovulatory maturation, only one of which is the resumption of meiosis in the oocyte.

FSH is equally effective in inducing meiotic maturation. But the levels of circulating FSH would normally be insufficient in the normal oestrous cycle of rodents without the synergistic action of LH. Studies of isolated mouse ovaries of rat Graafian follicles treated in vitro, have clearly revealed that both FSH and LH can cause pre-ovulatory maturation of the follicle, although in the absence of these supplements to the culture medium resumption of meiosis in oocytes is a rare event (Baker and Neal 1972, Lindner et al. 1974). Indeed, FSH is more effective than LH (Neal and Baker 1975). FSH and LH can also induce resumption of meiosis in rabbit follicular oocytes in organ culture (Tsafriri 1978a, Baker 1979). In contrast to this situation neither FSH nor LH appears to induce meiotic maturation in isolated follicles derived from pig, ox, sheep, human and rhesus, suggesting that the control of oocyte maturation in them may differ slightly from that in rodents and lagomorphs (Tsafriri 1978a, 1983, Baker and Hunter 1978, Tsafriri et al. 1982a, b). It appears likely, therefore, that the physiological requirements for oocyte maturation in these species may be somewhat different from those in rodents and lagomorphs (see also Thibault et al. 1975). In porcine and ovine Graafian follicles maintained in organ culture meiotic maturation of oocytes can be induced if FSH, LH and oestradiol-17β are injected simultaneously (Moor and Trounson 1977, Baker and Hunter 1978). Whether it is the FSH or LH which triggers the resumption of meiosis within the oocyte remains uncertain, although it is tempting to draw parallels with the results for mouse follicles discussed above. Further studies are recommended to determine the apparent contrasting behaviour of follicle-enclosed oocytes of different mammalian species.

The timing of maturational changes in liberated rat oocytes and follicle enclosed oocytes differs (Tsafriri 1978a, 1983). Puromycin, at dose levels which inhibit protein synthesis, does not stop spontaneous GVB in denuded oocytes, but does inhibit LH-induced GVB in rat follicle-enclosed oocytes, suggesting that the liberated oocyte escapes meiotic arrest and skips some of the regulatory steps involved in the normal hormone-induced maturation (see Lindner et al. 1980, Tsafriri et al. 1982a, b, Tsafriri 1983). For the rat, it is well documented that LH and not FSH is the oocyte maturation and ovulation-inducing hormone. This is also supported by the fact that treatment of LH with antiserum to the β-subunit of LH abolishes its effect on oocyte maturation and ovulation, whereas the influence of FSH remains unaffected by this antiserum (Yang and Papkoff 1973, Lindner et al. 1974).

It is interesting to mention here that the most significant alteration in the composition of human follicular fluid at the time of the LH surge is the significant and sharp decrease in the concentration of prolactin (McNatty et al. 1975). In this regard, Baker and Hunter (1978) have observed that an antiserum to prolactin when added to porcine Graafian follicles in organ culture apparently induces meiotic maturation in the oocyte. Baker and Hunter (1978) believe that it is doubtful whether prolactin itself plays the role of an inhibitor of oocyte maturation, or of the action of gonadotrophic hormones (see also Baker et al. 1977). But it may prevent the availability of LH-binding sites or induce the binding to protein of intrafollicular steroids. On the other hand, it is possible that prolactin plays a purely passive role, its elimination from the follicular compartment normally reflecting an alteration in the permeability of the basement lamina of the follicle, apparently induced by LH. Baker and Hunter (1978) have further stated that even if prolactin is not in itself an inhibitor of oocyte maturation, it may be involved in the regulation of the inhibitor substance believed to be the product of granulosa cells (Tsafriri and Channing 1975a, see also Chap. III). This inhibitor is formed in the immature follicle, whereas another protein (antagonist) which neutralizes the effect of the inhibitor is apparently formed in response to gonadotrophins at their surge levels. This suggestion and the possible involvement of prolactin and/or LH in the control of the process needs to be confirmed further by means of carefully controlled experiments in vivo and in vitro.

Mandelbaum et al. (1977) have demonstrated a new aspect of oocyte maturation for the hamster. When nuclear maturation of the oocyte occurs in vitro the zona pellucida is not penetrable by spermatozoa. In vivo this property is acquired 6 h or more after hCG injection. When pre-ovulatory follicles are cultured in vitro in presence of gonadotrophins, complete oocyte maturation is obtained and fertilization occurs normally. These observations suggest that gonadotrophins (LH/hCG) may also cause some molecular changes in the zona pellucida during ovum maturation, which facilitate the penetration of spermatozoa through it. The precise mechanism of these changes must be determined in future studies.

Erickson and Ryan (1976) observed that oocytes from pre-antral follicles obtained from hypophysectomized rats are capable of resuming meiosis, but maturation was blocked at metaphase I in the oocytes derived from hypophysectomized rats (see also Lindner et al. 1980, Tsafriri et al. 1982a, b, Tsafriri 1983). In contrast to this, oocytes from mature, intact rats completed the first meiotic division. These results have further confirmed that hypophyseal hormones play an important role in the acquisition of capability to complete the first meiotic division (see also Tsafriri 1983). But the attainment of maturation of competence by the oocyte has not been correlated with the physiological differentiation of granulosa cells. An attempt to define the hypophyseal hormone required for completion of the first meiotic division was not made. Smith and Tenney (1979) observed no difference in frequency of maturation of oocytes obtained from mice hypophysectomized for 2 weeks compared to those from sham-operated or untreated (control) animals of the same age. By 7 weeks the incidence of polar body formation in vitro is significantly reduced. The number of oocytes, which remain meiotically inactive in culture, is increased at 7 and 12 weeks after hypophysectomy. This decrease in spontaneous oocyte maturation in vitro can be partly overcome by administering exogenous PMSG, oestradiol-17β or PMSG oestradiol-17β but not progesterone or hCG to hypophysectomized mice, suggesting the influence of PMSG and oestradiol-17β on oocyte maturation.

Smith et al. (1978b) studied the influence of season and age on maturation in vitro of rhesus monkey oocytes. The oocytes obtained from antral follicles of adult and adolescent monkeys during the annual breeding season extrude polar bodies in vitro at significantly higher rates (50%–60%) than oocytes from animals of similar age during the nonbreeding season (20%–30%). The proportion of oocytes degenerating in culture is greatest in groups where maturation is highest. Wassarman et al. (1979a) also observed that the ability to resume meiosis "meiotic competence" is acquired at a specific stage of oocyte growth in the juvenile mouse and that the ability to complete meiotic maturation is acquired subsequently (see also Sorensen and Wassarman 1976, Wassarman et al. 1976). These studies also appear to provide indirect evidence for the influence of gonadotrophins on oocyte maturations, as the levels of their secretion are influenced by season as well as by age of the animal (see Lindner et al. 1980, Tsafriri et al. 1982a, b, Tsafriri 1983). The isolated oocytes resume meiosis in culture, i.e. acquire meiotic competence only after reaching certain age, which is variable in different species of mammals (Tsafriri 1983). The capability of oocytes to mature is acquired gradually. Tsafriri (1983) has suggested that the ability to resume meiosis and to undergo GVB is attained earlier than the ability to reach metaphase II. Therefore, GVB is usually taken as the criterion for the acquisition of meiotic competence.

2. Cyclic Nucleotides, Prostaglandins and Hypothalamic Gonadotrophin – Releasing Hormone (GnRH)

The results of several studies have indicated that cAMP is involved in the regulation of meiotic maturation in amphibian oocytes (Masui and Clarke 1979, Schorderet-Slatkine et al. 1982). The involvement of cAMP in the regulation of meiosis in mammalian oocytes is less clear (Schultz et al. 1983). Resumption of meiosis can be triggered in rat follicles explanted on the day of pro-oestrus by LH, hCG and immunochemically pure FSH, as already discussed. A number of substances can also mimic the action of gonadotrophins (PGE_2, $PGF_{2\alpha}$, cAMP. etc. but not prolactin) (Lindner et al. 1974, Neal et al. 1975, 1976, Peters and McNatty 1980). The lack of an in vitro action of gonadotrophins in ungulates and primates may be due to an inadequate number of binding sites, since PGE_2 [which like LH acts on adenylate cyclase (AC) system] can induce the resumption of meiosis in follicle-enclosed porcine oocytes (Baker and Hunter 1978). However, cAMP itself is without effect and it is more difficult to explain the action of $PGF_{2\alpha}$ (Baker and Hunter 1978). All of these hormones or substances used induce an immediate rise in cAMP accumulation in vitro (see Moor and Heslop 1981, Tsafriri 1983), some of these also induce a somewhat later increase in progesterone synthesis (see Tsafriri 1978 a). When dibutyryl cAMP (db cAMP) is injected directly into the follicle antrum, meiosis is also induced. A short-term exposure of follicle to 8-bromo-cAMP (Hillensjö et al. 1978), db cAMP or isobutyl methyl xanthine (IBMX) (Dekel and Beers 1980) also triggers the resumption of meiosis but the mode of their action needs to be determined. By contrast, the continuous presence of cAMP derivatives or several inhibitors of phosphodiesterase prevent the induction of meiosis by LH (Lindner et al. 1974, Hillensjö et al. 1978, Dekel et al. 1981, Tsafriri 1983). The spontaneous maturation of isolated oocytes from mice and rats is inhibited in the presence of cAMP derivatives or phosphodiesterase inhibitors (Cho et al. 1974, Nekola and Smith 1975, Hillensjö 1977, Hillensjö et al. 1978, Dekel and Beers 1978).

As follicular cAMP activity is increased by LH, the action of LH on the resumption of meiosis is believed to be mediated by cAMP (Tsafriri et al. 1972, Lindner et al. 1974, Amsterdam et al. 1975, Tsafriri 1978 a, 1983, Lindner et al. 1980, Moor and Heslop 1981, Tsafriri et al. 1982 a, b). cAMP does not act directly on the oocyte but by apparently overcoming the inhibitory influence that granulosa and cumulus oophorus cells have on the oocyte (Fig. 47) (Chang 1955, Tsafriri and Channing 1975 a, Hillensjö 1978, Hillensjö et al. 1979, Tsafriri 1978 a, 1983). Tsafriri (1978 a, 1979, 1983) has discussed the details of effects of cAMP on oocyte maturation (see also Lindner et al. 1980, Tsafriri et al. 1982 a, b, Schultz et al. 1983). The results of all these studies suggest that enhanced production of cAMP is involved in the mediation of the meiosis-inducing action of LH. But this response is limited to the somatic cell compartment of

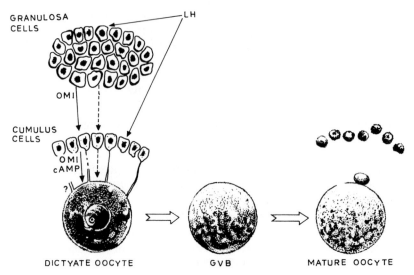

Fig. 47. Schematic representation of the control of meiotic maturation in mammalian oocytes. See text for discussion of views expressed in this regard (Redrawn from Tsafriri et al. 1982a)

the follicle, as the elevated levels of cAMP in the oocyte prevent the resumption of meiosis. It is suggested, therefore, that cAMP also serves as a physiological regulator of meiosis in mammalian oocytes (Lindner et al. 1974, Dekel and Beers 1978, Moor et al. 1980, Dekel et al. 1981, Schultz et al. 1983). The observation that addition of the adenylate cyclase activator, cholera enterotoxin (Cuatrecasas et al. 1975), prevents spontaneous maturation of rat oocytes cultured within their cumulus but not in denuded oocytes (Dekel and Beers 1980) supports the suggestion that cAMP produced by cumulus cells and transported into the oocytes blocks the resumption of meiosis. According to this hypothesis, LH and short exposure to db cAMP or phosphodiesterase inhibitors result in resumption of meiosis due to breakdown of cumulus–cell communication and termination of transfer of cAMP to the oocyte. However, Moor and Heslop (1981) have observed that there is no fall in oocyte cAMP concentration as an immediate response to the gonadotrophins but at 12–18 h after stimulation, the concentration in both the oocytes and follicle cells is considerably increased. No comparable increase is observed in intracellular cAMP in oocytes denuded of follicle cells before culture, even when both gonadotrophins and phosphodiesterase inhibitors are included in the medium. These results suggest that cellular interactions within the mammalian follicle are essential for the characteristic periods of increased cAMP in oocytes during maturation. The oocyte and cumulus cells also remain highly coupled to one another even though the oocyte has already resumed meiosis (Moor et al. 1981, Eppig 1982). Hubbard and Terranova (1982) have suggested that 8-Br-cyclic guanosine 5' phosphoric acid (GMP) may exert its inhibitory effect on oocyte maturation through the cumulus cells, whereas cAMP appears to inhibit oocyte maturation directly. The observed block of meiosis by inhibitors of phosphodiesterase may be related to their other effects on adenosine receptors, protein synthesis and calcium metabolism (Tsafriri 1983). Further investigations are needed to determine more precisely whether cAMP is involved in the effect of LH on resumption of meiosis and also to define the precise site of cAMP action. However, Schultz and Wassarman (1977a) proposed that a drop in mouse oocyte cAMP initiates resumption of meiosis by allowing the dephosphorylation of an inactive phosphoprotein to the active form, an activation that in some manner leads to GVB. The equilibrium between the inactive phosphorylated form (possibly the 28,000 MW phosphoprotein) and the active dephosphorylated protein was suggested to be regulated by levels of protein kinase and protein phosphatase. Recent studies by Schultz et al. (1983), which support this hypothesis, describe changes in oocyte phosphoprotein metabolism that precede GVB. The results show that a drop in oocyte cAMP and dephosphorylation and phosphorylation of specific phosphoprotein precede GVB and occur during the period of oogenesis in which oocytes become irreversibly committed to resuming meiosis. But further experimentation is necessary to clarify the roles of cAMP, cAMP metabolizing enzymes, cAMP-dependent and -independent protein kinases, protein phosphatases and changes in phosphoprotein metabolism in regulation of mammalian oocyte maturation. These problems may be approached by micro-injection of oocytes with compounds that affect cAMP levels and state of protein phosphorylation, and observing their effects on oocyte maturation.

Meiosis can also be induced in follicle-enclosed rat oocytes by factors which do not increase the production of cAMP in the follicle. GnRH or its agonistic analogue induces resumption of meiosis both in vivo (Ekholm et al. 1981a) and in vitro (Hillensjö and Le Maire 1980). Whereas a GnRH antagonist abolishes the meiosis-inducing action of GnRH or of its analogue, it does not block this effect of LH (Ekholm et al. 1982, Tsafriri 1983). These observations suggest that GnRH, or a similar ovarian factor, apparently does not mediate the effect of LH on meiosis. GnRH induces follicular prostaglandin synthesis (Clark et al. 1980), and PGE_2 is shown to induce resumption of meiosis (Tsafriri et al. 1972). However, indomethacin, a potent prostaglandin synthetase inhibitor, blocks the effect of GnRH analogue on ovulation but not on oocyte maturation (Ekholm et al. 1981b), showing that induction of meiosis by GnRH is not mediated by follicular prostaglandin synthesis.

3. Follicular Fluid Inhibitors

The results of various studies as discussed above have suggested that the follicle, or some compartment thereof, exerts an inhibitory action on the resumption of meiosis. Pincus and Enzmann (1935) suggested that follicle (or granulosa) cells in mammals "supply to the ovum a substance or substances which directly inhibit nuclear maturation". This suggestion is supported by the in vitro studies of mammalian oocytes, which have indicated that a follicular factor derived from granulosa cells or cumulus cells and accumulated in follicular fluid (FFI) of rabbit, sheep, ox and human inhibits the resumption of meiosis within the large Graafian follicle(s) (Tsafriri 1983). The inhibitory effects of granulosa cells on oocyte maturation depends upon their concentration in culture. Granulosa cells collected from small follicles inhibit meiosis more effectively than cells from medium or large follicles (Tsafriri et al. 1982a, b, Tsafriri 1983). Extracts of granulosa cells also inhibit the resumption of meiosis. The factor produced by the granulosa cells and stored in the follicular fluid is called oocyte maturation inhibitor (OMI) (see Sect. F of Chap. III) which keeps the follicle meiosis in abeyance within the large Graafian follicle(s) (Fig. 47).

Both FFI and porcine FFI (PFFI) prevent the resumption of meiosis (GVB) of rat oocytes cultured with adherent cumulus cells (Tsafriri 1978 b, 1983, Tsafriri et al. 1982a, b). Co-culture of oocytes with freshly isolated rat granulosa cells do not affect the spontaneous resumption of maturation, but granulosa cells cultured for 24 h before adding the oocytes prevent maturation. Culture of oocytes in a medium in which granulosa cells have been cultured for 48 h leads to a similar inhibition of meiosis. Addition of LH (5 μg/ml) overcomes the inhibitory action of co-culture with granulosa cells conditioned medium and PFFI. These studies have revealed that (1) the inhibitory effect of follicular fluid (FFI) is on the maturation of oocytes, which does not appear to be species-specific (Tsafriri 1978a, b, 1979, 1983, Lindner et al. 1980, Tsafriri et al. 1982a, b), (2) the rat granulosa cells in culture produce a similar inhibitor of maturation, and (3) LH at least partially counteracts these inhibitory activities, lending support to the physiological role of OMI in the regulation of meiosis (Fig. 47). Channing et al. (1982) have observed that addition of FSH (1 μg/ml) or prolactin (1–10 μg/ml) to suspension cultures of porcine granulosa cells facilitates the accumulation of OMI in the medium, whereas testosterone or dihydro-testosterone reduces it.

Gwatkin and Andersen (1976) also observed that when pre-ovulatory hamster follicles are cultured in vitro, the oocytes within them remain at the germinal vesicle stage. This maturation arrest can be overcome by washing the follicles before cultivation or by the addition of LH to the medium. The ability of LH to partly overcome this inhibition further supports the view that OMI may be involved in the control of meiosis. The results of various studies as discussed by Tsafriri and co-workers (Lindner et al. 1980, Tsafriri et al. 1982a, b, Tsafriri 1983) suggest that OMI exerts its inhibitory action upon meiosis not directly on the oocyte, but through the mediation of cumulus cells (Fig. 47). The oocyte is connected by processes to granulosa cells in the cumulus layer (see Chap. III). But unequivocal evidence that OMI has a physiological role in vivo has not been obtained. However, Cameron et al. (1983) have recently observed that human follicular fluid from healthy mature Graafian follicles and from pathological ovarian cyst fluid is inhibitory to progesterone-induced meiotic maturation of oocytes from the South African clawed toad, *Xenopus laevis,* indicating the presence of OMI, which is inactivated when subjected to a boiling water-bath for 2 min. The OMI action is shown to be reversible in its inhibitory action. The fact that OMI can act directly on the oocyte is demonstrated by its inhibitory action on maturation in defolliculated oocytes. These findings also indicate that inhibitory action of human OMI is not species-specific.

In a series of studies, Channing and co-workers (see Channing 1979a, b for references) have also suggested that there are some inhibitors in the follicular fluid, which are responsible for keeping the oocyte from maturing and extruding the first polar body until time of ovulation (see also Tsafriri 1978a, b, 1979, 1983, Channing et al. 1982, Schwartz 1982, Tsafriri et al. 1982a, b). Tsafriri et al. (1976a) suggested that this inhibitor is a polypeptide with a molecular weight of 2000. Actually, very divergent views exist about its molecular weight, as already discussed in Chap. III. Stone et al. (1978) studied purification and reversible action of inhibitor of oocyte maturation from porcine follicular fluid, which was generally called the PFFI. The major inhibitory fraction had a MW of 2000. The various data suggest that OMI is probably a peptide (Channing et al. 1982, Tsafriri 1983), which would have a MW of 100, not 2 K or more (personal discussion with van Blerkom). Nevertheless, the purification of OMI is far from being completed. Even its existence is not proved definitely, as it may be an artifact of purification. This suggestion is supported by the observations of some investigators who have not been able to show OMI activity of follicular preparations. Granulosa cells do not inhibit maturation of cow, sow or ewe oocytes, whereas FFI contains such activity (Jagiello et al. 1977). Porcine follicular fluid is inactive, whereas co-culture with granulosa cells inhibits resumption of meiosis (Sato and Ishibashi 1977, Sato et al. 1982). Leibfried and First (1980a) observed that neither bovine follicular fluid nor granulosa cells affect the completion of the first meiotic division in vitro of bovine oocytes. Porcine follicular fluid has no effect on maturation of bovine oocytes in vitro. However, porcine follicular fluid and granulosa do not inhibit maturation of porcine oocytes. Co-culture of oocytes with follicle hemisection prevents meiosis and this is overcome by addition of LH (Leibfried and First 1980b). The recent results obtained by Fleming et al. (1983) indicate that native pig follicular fluid is unable to inhibit the initiation of maturation of rat oocytes in vitro. Zeilmaker (1978) observed that the presence of rat granulosa cells in oocyte cultures has no appreciable effect on GVB. He, therefore, concluded that there is no indication for the presence of a maturation-inhibiting factor produced by granulosa cells. Nekola and Smith (1975) also could not demonstrate the influence of follicle cells in vitro on maturation of mouse oocytes (polar body formation). The investigations using co-culture of oocytes with granulosa cells and with follicular fluid fall just short of providing unequivocal evidence for the physiological role of FFI inhibitor in the control of resumption of meiosis (Tsafriri 1978a, b, 1979). The varying findings are suggestive of some alternative factors as inhibitors of meiosis. Therefore, cAMP, steroids and oocyte–cumulus cell communication are implicated as regulators of meiosis in mammals in some studies. But the precise molecular mechanisms of their regulations are also not understood.

Hillensjö et al. (1979) demonstrated that cumulus cells appear to be required for the inhibitory action of OMI, since it only functions in cumulus-enclosed, not in denuded oocytes. In parallel to inhibiting meiosis, PFFI inhibits cumulus progesterone synthesis. It has been suggested that oocyte maturation inhibitor, inhibition of cumulus cell progesterone secretion, cumulus cell differentiation (partial luteinization) and oocyte maturation all reflect actions directed at keeping the oocyte–cumulus cell mass in an immature, "arrested" state.

It is still not known how granulosa cells prevent the resumption of maturation and whether this action is exerted solely through the inhibitor in follicular fluid (Tsafriri 1978a, b, 1983, Lindner et al. 1980, Tsafriri et al. 1982a, b). The site of action of LH in the ovum maturation is also not clear. LH effect may be exerted by a direct action on the oocyte, indirectly through the mediation of granulosa cells, or by a direct interaction with follicular fluid inhibitor (Tsafriri 1978b). However, there is no conclusive evidence for a specific interaction of LH by granulosa cells with the oocyte or with FFI. Further study is required to define the mechanism by which the follicular inhibitor exerts its action on meiosis, and the mode of action of LH in overcoming this effect. Even in the presence of LH, PFFI causes a delay in GVB, which is directly related to the dose of inhibitor used. It appears that the effect of LH on resumption of meiosis is mediated by granulosa cells. This may explain the limited effect of LH in overcoming the inhibitory action of granulosa cells conditioned medium and the delay in GVB in the presence of PFFI and of LH. Tsafriri (1979) suggested that the dependence of the meiosis-inducing action of LH upon the presence of granulosa cells is compatible with the notion that this effect of LH on maturation is exerted through the mediation of granulosa cells. In vivo and in vitro studies may throw some further light on these issues. Leibfried and First (1980 b) observed that follicle wall segments maintain dictyate arrest in oocytes, an action which is specific to the follicle hemisections and not shown by control sections. LH induces meiotic resumption in this in vitro system. The cumulus oophorus is necessary to prevent oocyte maturation in vitro when follicle wall is present. Tsafriri (1978 b) suggested that it is possible that the cumulus cells included in the culture are unable to act as substitutes for the granulosa cells of the whole follicle. This inability may be either due to the low number of cells included in the culture or due to the lesser ability of cumulus cells, compared to mural granulosa cells to bind LH (Amsterdam et al. 1975). Schultz et al. (1983) have suggested that LH elevates follicle cAMP, which appears to be necessary for granulosa cell luteinization and cumulus cell mucification and expansion. LH may also result in inactivation of the OMI, causing a decrease in oocyte cAMP and resumption of meiosis.

All these findings, as discussed here, do not provide conclusive evidence for the physiological role of the follicular factor derived from granulosa cells in the control of maturation, and do not define whether this factor is acting directly on the meiotic process or through the mediation of some other physiologically coupled process(es) involved in oocyte maturation. The complete removal of the inhibitory effect of granulosa cells co-cultured with oocytes by the addition of LH resembles closely the behaviour of rat follicle-enclosed oocytes (Tsafriri 1978 b). Further investigations with co-cultures of rat granulosa cells and oocytes will help us in determining the definition of the primary site(s) and mechanism(s) involved in both, the inhibition of maturation by granulosa cells and the induction of meiosis by LH.

The results of various studies have, however, conclusively shown that the capability of oocytes to mature is acquired at a certain stage of follicular maturation (Tsafriri 1978 a, b, 1983, Tsafriri et al. 1982 a, b). The various in vitro experiments have suggested that, in addition to the well-established changes in follicular sensitivity and responsiveness to gonadotrophins during follicular development, there is a concomitant reduction in the sensitivity of oocytes to the inhibitor of maturation. Tsafriri (1978 b) has demonstrated that (1) inhibitor of maturation present in porcine follicular fluid is not species-specific and that rat oocytes can also be employed as an assay system for further purification and characterization of this factor; (2) the sensitivity of rat oocytes towards the inhibitor changes during follicular development; (3) rat granulosa cell cultures accumulate a similar inhibitor of oocyte maturation; (4) LH completely removes the inhibitory action of rat granulosa cells when added to co-cultures of oocytes and granulosa cells, and partially when added to a medium containing PFFI or to granulosa cells-conditioned medium. Purification of OMI to homogeneity in future may allow the assessment of its exact role in the regulation of meiosis in mammals. However, the synchrony of morphological and physiological maturation of the follicle differs among species of mammals. The biochemical maturation of the granulosa cells and capability of the oocyte to mature appear to be independent (Sorensen and Wassarman 1976).

4. Steroid Hormones

The involvement of steroids in the maturation of mammalian oocytes is the subject of much controversy (Lindner et al. 1980, Tsafriri et al. 1982a, b, Fukui et al. 1982, Racowsky and McGaughey 1983, Tsafriri 1983). It has been postulated that steroids are not involved in meiotic maturation as LH initiates meiosis in follicles even after the total inhibition of follicular steroidogenesis (Lindner et al. 1974; Lieberman et al. 1976, Tsafriri et al. 1982a, b, Tsafriri 1983). In contrast to this suggestion, other workers have reported that steroid hormones stimulate the rate of meiosis (Robertson and Baker 1969b, Bae and Foote 1975, Lindner et al. 1980, Tsafriri et al. 1982a, b, Fukni et al. 1982, Tsafriri 1983) and decrease the incidence of chromosomal abnormality at telophase I and metaphase II (McGaughey 1977b). In addition, steroids have been involved, together with gonadotrophins, in the induction of cytoplasmic maturation in human, rodent and sheep oocytes (see Thibault 1977, Moor and Warnes 1977, Lindner et al. 1980, Tsafriri et al. 1982a, b, Tsafriri 1983). Tsafriri (1983) has suggested that action of FSH (or PMSG) to increase the number and the percentage of competent oocytes is, at least partly, mediated by ovarian oestrogen production in vivo.

The results of some studies have suggested that the addition of steroid hormones to the culture medium may cause physiological maturation of liberated oocytes (Lindner et al. 1980, Tsafriri et al. 1982a, b, Tsafriri 1983). It is often suggested that steroid hormones are the intermediates affecting the oocyte or its cumulus oophorus (Schuetz 1977), although there is evidence that steroids do not have to be produced in response to the LH surge since inhibitors of steroidogenesis, such as aminoglutethimide and cyanoketone, do not block LH-induced oocyte maturation (Lindner et al. 1974, 1980). Therefore, the possibility remains that the high levels of steroids in the follicle prior to the LH surge are sufficient to act as intermediates without the necessity for further steroid synthesis (Baker et al. 1977). The precise role of follicular oestrogen is also obscure. It is possible that it is an intermediate and augments the action of the FSH or LH in the culture system (Baker and Hunter 1978). There is also the possibility that it may compensate for an inadequate synthesis of steroid by the follicle in vitro. Exogenous oestradiol has the capacity to arrest meiosis of either intact or denuded oocytes in vitro, but this capacity can only be expressed if no exogenous protein(s) are present (Racowsky and McGaughey 1983).

According to Moor and Warnes (1977), two distinct phases occur during the maturation of sheep oocytes: (1) an inductive phase of about 6–8 h duration which is follicle-dependent, and (2) a synthetic phase which is not dependent upon the follicle. Moor (1978) suggested that an inductive period of 6–8 h within the follicle after gonadotrophin treatment is a prerequisite for complete maturation of oocytes in sheep. Steroid function during the inductive phase is characterized by (1) elevated levels of oestrogen (in vitro: 21.0 n mol/ml follicular fluid; in vivo: 3.1 n mol/ml), (2) intermediate levels of androgen (7.0 and 3.5 n mol/ml in vitro and in vivo respectively) and (3) relatively low levels of progesterone (0.4 and 0.8 n mol/ml in vitro and in vivo respectively). Steroid support is not needed for the breakdown of the germinal vesicle and the formation of the first metaphase plate (Lindner et al. 1980, Tsafriri et al. 1982a, b, Tsafriri 1983). However, substantial levels of steroid, especially oestrogen, are essential for cytoplasmic maturation. Inadequate steroid support during maturation results in anomalies in fertilization, delayed cleavage and an almost total failure to undergo differentiation and blastocyst formation in rodents (Tsafriri 1983).

The role of androgens and progestins in the maturation of mammalian oocytes is not clear. Steroids of both types have been involved in the synthesis of the putative male pronucleus factor (MPFG) in the oocytes from rabbits and humans (Soupart 1973, Thibault et al. 1975). This factor has never been isolated, and thus only assumed to exist. MPFG is believed to be present in the cytoplasm as the oocyte approaches ovulation. The ability of the cytoplasm to decondense spermatozoal chromatin totally, and to support formation of the male pronuclear membrane is acquired progressively between GVB and metaphase II (van Blerkom 1984). Van Blerkom (1984) has suggested that post-translationally modified proteins released from the oocyte nucleus at GVB, or newly synthesized or activated proteins (or both) involved in post GVB protein modification may be related to the acquisition by the cytoplasm of the ability to decondense sperm chromatin and support pronuclear formation. Soupart (1975) has reported that male pronucleus formation occurs in human oocytes matured in vitro after sequential treatment with oestradiol and 17β-hydroxy progesterone. Some degree of pronucleus formation has also been achieved when a mixture of gonadotrophins, prolactin, oestradiol, and testosterone is added to rabbit oocytes in culture (Thibault et al. 1976). The spontaneous maturation of liberated oocytes occurs, in most species of mammals, in morphologically normal secondary oocytes, but a very low fertilization rate has been reported for such mature oocytes (Tsafriri 1978a, 1983, Lindner et al. 1980, Tsafriri et al. 1982a, b). The absence of MPGF is believed to be related to the failure of development of the male pronucleus, which is mostly responsible for low fertilization rate. But the results of studies on guinea-pig (Yanagimachi 1974) and rat (Niwa and Chang 1975) have suggested that in some species the failure of in-vitro-matured oocytes to undergo normal fertilization is not necessarily expressed only by the lack of putative MPGF, but also by some other factors.

In some studies steroids, especially progesterone, are considered to be a necessary intermediate in LH-induced oocyte maturation in rodents (Tsafriri 1978a, 1983, Lindner et al. 1980, Tsafriri et al. 1982a, b). Readhead et al. (1979) presented data showing that oocyte maturation, whether "spontaneous" or LH-induced, is always associated with incrased levels of steroids in the medium, which are believed to play a role in initiating the resumption of meiosis in rat oocytes. Schuetz (1979) reported that large quantities of progesterone, smaller increases of prostaglandin, and unchanged levels of oestradiol are present in media incubated with cumulus masses, indicating that cumulus cells secrete steroid hormones. But it needs to be determined whether these are the only important molecules or whether other products or activities of the cumulus cell–oocyte complex perform any significant role in the meiotic process.

Progesterone facilitates maturation of denuded rabbit and bovine oocytes (Tsafriri 1978a). But the facilitatory effect of progesterone has not been seen on maturation on human and porcine liberated oocytes. Addition of steroids in the culture medium do not induce the maturation of bovine, porcine and rat follicle-enclosed oocytes (Tsafriri 1978a, 1983). Experiments involving the addition of inhibitors of steroid synthesis (amino-glutethimide or cyanoketone) at the time of LH treatment do not prevent the maturation of the oocyte (Lindner et al. 1974, Tsafriri 1978a, 1983, Lindner et al. 1980, Tsafriri et al. 1982a, b). A dissociation of the steroidogenic and meiosis-inducing actions of LH is seen in vitro under two additional experimental conditions: (1) during selective inhibition of RNA synthesis by actinomycin D and (2) during inhibition of aerobic glycolysis by iodoacetate. In both the experiments, there occurs the inhibition of progesterone accumulation in the medium, but ovum maturation is not impaired (see Tsafriri 1978a, 1983, Lindner et al. 1980, Tsafriri et al. 1982a, b). The results of all these studies have suggested that the gonadotrophic stimulus for resumption of meiosis is not mediated by steroid synthesis. Since the mammalian oocyte is exposed within the follicle to high levels of steroids prior to the pre-ovulatory surge, this previous exposure to steroids is probably essential for its ability to respond to the meiosis-inducing action of gonadotrophins. But this suggestion needs to be confirmed by further studies (Tsafriri 1983).

Smith et al. (1978a) studied the effects of five chemically defined media and progesterone on maturation in vitro of rabbit oocytes from Graafian follicles of different sizes. In all media, oocytes from large antral follicles (1–1.5 mm diameter) show the highest incidence of meiotic activity, followed by those from follicles of medium size (0.5 mm). Most oocytes from small follicles (0.15–0.25 mm) do not resume meiosis in culture. The addition of glutamine to a standard medium for ovum culture significantly improves maturation of oocytes from medium-sized follicles but does not affect those from large or small follicles. When polyvinylpyrrolidene is substituted for bovine serum albumin, maturation of oocytes from late and medium-sized follicles is reduced. Progesterone at a concentration of 10 μM does not affect maturation, but 100 μM progesterone blocks germinal vesicle breakdown in oocytes from medium-sized follicles and reduces both GVB and polar body formation in oocytes from large follicles. This effect is reversible. A partial inhibition of GVB is observed when mouse oocytes are cultured with progesterone (6.4–12.8 μM), androstenedione (20–40 μM) and testosterone (40–80 μM), whereas oestrone, oestradiol 17β (10–30 μM) and dihydrotestosterone (1–8 μM) are without effect on GVB (Smith and Tenney 1979). Eppig and Koide (1978) also observed that only progesterone (36.7 μM), but not oestradiol, inhibits GVB in mouse oocytes. In porcine oocytes oestradiol 17β (3.67–36.7 μM) is the only steroid that significantly and reversibly inhibits the resumption of meiosis in culture (McGaughey 1977b, Richter and McGaughey 1979). Rice and McGaughey (1981) have shown a synergistic inhibition of the maturation of porcine oocytes by db cAMP and testosterone.

Gonadotrophins, which induce the resumption of meiosis, also induce a rise in follicular progesterone synthesis (Richards 1980, Tsafriri 1983). Therefore, it is difficult to assume that progesterone can serve as a physiological inhibitor of meiosis in vivo. On the other hand, the pre-ovulatory surge of gonadotrophins changes follicular steroidogenesis from production of predominantly oestrogen to progesterone as the main product, as will be discussed in Chap. VI. It appears that the pre-ovulatory surge of gonadotrophins induces not only the resumption of meiosis but also a rise in follicular steroidogenesis, including a transient rise in oestrogen production. Oestrogen remains in rat and hamster elevated at the time when GVB takes place, or in pig and sheep when the oocyte becomes committed to mature, even though GVB takes place later (see Tsafriri 1983). These observations argue against the suggested role of oestrogen as a follicular inhibitor of the resumption of meiosis. Although the available evidence for a regulatory role of steroids in oocyte maturation is unconvincing, their modulatory role in meiotic process cannot be overlooked. Bar-Ami et al. (1982) have demonstrated the role of FSH and oestradiol 17β in the development of meiotic competence in rat oocytes (see also Bar-Ami and Tsafriri 1981). The experiments of McGaughey (1977a) suggest that the incidence of chromosomal abnormality in pig oocytes during maturation is significantly reduced by the inclusion of progesterone and oestrogen in the culture medium. Steroids, especially oestradiol, are the most important factor for bovine oocyte maturation rather than gonadotrophins and culture media, and the effect of the hormones differs in the media tested (Fukni et al. 1982). However, Racowsky and McGaughey (1983) have observed no significant effect of oestradiol on the onset of maturation of either intact or denuded porcine oocytes that were cultured in medium containing either BSA or serum. The steroid significantly inhibited the release from meiotic arrrest of both types of oocytes cultured in medium supplemented with dextran. The various findings in regard to the effects of steroids on ovum maturation, as discussed here, are controversial. Therefore, further studies are needed to determine more precisely that steroids may facilitate changes not only in the cytoplasm but also in the nucleus during the later stages of oocyte maturation.

5. Ions and Electrolytes

The effects of ions and electrolytes on the maturation of oocytes form the subject of some studies. But their roles need to be determined more precisely at the molecular level. Leibfried and First (1979b) observed that lowering of external Ca^{2+} levels singly has no effect on either meiotic resumption or completion of the first meiotic division in bovine follicular oocytes. Lowering of external Mg^{2+} concentration alone, although having no effect on meiotic resumption in vitro when Ca^{2+} is present, does interfere with the completion of the first meiotic division. The result is arrest of oocyte maturation between germinal vesicle breakdown and formation of the first metaphase plate. Tsafriri and Bar-Ami (1978) studied the role of divalent cation in the resumption of meiosis of rat oocytes. Neither the ionophore nor Ca^{2+}-deficient medium interferes with the spontaneous maturation of isolated oocytes. Redistribution of divalent cations is suggested to participate in the physiological control of meiosis in mammalian oocytes. Tsafriri (1979) suggested that divalent cations, possibly Ca^{2+}, appear to be involved in regulating hormonal mechanisms of meiotic maturation of mammalian oocytes. Iwamatsu et al. (1979) studied the effects of Ca^{2+} and Mg^{2+} on follicle cell-free mouse oocytes during in vitro spontaneous maturation. The change in the resistance of oocytes to a brief exposure to a Ca^{2+}-Mg^{2+}-free medium suggested that the divalent cations change from a diffusible to a bound state and accumulate within the oocyte until the first meiotic metaphase. Further investigations are needed to assess the importance of alterations in intracellular distribution and transmembrane movement of various cations in the regulation of meiosis.

McGaughey (1977a) observed that denuded (granulosa cells removed) oocytes mature equally well in media containing either bovine serum albumin or dextran T70, but at a reduced incidence in media containing polyvinylpyrrolidone. They mature at an optimal incidence in medium with a total calculated osmolarity of 285 mosmol.

Bae and Foote (1980) have studied maturation of rabbit follicular oocytes in a defined medium of varied osmolarity. The proportions of oocytes that matured to meiosis II with polar body formation were 64%, 68% and 65% in media of 250, 290 and 310 mosmol respectively. Powers (1982) has studied changes in mouse oocyte membrane potential and permeability during meiotic maturation. The membrane potential depolarization appears to be caused largely by a decrease in K permeability and there appears to be a linkage between depolarization and GVB. But membrane depolarization by itself is not sufficient to GVBD. Biggers and Powers (1979) suggested that as an oocyte is freed from the Graafian follicle and is placed in foreign medium it experiences "stresses" than can modify the potential of the oolemma. Different responses to various media explain why some authors report follicular fluids or cumulus cells to be inhibitory (Chang 1955, Edwards 1962, Tsafriri and Channing 1975a, b), whereas others report them to have a stimulatory effect (Cho et al. 1971, · Hunter et al. 1972, Robertson and Baker 1969a). Further studies should be continued to determine more precisely the effects of follicular fluids and cumulus cells on the control of oocyte maturation.

6. Miscellaneous Factors

Besides hormones, nucleotides, follicular fluid inhibitors, electrolytes, ions, etc., the effects of other factors on oocyte maturation have also been studied. Shea et al. (1976a) observed that rabbit follicular oocytes mature in vitro more consistently in a medium containing 10% rabbit serum (82%) than when 0.3% bovine serum albumin is substituted for serum (54%). The first maturation division occurs within 10 h both in vivo after stimulation by hCG in vitro. Of 27 oocytes incubated in vitro, transferred to mated recipients and recovered 1 day later, 59% show evidence of fertilization. Development occurs up to the blastocyst stage, but no live young are obtained. Bachvrova et al. (1980) observed that naked growing mouse oocytes, free of attached follicular cells, when cultured in minimaly essential medium plus 10% foetal calf serum over the developing monolayer of ovarian cells, increase in diameter for a period of at least 1 week at 1.6 μm/day, or 70% of the in vivo growth rate, and show normal ultrastructure of ooplasmic organelles. During the second week, about 35% undergo spontaneous fragmentation, and at least 10%–12% resume meiotic maturation. Free oocytes with attached follicle cells grow to a larger average diameter than naked oocytes, and incorporate [^{3}H]-leucine by 75% more rapidly. Naked oocytes cultured in the absence of ovarian cells do not grow, and die when pyruvate is omitted from the medium. Naked oocytes do not grow when cultured over primary mouse fibroblasts, L-cells, hamster ovary cells or hepatoma cells. Naked oocytes separated from the monolayer of mouse ovarian cells by 0.7 mm layer of agar grow at 60% of the rate of control oocytes.

  Shea et al. (1976b) also studied maturation in vitro and subsequent penetrability of bovine follicular oocytes for determining optimum conditions for the resumption of meiosis. They incubated 450 follicular oocytes for 0 to 36 h in Ham's medium F-10 with 10% foetal calf serum. At recovery, one oocyte was binucleate, 4% were degenerative, and 8% had undergone germinal vesicle breakdown. After 28 h in culture, 60% of the oocytes reach metaphase II, but further incubation did not increase the proportion that mature. No significant difference is seen in the proportion of oocytes that mature in vitro from follicles recovered either 1 h (78%) and 2 h (72%) after the death of donors. Bovine oocytes mature in media with pH ranging from 6.70–7.59. The highest proportion (69%) reach metaphase II in media at pH 7.00 to

7.29. Motlik et al. (1978a) observed that in bovine oocytes the germinal vesicle breakdown is completed in 30.4% of the oocytes after 5 h of cultivation, and in 92.5% after 6 h. The time necessary to recover the rabbit follicular oocytes was found to be important (Shea et al. 1976a). When the oocytes were in culture within 30 min of death of the donor, 67% reached second metaphase, when the time from death to oocyte recovery from the follicle was extended to 2 h, none reached metaphase.

Meinecke (1977) made a comparative study of maturation of the nucleus and incidence of degeneration in the follicular ova from cattle, pig and sheep, which were cultured in TC 199 under an oil film, Porcine ova require about 36 h and ovine ova about 24 h to reach metaphase II. The number of degenerate oocytes is increased with the time for which they are cultured, 10%–30% of the ova show signs of degeneration, a further 20% become metabolically inactive after an incubation time of 48 h. Smith and Tenney (1978) observed that maintenance of mouse ovaries in ice-cold medium for 1, 2 and 5 h before removing the oocytes does not alter the incidence of spontaneous maturation or increase the number of oocytes degenerating, fragmenting or forming two equal blastomeres in culture.

Wassarman et al. (1976, 1979a) investigated the meiotic maturation of mouse oocytes in vitro with special reference to inhibition of maturation at specific stages of nuclear progression in the presence of several drugs. The germinal vesicle breakdown is inhibited by db cAMP and the nuclear envelope becomes greatly convoluted and condensation of chromatin is initiated but aborts at a stage short of compact bivalents. Similarly, chloroquine also prevents GVB. The inhibitory effects of db cAMP and chloroquine suggest that GVB may occur via a cAMP-controlled protease activation mechanism, analogous to that proposed for a variety of polypeptide hormone-mediated biological phenomena. GVB and chromatin condensation occur in an apparently normal way in the presence of puromycin, colcemid or cytochalasin B. Nuclear progression is blocked at the circular bivalent stage when oocytes are cultured continuously in the presence of puromycin or colcemid, but oocytes cultured in the presence of cytochalasin B proceed to the first metaphase, form an apparently normal spindle and arrest. Extrusion of a polar body is inhibited by these drugs. These observations suggest that dissolution of the mouse oocyte germinal vesicle and condensation of chromatin are probably not dependent either upon concomitant protein synthesis or upon microtubule assembly. The complete condensation of chromatin into compact bivalents appears to require breakdown of the germinal vesicle, and the failure of homologous chromosomes to separate after normal alignment on the meiotic spindle in the presence of cytochalasin B suggests that microfilaments may be involved in nuclear progression at this stage of maturation. Cytokinesis in the form of polar body formation, is apparently blocked when any one of the earlier events of maturation fails to take place. The inhibitory effects of these drugs on meiotic events are reversible to a varying degree depending upon the duration of exposure to the particular drug.

Other drugs, which also induce resumption of meiosis in follicular enclosed oocytes, include clomphene citrate (0.01–0.5 µM), tetrahydrocannabinol (100–200 µM), the major psychoactive ingradient of marihuana and cannabidiol (100–200 µM), and one of its nonpsychoactive constituents (see Tsafriri 1983). The mechanism by which these drugs act for inducing the resumption of meiosis remains to be determined. But they can be used as pharmacological tools for the study of oocyte maturation.

E. RNA and Protein Synthesis and Other Metabolic Changes

In recent years, many studies have been made of RNA and protein synthesis and other metabolic changes duing ovum maturation both in vivo and in vitro. The results of these studies will be discussed here.

1. RNA Synthesis

The study of RNA and protein synthesis during ovum maturation forms the subject of numerous recent studies (Tsafriri 1978a, van Blerkom 1980, 1981a, b, 1983, 1984); some of these studies have shown that RNA snythesis continues in the fully grown oocytes within antral and Graafian follicles, as already discussed in Chap. III. Rodman and Bachvarova (1976) observed that in the mouse synthesis of RNA by the oocyte occurs during the last 7 days of the dictyate stage that precedes ovulation (an in-vivo study). In the mouse, detectable RNA synthesis terminates at germinal vesicle breakdown (GVB) (Rodman and Bachvarova 1976, Wassarman and Letourneau, 1976a, Moore and Lintern-Moore 1978). Further, Wassarman and Letourneau (1976a) reported that RNA synthesis occurs during meiotic maturation of mouse oocytes within intact follicles and in an in vitro culture. Wolgemuth and Jagiello (1979) have confirmed that RNA synthesis occurs in in vitro throughout meiotic maturation of all species (mouse, cow and pig) investigated. They have further suggested a possible interaction between the granulosa cells and the oocyte in this process.

Recent ultrastructural studies of chromatin in pig oocytes during transition from germinal vesicle configuration through the first meiotic division by McGaughey (1983b) have shown some degree of transcriptional activity. But the nature of this RNA is not known. It is still to be determined more precisely whether the nascent fibrils on oocyte chromatin are newly synthesized or pre-existing. In any case, the level of transcriptional activity shown by electron microscopy is very low and probably does not contribute much to the extensive translational changes that occur during meiotic maturation in the pig. Failure to demonstrate transcriptional activity in mouse oocytes by light microscope autoradiography (Wassarman and Letourneau 1976a) does not exclude the possibility of extremely low-level activity detectable by electron micro-scopy. It is doubtful that transcription at such a level could support the magnitude of translational changes demonstrated in this species during the 36 h of development from resumed meiosis to the late one-cell early two-cell stage (van Blerkom 1984). It is still to be determined more precisely whether the demonstrated synthesis of RNA prior to GVB is necessary for the meiotic process or for later stages of embryonic development (van Blerkom 1981a, b, 1983, 1984). Therefore, the effects of actinomy-cin D demonstrated in several studies (Tsafriri 1983a) may be a result either of its specific inhibitory action on RNA synthesis or of some other unrelated effect. Crozet and Szöllösi (1980) have observed that actinomycin D at 4 μg/ml and α-amanitin at 50 μg/ml (inhibitors of RNA synthesis) by themselves do not inhibit nuclear progression: germinal vesicle breakdown and chromosome condensation, suggesting that RNA synthesis is apparently not necessary for the initiation of meiosis resumption. However, Osborn and Moor (1983) have suggested that concurrent transcription of mRNA may be necessary for normal ontogenetic changes in protein synthesis in sheep during pre-ovulatory resumption of meiosis.

Recent studies on mouse oocytes have suggested the presence of two populations of mRNA: (1) RNA templates which code for proteins synthesized in common by oocytes and newly fertilized eggs and (2) RNA templates that are differentially activated and translated during the period of resumed meiosis and early post-fertilization (see van Blerkom 1981 a, b, 1983, 1984). However, the proportion of previously untranslated mRNA that is contained in the pool of mRNA utilized during resumed meiosis and inherited and translated by the newly fertilized egg is unknown.

2. Protein Synthesis

Very divergent views have been expressed about the protein synthesis and its relationship to the maturation process of ova in different species of mammals (van Blerkom, 1981 a, b, 1983, 1984). Specific polypeptide syntheses are believed to accompany and define each stage of resumed meiosis. Stage-specific changes in the profile of polypeptide synthesis are also indicative of cytoplasmic maturation (Tsafriri 1978 a, van Blerkom 1980, 1981 a, b, 1983, 1984).

The uptake and incorporation of labelled valine into the trichloroacetic acid-insoluble fraction of maturing oocytes have been demonstrated (Stern et al. 1972). The presence of labelled precursor does not interfere with normal maturation of the oocytes. Cross and Brinster (1974) did not observe any difference in leucine uptake and incorporation at three stages of maturation of mouse oocytes, or between small and large dictyate oocytes. Moor and Smith (1979) made a comparative study of amino acid transport in mammalian oocytes. Amino acid fluxes in functional germinal vesicle oocytes are similar at all stages of development studied. An increase in Vmax but not Km during meiotic maturation causes a doubling of amino acid uptake in metaphase II oocytes. The increased fluxes are under gonadotrophic regulation and are independent of nuclear maturation. Amino acid uptake by mouse oocytes is approximately half that measured in sheep oocytes. The transport of amino acid (methionine) into unfertilized and fertilized mouse eggs is believed to involve active transport mechanisms with similar Vmax, Km, substrate specificity, and independently from Na^+ (Holmberg and Johnson 1979). An exchange diffusion system with an amino acid specificity similar to the uptake system is also observed in both unfertilized and fertilized eggs. An estimate of 6.5 fmol (femtomol) is made for the size of the total internal pool of exchangeable amino acids.

Recently, various studies using polyacrylamide gel electrophoresis, have demonstrated a change in the pattern of the proteins synthesized by maturing oocytes against a background of arrested transcription (Fig. 48) (van Blerkom 1980, 1981 a, b, 1983, 1984). This relatively brief period of pre-ovulatory meiotic maturation is believed to be accompanied by major quantitative and qualitative changes in protein synthesis (Fig. 48). For example, in the mouse, the absolute rate of ribosomal protein synthesis decrease by about 40% as compared to a 23% decrease in the rate of total protein synthesis during the same period (LaMarca and Wassarman 1979). Qualitative alterations in the pattern of protein synthesis have been reported for the mouse (Schultz and Wassarman 1977a, van Blerkom 1980, 1984, Richter and McGaughey 1982), rabbit (van Blerkom and McGaughey 1978a, b), pig (McGaughey and van Blerkom 1977, McGaughey et al. 1979) and sheep (Warnes et al. 1977). Each stage of meiotic maturation appears to be accompanied by specific qualitative and quantitative

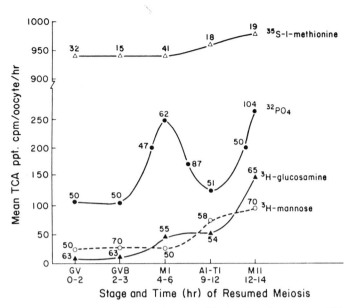

Fig. 48. Relative rates of incorporation of ^{35}S-1 methionine, ^{32}PO4, 3H-glucosamine ^{3}H-mannose into acid precipitable counts per minute/oocyte. Incorporation values were obtained from measurements and individual oocytes. The number of oocytes for each point is shown on the figure. In the presence of relatively constant rates of protein synthesis (indicated by methionine incorporation) the relative rates of phosphate glucosamine and mannose incorporation change independently (Redrawn from van Blerkom 1983)

alterations of protein synthesis (Fig. 48). Presumably, a programme of stage-specific protein synthesis is supported by preformed mRNA and is initiated with the resumption of meiotic maturation, as there are undetectable or extremely low levels of mRNA synthesis (van Blerkom 1984). The demonstrable synthesis of some proteine is short-lived or stage-specific, while others, synthesized at a specific stage, continue to be produced for a variable period of time (semi-constitutive to constitutive). A very small number of proteins is apparently synthesized nonsequentially during meiotic maturation, i.e. a particular protein may be found in the patterns of oocytes of GVB and not again until arrest of meiosis at metaphase II, or perhaps not until the early post-fertilization stages (van Blerkom 1983, 1984). GVB is one logical point for a programme of protein expression to be initiated, because mixing of nucleoplasm and cytoplasm occurs at this stage.

Golbus (1976) observed that mouse oocytes synthesize proteins actively at the germinal vesicle, metaphase I, metaphase II and pronuclear (6 h post fertilization) stages. The qualitative pattern of protein components being synthesized in vitro is changed thoughout maturation and fertilization. Oocytes are arrested at metaphase I by 0.01 μg/ml cyclohexamide or actinomycin D, then progress to a metaphase II pattern in spite of the nuclear maturation arrest, indicating a dissociation between meiotic maturation and the changes in the pattern of proteins synthesized at different stages of maturation (see also Golbus and Stein 1976).

Schultz and Wassarman (1977a) and Schultz et al. (1978b) reported specific changes in the pattern of protein synthesis during meiotic maturation of mouse oocytes in vitro which have either failed to undergo GVB or have been arrested at the GV stage by culture in the presence of db cAMP. Fluorograms of [^{35}S] methionine-labelled oocyte proteins have shown that meiotic progression from dictyate to metaphase II (meiotic maturation) is closely accompanied by marked qualitative changes in the pattern of proteins synthesized by oocytes. Their results have also shown that stage-specific changes in protein synthesis do not occur in mouse oocytes that do not undergo GVB. These workers have concluded that mixing of the oocyte's nucleoplasm and cytoplasm may trigger many of the alterations in protein synthesis that accompany meiotic maturation. According to Schultz et al. (1978b), the absolute rate of protein synthesis in the mouse is decreased from 43 to 31 pg/h per oocyte during meiotic progression from dictyate to metaphase II (meiotic maturation), while the size of the intracellular free methionine pool is decreased from 61 to 35 fmol per oocyte during the same period. Comparable measurements made on ovulated mouse oocytes that have undergone meiotic maturation in vivo strongly suggest that the decrease in the absolute rate of protein seen during meiotic maturation in vitro is physiologically significant. An alternative method that depends upon differential expansion of the oocyte's endogenous methionine pool determines absolute rates of protein synthesis. The oocyte's free methionine pool is not compartmentalized. The results of these studies have indicated that there occur two major kinds of changes in protein synthesis: (1) the appearance or disappearance of particular proteins and (2) an increase or decrease in the relative rates of synthesis of particular proteins (see also van Blerkom 1984). However, the overall qualitative pattern of protein synthesis remains remarkably constant throughout oocyte growth, but not during meiotic maturation (van Blerkom 1984) (see Wassarman et al. 1979a). Suzuki et al. (1981) have observed that nuclear DNA contents of human oocytes are diminished to approximately one half, but cytoplasmic protein contents measured by cytofluorometry remain unchanged after culture in spite of various ultrastructural changes that occur during the oocyte maturation in vitro, as already discussed in Sect. A of this chapter.

Wassarman et al. (1979a) observed that isolated fully grown mouse oocytes, arrested in dictyate of the first meiotic prophase synthesize a protein with apparent MW of 28,000. This is a germinal vesicle-associated protein (GVAP), as it appears to be at least 1000-fold more concentrated in the germinal vesicle than in the cytoplasm of the oocyte. The synthesis and phosphorylation of GVAP are apparently terminated at a time which coincides with GVB during spontaneous meiotic maturation of mouse oocytes in vitro. GVAP appears to be an example of protein that is selectively sequestered in the germinal vesicle of the oocyte during oogenesis and whose synthesis and modification are dependent upon the presence of an intact germinal veside.

Virtually all of the protein changes seen occur subsequent to the breakdown of the oocyte's germinal vesicle, but are not dependent upon the occurrence of other morphological events, such as spindle formation or polar body emission, as inhibition of spindle formation by Colcemid or of polar body extrusion by cytochalasin B does not affect protein synthesis (Schultz and Wassarman 1977b). These alterations in protein synthesis also do not occur in oocytes that fail to undergo breakdown of germinal vesicles spontaneously, or in oocytes arrested at the germinal vesicle stage by db cAMP. The addition of db cAMP to the culture medium after GVB has occurred does not prevent the changes in protein synthesis associated with meiotic maturation from taking place (see Wassarman et al. 1979 a). But Richter and McGaughey (1982) obtained quite different results. All these observations suggest that mixing of the oocyte's nucleoplasm and cytoplasm must occur to trigger many of the changes in the pattern of protein synthesis that accompany meiotic maturation of mouse oocytes in vitro (see also van Blerkom 1983, 1984). Wassarman et al. (1979 a) also concluded that although minor alterations in protein synthesis can be found during the first 5 h of culture, the major alterations associated with meiotic maturation in the mouse occur during the 5–10 h period following GVB and are then accentuated during the 10–15 h period (Schultz and Wassarman 1977 a). Inhibitors of protein synthesis, puromycin, or cycloheximide do not prevent GVB in mouse liberated oocytes; however, meiosis is arrested at circular bivalent stage (Stern et al. 1972, Golbus and Stein 1976, Wassarman and Letourneau 1976 b). Wassarman et al. (1979 a) observed that protein synthesis is necessary for a brief period just following GVB for meiotic maturation to proceed beyond the circular bivalent stage (see also Tsafriri 1979). By contrast, LH-induced maturation of rat follicle-enclosed oocytes is blocked by cycloheximide or puromycin (Tsafriri et al. 1973 b, Lindner et al. 1974). This discrepancy in the behaviour of the denuded and follicle-enclosed oocyte may show that LH exerts its action on the resumption of meiosis through a protein synthesized in granulosa cells. Although alterations in the pattern of protein synthesis must take place for nuclear progression to proceed beyond GVB, these alterations are not dependent on morphological events subsequent to GVB.

The precise nature and role of the proteins synthesized during meiotic maturation are not known. One of the proteins synthesized shows several of the characteristics of the lysine-rich histone F_1 (Wassarman and Letourneau 1976 b). The failure of nuclear progression to proceed beyond the circular bivalent stage does not prevent those changes in the pattern of protein synthesis that characterize meiotic maturation from taking place. These studies have also indicated that there may be basic differences in the control of meiotic maturation in oocytes isolated from mammalian as compared to nonmammalian animal species (Wassarman et al. 1979 a).

Balakier and Dzolowska (1977) observed that germinal vesicle breakdown and nuclear maturation in mouse oocytes is apparently induced by a cytoplasmic factor which is produced or unmasked independently of the nucleus. For these studies, anucleate and nucleate fragments of oocytes have been used. Schultz et al. (1978 a) found that nucleate oocyte fragments of mouse resume meiosis in vitro progressing from dictyate of the first meiotic prophase to metaphase II (meiotic maturation), and show some of the changes in protein synthesis normally associated with meiotic maturation of mouse oocytes. The anucleate fragments of the oocyte also undergo certain of the changes in protein synthesis associated with meiotic maturation despite the absence of nuclear progression. These findings have suggested that the acquisition of meiotic competence (i.e. the ability to undergo meiotic maturation) during growth of the mammalian oocyte is apparently due to changes in the quality, rather than the quantity, of cytoplasm, and the reprogramming of protein synthesis during meiotic maturation appears to be directed by RNA templates (mRNA) already present in the cytoplasm of the oocyte (Wassarman et al. 1979 a). However, the extent to which stage-specific proteins characteristic of meiotic maturation are derived from preformed, cytoplasmic RNA templates needs further investigation. These results would also suggest that the "trigger" for programmed alterations in protein synthesis may be initiated during the GV stage rather than subsequent to GVB. The recent findings of van Blerkom (1981 b, 1983, 1984) have indicated that a programme of protein synthesis and expression is initiated at some point between receipt of signal to resume meiosis and GVB and continues to influence translational patterns for at least an additional 36 h after removal of the nucleus. These preliminary results need to be extended and confirmed. Schultz and Wassarman (1977 a) observed that some of the proteins characteristic of different stages of meiotic maturation appeared in anucleate compartments derived from germinal vesicle stage oocytes exposed to cytochalasin D at the initiation of resumed meiosis. These results indicated the presence of a pool of translationally inactive, preformed mRNA that is differentially activated during resumption of arrested meiosis. The presence of preformed mRNA is also suggested by the studies of Petzoldt et al. (1980), who have observed that mouse eggs enucleated shortly after syngamy (fusion of male and female pronucleus) retain the capacity to express "on schedule", stage-related proteins charateristic of early cleavage. Balakier (1978) studied induction of maturation in small oocytes from sexually immature mice by fusion with meiotic or mitotic cells. The findings of this study have shown that the incompetence of these oocytes to initiate maturation appears to be due to the absence or the insufficient level of a maturation-promoting factor in their cytoplasm.

It has been shown that the pattern of protein synthesis seen with denuded oocytes is similar to that seen with oocytes cultured with surrounding cumulus cells and that alterations in protein synthesis observed during meiotic maturation cannot be attributed to post-translational modification of proteins synthesized prior to GVB (Schultz et al. 1978 a). However, van Blerkom (1981 b, 1983, 1984) has suggested that regulation of protein expression in the mouse may involve an intrinsic programme of mRNA activation or post-translational modification, or both, during meiotic maturation as well as during the early post-fertilization period (see also Pratt et al. 1983). The ability of mouse oocytes to modify post-translationally proteins has been demonstrated in the studies of Wassarman et al. (1979 a), who have observed protein phosphorylation during the GV stage, and by McGaughey (1983 a), who has described extensive phosphorylation throughout the entire period of meiotic maturation. But the extent to which stage-specific changes in protein synthesis during meiotic maturation occur as a result of post-translational modification of precursor proteins remains to be established. However, post-translational modifications including phosphorylation and glycosylation could alter the charge or molecular weight, or both, of specific polypeptides (van Blerkom 1984). Post-translational modifications could, therefore, generate families of proteins with identical primary structure but different electrophoretic mobilities. The functions of protein modification during pre-ovulatory resumption of arrested meiosis in the mammal are at present unknown. However, it is possible that ontogenic shifts in protein modification may be related to major and rapid nuclear cytoplasmic and plasma membrane changes that accompany each stage of this terminal phase of oogenesis, as already discussed in Sect. A of this chapter. The extent to which protein modifications are involved in the nuclear, cytoplasmic and plasma membrane events of resumed meiosis remains to be determined. However, several studies do provide important insights in defining possible relationships between protein modification and specific processes of pre-ovulatory oocyte maturation in the mammal (see van Blerkom 1984).

McGaughey and van Blerkom (1977) using high resolution, two-dimensional, polyacrylamide gel electrophoresis have studied the changes in the pattern of polypeptide synthesis during the in vitro maturation of the porcine oocytes, which not only mature but also synthesize a complex pattern of polypeptides throughout maturation. The appearance, disappearance and significant change in the intensity of molecular markers demonstrated are temporally related with the sequential stages of nuclear meiotic maturation. Collectively, the evidence has indicated the presence of a developmental programme during oocyte maturation, as also recently emphasized by van Blerkom (1981a, b, 1983, 1984).

McGaughey et al. (1979) studied germinal vesicle cytogenetic configurations and patterns of polypeptide synthesis of porcine oocytes from antral follicles of different sizes as related to their competency for spontaneous maturation. A high incidence of gross and cytogenetic degeneration among oocytes from small antral follicles is observed, as compared with those from medium or large follicles. The patterns of polypeptide synthesis differ greatly for normal oocytes from different follicular classes, and the patterns for oocytes from medium and large follicles are more similar to each other than to patterns for oocytes from small follicles. The normal oocytes from large follicles complete maturation in vitro (i.e. undergo the first meiotic division) at a significantly higher incidence (55%) than do oocytes from small (11%–20%) or medium (16%) follicles. A high proportion of oocytes from small antral follicles are evidently atretic. These findings further support the concept that a developmental programme apparently controls the molecular and cytogenetic changes occurring in porcine oocytes during follicular growth, which appear to be highly correlated with the acquisition of competency to complete maturation in vitro and possibly also are required for normal fertilization and embryogenesis, as recently discussed in detail by van Blerkom (1981a, b, 1983).

Van Blerkom and McGaughey (1978a) observed that both qualitative and quantitative changes occur in the pattern of polypeptide synthesis during maturation of rabbit oocyte in vitro and in vivo, which are directly comparable. Each stage of maturation is associated with the synthesis of specific polypeptides in the autoradiographic patterns. The major differences found between oocytes matured under these two conditions are that several polypeptides fail to appear in in vitro matured oocytes at the time they are detected in vivo and that synthesis of some polypeptides is prolonged in vitro compared to in vivo matured oocytes. Peluso and Hutz (1980) studied the effect of age on the ability of the oocyte to resume meiosis in vitro and to incorporate [^{2}H]-uridine and [^{3}H]-leucine into RNA and protein, respectively. In comparison with the mature control oocyte, the nucleolus in the aged oocyte tends to be retained although the rate of GVB is not changed. The incorporation of [^{3}H]-uridine is decreased, whereas [^{3}H]-leucine incorporation is not impaired. These results suggest that the inability of the aged oocyte to synthesize RNA may be responsible for its inability to complete meiotic maturation in vitro.

Warnes et al. (1977), using electrophoretic separation in one dimension on polyacrylamide SDS gels, studied changes in protein synthesis during maturation of sheep oocytes in vivo and in vitro. No change is observed in protein synthetic pattern of oocytes removed before or up to 6 h after the release of LH in vivo. By 9 h intermediate changes occur in at least 13 separate protein bands. Marked alterations in the synthesis of some proteins become apparent 15 h after LH; formation of proteins in five of the original bands is either reduced or not demonstrable, while new synthesis becomes evident from the appearance of seven additional bands. The pattern of proteins produced by oocytes cultured within the follicle corresponds closely to that demonstrated in vivo: alterations in synthesis are initiated about 9 h after addition of gonadotrophin and are completed by 15 h. Oocytes cultured outside the follicle in a gonadotrophin-containing medium do not show a change in protein synthesis and at 15 h only proteins produced during the early stages of maturation are being synthesized. The change in protein synthesis observed by Warnes et al. (1977) does not appear, therefore, to be necessary for the completion of nuclear maturation. The recent study of protein synthesis in ovine oocytes by Osborn and Moor (1983) shows that concurrent transcription of mRNA may be necessary for normal ontogenic changes in protein synthesis observed in this species during pre-ovulatory resumption of meiosis. These authors observed that addition of alpha-amanitin to extra-follicular, cumulus-enclosed ovine oocytes at explantation inhibits meiotic maturation and prevents many of the changes in protein synthesis that normally accompany maturation. These inhibitory effects are considerably reduced by delaying addition of inhibitor or by denuding oocytes of all associated cumulus cells at the outset of culture. However, it needs to be determined more precisely whether this transcriptional event(s) involves the cumulus cells or the oocyte, or both. While important species variations may exist, available evidence for the mouse, pig, and rabbit supports the notion that a programme or schedule of stage-related protein expression starts during the initial stages of pre-ovulatory resumption of arrested meiosis in the oocyte (terminal events of oogenesis) and continues to function in the absence of genomic input through the early post-fertilization period (van Blerkom 1984).

From the results of various studies discussed above, it is apparent that reproducible "stage-specific" alterations in protein synthesis and expression accompany each stage of resumed meiosis, with the duration of detectable synthesis of many species of proteins observed to be only a few hours (McGaughey and van Blerkom 1977, van Blerkom 1977, 1981 a, 1983, 1984, Schultz and Wassarman 1977 a, Warnes et al. 1977, Tsafriri 1978 a, van Blerkom and McGaughey 1978 a, Lopo and Calarco 1982, McGaughey 1983). These alterations in protein expression also occur in the absence of detectable transcription (Rodman and Bachvarova 1976, Wassarman and Letourneau 1976 a, Moore and Lintern-Moore 1978, van Blerkom and McGaughey 1978 a, van Blerkom 1984). While the identification of most stage-specific polypeptides needs to be determined more precisely, it is generally believed that such alterations are most probably involved in the process of resumed meiosis and possibly in the preparation of the oocyte for fertilization as well. In this regard, it is of interest to mention that ribosomal proteins and the contractile proteins (actin and tubulin) are translated throughout the period of resumed meiosis (van Blerkom 1980, 1981 a, b, 1983, 1984). The translation of contractile proteins may be related to cellular requirements for chromosomal maturation. But the continued formation of ribosomal proteins in the absence of detectable ribosomal RNA synthesis is somewhat unexpected, as significant stores of translationally inactive ribosomes are present in the mature mouse oocyte (van Blerkom 1983). Studies of protein synthesis during oogenesis (discussed in Chap. III), resumed meiosis (as already described here) and early pre-implantation embryogenesis reveal the presence of a translational continuum extending from oocyte through pre-implantation stages (Schultz and Wassarman 1977 b, Schultz et al. 1978 a, van Blerkom and McGaughey 1978 a, b, van Blerkom 1979, 1980, 1981 a, b, 1983, 1984, McGaughey 1983 a). The results of all these studies also indicate that a schedule or programme of translation is apparently initiated during the resumption of meiosis, most probably at or after the breakdown of the germinal vesicle, and continues to some extent in the oocyte regardless of whether or not ovulation and/or fertilization has taken place. This suggestion is supported by the fact that some of the proteins characteristic of specific stages of meiotic maturation (Schultz et al. 1978 a) are expressed approximately "on schedule" in the anuclear compartments cultured in vitro for varying lengths of time (see van Blerkom 1983, 1984). Fertilization-independent, time-dependent protein synthesis might be expected to be supported by preformed, oocyte RNA templates because transcription is barely observable in post-ovulatory mouse (Young et al. 1978) and rabbit oocytes (van Blerkom 1979). The change in protein pattern may be related to cytoplasmic maturation essential for normal fertilization and subsequent development. Investigations should be continued to determine the alterations in macromolecular synthesis and their exact function in the maturation of both isolated and follicle-enclosed oocytes.

3. Metabolic Changes

It has been observed that LH can stimulate the lactate formation in the isolated pre-ovulatory follicle in vitro (Nilsson 1974, Tsafriri et al. 1976 c). Furthermore, following the ovulatory LH surge, explanted rat follicles reveal a markedly increased lactate production. At the same time a decrease of the respiration by isolated cumulus cells occurs, as already described (Hillensjö 1976, Dekel et al. 1976, Ahrén et al. 1978). These acute alterations in follicular energy metabolism are indicative of a specific role for the LH effects on ovarian carbohydrate metabolism in relation to the ovulatory process. This also suggests one possibility, namely that LH by stimulating follicular glycolysis alters the intrafollicular environment to suit the maturing oocyte. One of the first experimental studies with isolated oocytes was the demonstration that pyruvate and oxaloacetate are necessary for meiotic maturation of the isolated mouse oocytes, which have not been hormonally stimulated, and that glucose will only support the spontaneous maturation if the cumulus cells, which convert glucose into pyruvate for oocyte, are also present in the culture system (Biggers et al. 1967). A lack of energy substrate should, therefore, be considered in this case as a condition that inhibits maturation. Rat oocytes mature in media containing lactate, and to a certain extent even in media without energy substrate (Zeilmaker and Verhamme 1974). Lactate is produced in the medium of explanted follicles after the addition of LH (Hillensjö 1976), presumably by an increase in glycolytic activity in the follicle. Since lactate forms a substrate for rat oocytes and also mouse oocytes, provided enough NAD is present, it is possible that a rise in follicular lactate concentration is responsible for the induction of oocyte maturation. Zeilmaker and Verhamme (1977) determined the follicular lactate concentrations prior to, during, and after GVB in the rat oocyte. Zeilmaker (1978) detected high concentrations of lactate in pre-ovulatory follicles of the rat ovary before, as well as after, the time of the expected LH surge. Explanted oocytes obtained from prepubertal rats and surrounded by cumulus cells matured in the presence of 20 mM lactate. Lactate production by the follicle under the influence of LH cannot be considered as a factor initiating maturation.

Glucose-phosphenolpyruvate and lactate can supply the energy for mouse oocyte maturation only if follicular cells, which produce pyruvate from these substrates, are present in the culture (Donahue and Stern 1968). The dependance of the resumption of maturation on pyruvate led to the suggestion that the pre-ovulatory LH surge induces maturation by the supply of the essential nutrient for the oocyte by granulosa cells (Biggers 1972). The results of various studies suggest that LH-induced glycolysis does not appear to be the mechanism by which LH triggers the resumption of meiosis in all mammalian species (see Tsafriri 1978a). This problem needs to be investigated. However, recent results obtained by Fajer (1983) suggest that metabolism of glucose is changed significantly at the time of the onset of meiosis in the hamster ovary, as evidenced by a large drop in the CO_2 evolution from labelled glucose, which could be attributed to changes in the metabolic patterns in the germ cells. The reduced glycolytic metabolism is well established in the oocyte at the diplotene stage of the first meiotic division.

Yoon and Cho (1976) studied the glycogen content of the oocytes at the various stages of meiotic division induced during culture. The intensity of PAS-positive reaction gradually decreases as the meiotic resumption progresses. The amount of glycogen is also decreased in the degenerated ova. Glycogen consumption appears to be essential for meiotic resumption. Its loss while the germinal vesicle is intact seems to lead to degeneration. The endogenous glycogen is believed to be important for the support of meiosis.

Chang (1955) attributed the maturation of liberated oocytes to an increase in oxygen tension in the medium. The mouse and rat oocytes do not resume maturation in the absence of oxygen (Zeilmaker et al. 1972, Zeilmaker and Verhamme 1974, Zeilmaker 1978). Similarly, the maturation of hamster oocytes is blocked at "Chromatin II" stage in the absence of oxygen (Gwatkin and Haidri 1974). Although these studies suggest the involvement of oxidative phosphorylation in the process of oocyte maturation, they do not establish that oxygen is the limiting factor which prevents the resumption of meiosis within the follicle, and that LH initiates maturation by alteration in oxygen tension. A small slit in the follicle wall, presumably following access of oxygen, induces maturation in a limited number of cases. In a preliminary study the oxygen consumption by mechanically denuded oocytes obtained from antral follicles has been analyzed by micro-diver technique (Hillensjö et al. 1975). Certain energy substrates which are known to support mouse oocyte maturation in vitro, such as pyruvate and oxaloacetate (Biggers et al. 1967), when added to the incubation medium increase the oocyte respiration (Ahrén et al. 1978). It has also been observed that the inhibitory effect of granulosa cells on oocyte maturation cannot be eliminated by enriching the culture atmosphere with oxygen (Tsafriri 1978a). It is also of great interest to study oocyte oxygen consumtion in relation to the resumption of meiosis. When oocytes are induced to mature in vivo by LH, the oxygen consumption of isolated and denuded oocytes is increased as compared to oocytes in the dictyate stage (Ahrén et al. 1978). In these studies, carried out with the microspectrophotometric technique, the variation between oocytes within the same experimental group is small allowing a detailed estimation. An increase in oxygen consumption is observed after GVB–3 to 4 h after the LH injection. The oxygen consumption by the denuded maturing rat oocytes is also increased at the time of extrusion of the first polar body (Megnusson et al. 1977, Ahrén et al. 1978). This increase in oxygen consumption occurs when maturation is induced by LH in vivo or by culture of liberated oocytes with adherent cumulus cells. When the oocytes are induced to mature spontaneously by culture in the medium containing pyruvate, a similar rise in oxygen consumption is seen (Ahrén et al. 1978). The gradual increase in respiration follows the morphological alterations of the nuclear maturation when the resumption of meiosis is prevented by the addition of db cAMP into the culture medium, and oxygen consumption remains low, as in control, noncultured, oocytes (Tsafriri 1978a). The increase in oxygen use of the denuded oocyte contrasts with the depression by LH of oxygen utilization by oocyte–cumulus cell complexes (Dekel et al. 1976). It is not known whether this decrease in cumulus cell respiration is necessary for oocyte maturation. However, oocyte oxygen consumption is significantly increased after the onset of nuclear maturation (see Ahrén et al. 1978). It appears that the GVB per se triggers the increase in oocyte respiration and that no direct hormonal control of this metabolic event exists. This assumption is supported by the finding that cAMP and db cAMP, which prevent the spontaneous maturation, also prevent the rise in oxygen consumption normally induced by culture (Magnusson and Hillensjö 1977). cAMP appears to be involved indirectly in the resumption of oocyte meiosis, as already discussed in this chapter (Tsafriri et al. 1972, Hillensjö et al. 1978, Tsafriri 1983), but it does not act as a stimulatory agent directly on the oocyte.

Magnusson et al. (1977) investigated oxygen consumption of maturing rat oocytes. The oocytes were either induced to mature in vivo by LH or matured spontaneously when cultured. In both cases, there is seen a progressive increase in respiration of the oocytes following the onset of morphological nuclear maturation. The results of this study indicate that the resumption of oocyte meiosis is apparently accompanied by significant increase in O_2 consumption and that this increase appears to be related to the meiotic process rather than to hormonal stimulation. Oxygen consumption by rat oocytes and cumulus cells during induced atresia has been studied recently by separating denuded oocytes and oocyte–cumulus complexes (Magnusson et al. 1983). Oxygen consumption by oocytes that have resumed meiosis (GVB) is higher than in oocytes with an intact germinal vesicle, a change similar to that observed in oocytes maturing in healthy follicles, and suggesting that the meiotic process in the atretic follicles is similar to that in normal ones. Oxygen consumption by the cumulus cells is not altered during pentobarbitone-induced atresia. Hypophysectomy led to a rapid and marked increase in cumulus oxygen consumption in cyclic rats, but there was no change in PMSG-treated young animals. Both pentobarbitone-treatment and PDX result in follicular atresia, but changes in cumulus respiration occur in PDX adult rats. These results suggest that an increase in cumulus respiration is not inherent in the atretic process. Further investigations should be made to determine the metabolic changes within the oocyte and the follicle associated with oocyte maturation and their involvement in the regulation of the meiotic process.

The activities of LDH and phosphoglucose isomerase (PGI) during oocyte maturation in the mouse are increased (Engel and Franke 1975). Since their activity increase can be induced in germinal vesicle stage oocytes by proteolytic treatment with trypsin and pronase, and since cycloheximide does not antagonize the trypsin-induced effect, inactive molecules of LDH and PGI have been assumed to occur in the mouse oocyte. Their activation in vivo appears to be affected by proteolytic enzymes which are liberated in the oocyte from lysosomes or other vesicles during oocyte maturation. A similar observation has also been made for the rabbit oocyte, in which during the progression from late prophase of the first meiotic division to metaphase of the second meiotic division, new LDH isozymes appear (Engel et al. 1975). The studies of Poznakhirkina et al. (1975) on the unfertilized rat egg suggest the presence of inactive enzyme molecules in the mammalian egg, as the rat ovum normally shows only LDHI but acquires additional LDH isozymes after treatment with the detergent Triton X-100. Peacock et al. (1978) observed that rabbit oocyte LDH isozyme pattern shifts dramatically at ovulation. Both ovarian oocytes and ovulated eggs show predominantly LDH-5, with minor amounts of LDH-4 showing no change at ovulation. When oocytes or eggs are not treated to remove follicle cells, all five LDH isozymes are observed. Isolated follicle cells also show three isozymes. Failure to remove follicle cells may cause altered LDH isozyme patterns. Jagiello et al. (1978) observed that specific inhibitors of serine proteases inhibit meiotic progression in mammalian oocytes, revealing involvement of these enzymes in the process of maturation.

An increase in O_2 consumption is observed following maturation (Magnusson and Hillensjö 1977, Magnusson et al. 1977), as already discussed. But it is not known whether increase in O_2 is related to mitochondrial activity, activation, etc., or whether this increase in O_2 consumption is too low to have metabolic importance for maturation. However, when the oocytes are cultured in the presence of db cAMP or cAMP, no change in respiration occurs during culture. These findings argue against the theory that cAMP acts as a direct mediator of the action of LH on the oocyte maturation, discussed earlier in this chapter. Alterations in oocyte energy metabolism are apparently closely related to the maturation process. No doubt, additional studies are needed to determine their relationship between morphological and physiological maturation in various species of mammals and the specific requirement for hormonal and/or other stimuli for each of these components of oocyte maturation. Tsafriri (1978a, 1979, 1983), after reviewing the results of various studies, has suggested the involvement of at least three factors in the regulation of the resumption of meiosis: (1) the presence of a follicular inhibitor, (2) a change of oocyte sensitivity, and (3) alterations triggered by LH. The exact function and mode of action of each of these factors need to be determined more precisely. Our knowledge of the mechanism involved in the maturation of oocytes is still fragmentary and thus further investigations, especially at the subcellular and molecular levels, are recommended (see also Tsafriri 1983, van Blerkom 1983, 1984).

Chapter V
Ovulation

It is well established that ovulation in mammals occurs as a result of an endogenous pre-ovulatory LH surge. Gonadotrophins, such as hCG and PMSG, which have LH activity, also induce ovulation. The ovulatory process comprises alterations in the steroidogenesis, resumption of oocyte meiosis (discussed in Chaps. IV and VI), morphological and biochemical transformation of the cumulus oophorus (described in Chap. IV), luteinization of the granulosa cells (discussed in Chap. VI), and finally follicular rupture and release of the mature ovum, to be discussed in this chapter. The results of recent experimental studies have indicated that the processes of follicular rupture, ovum maturation and ovarian steroid production can proceed independently of one another (Bongiovanni et al. 1983, Testart 1983). Rondell (1974) has discussed the role of steroid and gonadotropic hormones in the process of ovulation at the cellular level (see also Schwartz 1974). During ovulation, the granulosa, theca interna and surface epithelium undergo major biophysical and biochemical changes (Lipner 1973, Espey 1974, 1978a, b, Guraya 1974a, 1979c), which will be discussed here in order to assess the increase in our kowledge about the cellular, subcellular and molecular aspects of the mechanism of ovulation.

The mature follicle in the ovaries of most mammals forms a spherical structure which is generally embedded within the stromal tissue of the ovary, but can be observed from the surface as a translucent vesicle (Fig. 49). For determining the details of alterations in the follicle wall during the ovulatory process, rabbit has been used

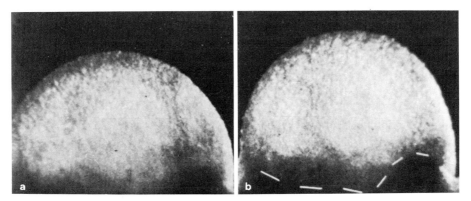

Fig. 49a, b. Photographs of a hamster follicle taken during in vitro ovulation. **a** 5 min before rupture; **b** About 2 min before rupture. The *dotted line* outlines the dark layer at the base of the follicle which rises during the final minutes before rupture (From Martin and Talbot 1981a)

extensively. The wall of its follicle consists of three layers: (1) the theca externa forming the fibroblastic layer; (2) the theca interna composed of spindle-shaped cells and hypertrophied polygonal vacuolated cells which are highly vascularized; and (3) a layer of granulosa cells, the membrana granulosa, which is sharply separated from the outer layers by the basement membrane or membrana propria (Fig. 50 a) (Espey et al. 1981). Similar layers have also been reported for other mammalian species (Fig. 26) (Mossman and Duke 1973). Their structure and function during the growth of the follicle has already been discussed in Chap. III. At its point of contact with the surface of the ovary the emerging follicle is also surrounded by the tunica albuginea, which is composed of connective tissue (Fig. 50 a). Outside the tunica albuginea is the surface epithelium, consisting of spindle, cuboidal cells (see Bjërsing and Cajander 1974 a). Rosenbaur et al. (1977) have made scanning electron microscopic studies on the course of noninduced ovulation in rats with normal cyclic behaviour. Such studies of the ovarian surface during the various phases of the cycle permit description of the morphology of ovulation. No signs of incipient ovulation are seen in pro-oestrus. The entire process of ovulation in the rat occurs in the oestrus. The epithelization of the ovulatory lesion starts in metoestrus and is always completed in dioestrus. Renaud et al. (1980), using echography, have demonstrated the physical event of ovulation during 18 menstrual cycles in ten women. The average follicular growth measured from the time of echographic visualization was 0.3 cm/day. The diameter of the follicle measured the day before disappearance was $2.7 + 0.3$ cm. These results have also been correlated with alterations in serum hormone (LH, 17β-oestradiol, progesterone) levels, basal body temperature curves and cervical mucus changes. In all cases, echography has revealed accurately the time of ovulation. Janson et al. (1982) have described methodology and pattern of steroidogenesis for the study of ovulation in the isolated perfused rabbit ovary. This experimental model may be useful in the further study of the physiology, biochemistry and anatomy of ovulation. Löfman et al. (1982), using photographic and cinematographic techniques, have evaluated the gross anatomy of ovulation in the perfused rabbit ovary. The use of infrared film has made it possible to visualize intrafollicular events prior to rupture.

   The first macroscopic sign that ovulation is about to occur is the development of the macula pellucida or stigma (Fig. 50 a, b) (Guraya 1971 b, 1979 c). This is also supported by the results of scanning and transmission electron microscopy (Nilsson and Munshi 1973, Motta and van Blerkom 1975, 1980, Bjërsing and Cajander 1974 b, c, Cajander 1976, Espey et al. 1981, Mori and Uchida 1981, Tsujimoto et al. 1982, Martin and Talbot 1981 a, b). The stigma in the Japanese long-fingered bat *(Miniopterus schreibersii)* is formed simultaneously with the expulsion of the first polar body and with the subsequent bleeding from the capillary lumina of the theca interna (Mori and Uchida 1981). In its size and form, stigma varies greatly in different mammalian species (Guraya 1974 a). Ovulatory stigma is avascular (Espey 1967, Motta et al. 1971, Gillet et al. 1980). On the basis of morphological, histochemical and biochemical observations, various theories have been put forward to explain the process of mammalian ovulation (for reviews see Blandau 1966, Guraya 1974 a, 1979 c, Espey 1974, 1978 a, b, 1980). The intrafollicular pressure (Espey and Lipner 1963, Rondell 1964, 1970, Talbot 1983), contraction of smooth muscle cells (O'Shea 1970 b, Osvaldo-Decima 1970, Fumagalli et al. 1971, Lipner 1973, Espey 1978 a, b, Vir-utasmasen et al. 1976, DiDio et al. 1980, Martin and Talbot 1981 a, b), enzymatic

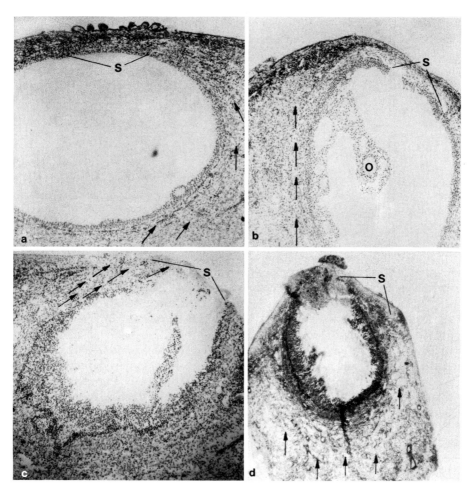

Fig. 50 a–d. Photomicrographs showing morphological and histochemical changes in the wall of rabbit follicle during ovulation. **a** Pre-ovulatory follicle, showing very little Feulgen-positive material in the surrounding stroma *(arrows)*, though such material is still quite conspicuous in the stigmal zone *(S)*. The follicle was removed 6 h 15 min after the injection of 100 i.u. hCG; **b** Follicle during pre-ovulatory swelling. There is a decrease in the staining of the nuclei of the surrounding stromal cells in the stigmal area *(S)*. The follicle was removed 8 h after the injection of 100 i.u. hCG; **c** Portion of newly ruptured follicle, showing further disappearance of Feulgen-positive substance from the area of stigma *(arrows)*. The follicle was removed 10 h after the injection of 100 i.u. hCG; **d** Portion of newly ruptured follicle, showing the absence of basophilic substance from the surrounding stroma *(arrows)* as well as from the area of stigma *(S)*. The basophilic substances are conspicuous in the cytoplasm of the granulosa and theca interna cells. The follicle was removed 10 h 40 min after the injection of 100 i.u. hCG (From Guraya 1971 b)

digestion (Espey and Lipner 1965, Guraya 1971 b, 1979 c, Lipner 1973, Espey 1974, 1978 a, Espey and Coons 1976, Bjërsing 1977, Strickland and Beers 1979), vascular changes (Burr and Davies 1951, Motta et al. 1971; other references in Lipner 1973, Reed et al. 1979, Makinoda 1980), nervous control (Bahr et al. 1974, Walles et al. 1976,

1977, Burden and Lawrence 1980) and neuromuscular mechanism (Weiner et al. 1977, Sjöberg et al. 1979, Owman et al. 1980, Morimoto et al. 1981) are believed to play roles in ovulation. At present, no single phenomenon has received wide acceptance as being the sole cause of ovulation. A combination of these phenomena to a variable degree appears to be responsible for ovulation (Lipner 1973, Nowacki 1977, Martin and Talbot 1981 a, b). These concepts will be discussed here.

A. Intrafollicular Pressure and Vascular Changes

The various studies have revealed that just before ovulation the follicle receives the largest volume of blood per unit time and has capillaries which are more permeable than those in other follicles (Burr and Davies 1951, Moor et al. 1975, Gillet et al. 1980, Kanzaki 1981). According to Kanzaki et al. (1981), pathways of venous drainage from a wall of a large follicle, and arteriovenous shunts seen in the wall of such follicles, appear to exert an important effect on the follicular haemodynamics in the process of follicle rupture. Dynamics of ovarian blood supply to pre-ovulatory follicles of the ewe are studied by injecting radioactive microspheres into the ovarian artery (Murdoch et al. 1983). The supply of ovarian blood to the wall (apex and base) of follicles is elevated after the rise in LH is initially ascertained (0–12 h). The distribution of blood to the follicular wall begins to decline from 12 to 16 h, and subsequently continues to decrease until after ovulation has occurred. Makinoda (1980) has studied the haemodynamic and histological aspects of ovarian blood-flow during ovulation. He has suggested the presence of a congestion in ovarian blood vessels during the ovulatory period, which appears to play a significant role in the mechanism leading to ovulation. Kanzaki (1981) has also suggested that there is increased capillary permeability and congestion of bloodstream at the apical region shortly before ovulation. Okamura et al. (1981) have observed that with the follicle at the mid-proliferative phase (8 mm diameter) in the human, the capillary lumen is narrow and empty. Capillary vessels seen in the theca interna of the pre-ovulatory follicle are dilated. Their endothelial cells show indented nuclei and elongated thin cytoplasm with unusually irregular membranous protrusions. Increased capillary permeability due to active transport about the time of ovulation is indicated. Precise morphological changes in ovarian vasculature just prior to ovulation still need to be determined. However, the vast capillary network in the follicle wall is suggested to play an important role in the mechanism of follicle rupture through its increased permeability and possibly through collagenolytic activities of leukocytes (Espey 1980). The idea of leukocytes contributing to follicular collagenolytic activity needs to be extended and confirmed in future studies.

The increase in the size of the follicle about the time of ovulation may be due to ballooning of the apex, rather than to any enlargement of the base of the follicle (Espey, pers. commun.). Espey believes that this follicular enlargement arises primarily from a proteolytic "softening" of the entire follicle. And, in this loosened state, the follicle simply expands under the force of a low, but sufficient intrafollicular pressure of 15–20 mm Hg (Espey and Lipner 1963). The increased follicular volume can lead to follicle rupture through increased distensibility of the follicular wall, as a result of collagenase activity. Nigi (1977a) using laparoscopy has observed two types of follicular rupture in the Japanese macaque *(Macaca fuscata)*. Type I is "fast and explosive", as it is characterized by a rapid exudation of surface follicular fluid. In the type II rupture, which is "slow and oozing", the follicular fluid is exuded gradually through one of the reddish difuse areas, developing on the surface of the pre-ovulatory follicles. This variability may result partly from the manner in which the apex (stigma) breaks down. If there is a thin strand of theca externa at the end, then this strand may "pop", and thus rupture appears to be more instantaneous. The intrafollicular pressure, may also be major factor here (Espey pers. commun.). However, if manipulation of the mesovarian tissue has even slightly impaired blood-flow to the ovaries, then intrafollicular pressure may decrease significantly, and there is insufficient force for a "fast" rupture. Recent studies by Schroeder and Talbot (1982) have shown that intrafollicular pressure is decreased in hamster pre-ovulatory follicles during smooth muscle cell concentration in vitro; 5-hydroxytryptamine can cause contraction of smooth muscle cells (Talbot and Schroeder 1982). According to Talbot (1983), intrafollicular pressure promotes partial evacuation of the antrum during hamster ovulation in vitro, indicating a role for intrafollicular pressure in hamster ovulation.

Wildt et al. (1977) have made a laparoscopic study of cyclic ovarian morphology in the bitch, as also made by Nigi (1977 b) for the Japanese macaque. DeCrespigny et al. (1981) have made ultrasonic observation of the mechanism of human ovulation. No demonstrable changes are seen in the size and appearance of the follicle upto 7 h prior to its rupture. In one of the four subjects in whom follicular collapse is seen, the follicle empties completely within less than 1 min. In two of the other three subjects, there is an initial rapid fluid loss followed by a slower release of the remaining contents. The slow phase of follicular collapse is believed to be an important aspect in ovum release.

The rapid accumulation of fluid within the follicle is not accompanied by any marked change in the intrafollicular pressure, which remains almost equal to that in the blood capillaries (Espey and Lipner 1963, Blandau and Rumery 1963, Rondell 1964, Richards 1966, Virutasmasen et al. 1972, 1976). Since there is no appreciable change in the pressure, and as cell mitosis in the follicle has ceased following the pre-ovulatory LH surge (see Chaps. III, VI), the follicular enlargement (particularly stigma formation) may be due to a decrease in tensile strength of the follicle wall (Espey 1974). Lipner (1973) has also suggested that for the follicle to expand a gradual realignment or decreased attachment of collagen fibrils to each other and to fibroblasts takes place.

Stahler et al. (1977) have measured intrafollicular, intra-ovarian and intra-arterial hydrostatic pressures in vitro on human ovaries in the follicular ripening phase. In the vascular system and intra-ovarian tissues, pressure variations occur spontaneously. With the influence of epinephrine, norepinephrine, prostaglandin $F_{2\alpha}$ and oxytocin, the tonicity of the vascular system increases; the number of spontaneous contractions increases rather noticeably especially under the influence of prostaglandin $F_{2\alpha}$ and various catecholamines. Follicles immediately before ovulation do not show a pressure increase, because the increase of the higher follicle runs parallel to the increasing elasticity of the follicle walls, i.e. follicular elasticity "dampens" any pressure fluctuations. However, Espey (pers. commun.) believes that intrafollicular pressure is directly dependent on capillary hydrostatic pressure and the follicle pressure changes instantaneously with capillary pressure. Furthermore, the intrafollicular pressure responses are much more pronounced in follicles that are close to rupture (i.e. in follicles with greater "elasticity") than in follicles that are distant from ovulation. Bronson et al. (1979) have observed no increase in intrafollicular pressure at the time of ovulation in the pig, which, however, varies within follicles of the unstimulated ovary. They have suggested that the follicle wall apparently develops the ability to undergo stress relaxation during follicular maturation and this plays a role in regulating intrafollicular pressure.

B. Nervous, Muscular and Neuromuscular Controls

Both adrenergic (sympathetic) and cholinergic (parasympathetic) nerves are present in the mammalian ovary (Bahr et al. 1974). Adrenergic nerves are relatively more abundant (Fig. 34) than the cholinergic nerves. Nerve fibres along with the blood vessels enter the ovary. Catecholamine-containing nerves, which surround the blood vessels, are distributed within the stromal fibromuscular layer and the theca of follicles (see Chap. III). The nerves follow a similar pattern of distribution in the ovaries of different species of mammals. However, their distribution varies greatly with species. The norepinephrine content of ovaries in these species correlates with the density of adrenergic nerves (Jacobowitz and Wallach 1967). The results of various experimental studies, as discussed by Bahr et al. (1974), did not provide any definite answer in regard to the role of nerves in normal ovarian function including ovulation. However, the results of recent studies have suggested that nerves do play some role in ovulation and possibly other ovarian functions. The detection of smooth muscle tissue within the ovary (Figs. 32 and 33) suggests that autonomic nerves within the ovary may intensify ovarian smooth muscle contractility at the time of ovulation and assist in follicular rupture and expulsion of the ovum.

Cholinergic nerves and receptors mediating contraction of the bovine Graafian follicles are suggested to play a significant role in ovulation (Walles et al. 1976, Owman et al. 1980, Morimoto et al. 1981). Walles et al. (1977) have also made characterization of autonomic receptors in the smooth musculature of human Graafian follicles at the pre-ovulatory stage. Various types of autonomic receptors have been studied in isolated 2×10-mm-size strips from the protruding part of such follicles, as also recently reviewed by Owman et al. (1980). This local autonomic mechanism in the follicle of the human ovary is believed to be involved in follicular development and/or ovulation. The presence of perivascular nerves suggests that they may have a vasomotor function. This vasomotor control may involve a selective or trophic role during follicular development and maturation (see Bahr et al. 1974, see also Chap. III). As adrenergic nerves are important in vasomotor regulation, the catecholamines formed by adrenergic nerves may affect the ovulatory process through haemodynamic changes.

Weiner et al. (1977) have studied the influence of ovarian denervation and nerve stimulation on ovarian contractions in the rabbit and then discussed the significance of the contribution of a neuromuscular mechanism to the control of ovulation.

Smooth muscle cells are distributed throughout the ovary. They are either vascular or nonvascular. The nonvascular smooth muscle cells occur mainly in the cortical stroma. Okamura et al. (1972) reported bundles of muscle cells interspersed with collagenous connective tissue throughout the stroma admixed with theca cells in various species. As already discussed in Chap. III, smooth-muscle cells have also been demonstrated in the theca externa of follicles in different mammalian species by other workers (Figs. 32 and 33) (Osvaldo-Decima 1979, O'Shea 1973, Espey 1978 b, Amenta et al. 1979, Pendergrass and Talbot 1979, Martin and Talbot 1981 a, b; see other references in Sjöberg et al. 1979, DiDio et al. 1980, Owman et al. 1979, 1980), which may have contractile as well as collagen-synthetic functions. Pendergrass and Talbot (1979) have studied the distribution of contractile cells in the apex of the pre-ovulatory hamster follicle. A discrete layer of smooth muscle cells is present within the theca externa of maturing follicle. Bundles and clusters of filaments are also present in the surface epithelial cells of maturing hamster follicles throughout the pre-ovulatory period. A contractile role in ovulation is suggested for both the smooth-muscle and surface epithelial cells. Amenta et al. (1979), using the electron microscope, have demonstrated the presence of smooth-muscle cells in the ovaries of rabbit, cat, rat and mouse, which occur as isolated cells or as cells in irregular layers. Their relationship with connective tissue around developing, mature and atretic follicles is analyzed. The relationship between ovarian myocytes and ovulation is studied. Martin and Talbot (1981 a) have demonstrated the presence of cells with all the ultrastructural features of smooth muscle only in the basal hemisphere of the hamster follicle. In contrast, the theca externa in the top half of the follicle is composed of fibroblasts. These workers believe that the smooth-muscle cells present in abundance in the theca externa in the base of the follicle contract during the final minutes before rupture of the follicle. Further contraction of these smooth-muscle cells may also help in the collapse of the follicle wall after rupture. The results obtained by Martin and Talbot (1981 b), after treatment of pre-ovulatory hamster ovaries with six classes of drugs (lanthanides, calcium antagonists, local anesthetics, prostaglandins, cAMP modulators, and cytochalasin B) known to inhibit smooth-muscle contraction in other tissues, further support the suggestion that contraction of follicular smooth-muscle cells constricts the follicle and is required for ovulation. These results have also revealed that follicular smooth-muscle cells are activated by an influx of extracellular calcium and that prostaglandin $F_{2\alpha}$ may be involved in promoting this contraction. The observations of Walles et al. (1978) are consistent with the idea that blockage of any influx of extracellular Ca^{2+} into smooth muscle of rabbit follicles inhibits their contraction and subsequent ovulation.

The localization of actin and myosin in the elongated cells in the theca externa in the rat Graafian follicles has been demonstrated by immunofluorescence (Fig. 33) (Amsterdam et al. 1977, Walles et al. 1978, Sjöberg et al. 1979). All these findings have further revealed that follicular growth and maturation is closely accompanied by the development of a smooth-muscle layer in the theca. The physiological studies have demonstrated contractions of the ovary (Rocereto et al. 1969, Coutinho and Maia, 1972, Walles et al. 1974, Virutasmasen et al. 1976, Espey 1978 b) and the follicle wall (Lipner and Maxwell 1960, Walles et al. 1975 a, b, 1977, Sjöberg et al. 1979), which can be attributed to the presence of smooth-muscle cells. The presence of myosin-like protein in the ovary further supports the presence of contractile smooth muscle cells in the ovary (Moscarini and Amenta 1980).

Rocereto et al. (1969) observed that ovarian contractions are enhanced by alpha adrenergic stimulation and inhibited by beta stimulating adrenergic agents. The contraction of the smooth-muscle layer in response to catecholamines and/or prostaglandins is believed to play a role in the extrusion of the oocyte (see Bahr et al.

1974, DiDio et al. 1980). In general, smooth-muscle cells may be stimulated by (1) hormones, (2) nerves, (3) prostaglandins and (4) local changes in the muscle itself (Wallach et al. 1980). Although the functional role of follicular smooth-muscle cells is not clear (Espey 1978 b), the effects of hormones and nerves have recently been reviewed (Owman et al. 1979, 1980). Sjöberg et al. (1979) have suggested that norepinephrine, released from the adrenergic nerve terminals in the follicular wall of the ovary, can influence ovulatory functions by exerting an effect on the follicular smooth musculature via adrenergic receptors (see also Bahr et al. 1974). Owman et al. (1980) believe that stimulation of the sympathetic component of the follicular nerve plexus releases sufficient amounts of norepinephrine to induce a motor response in the follicle wall strong enough to influence the intrafollicular pressure (see also Owman et al. 1975). The sympathomimetic increase in tension of the follicle wall involves adrenergic receptors, whereas the B_2 type of adrenoceptors mediate a relaxation of the follicle wall. Interference with the intra-ovarian sympathetic mechanisms affects the incidence of ovulation under in vivo conditions. The presence of well-developed smooth-muscle cells in the follicle wall suggests that some kind of mechanical factor is involved. It is not necessary that such a factor should cause an increase in follicular pressure to bring about the rupture of the follicular wall and expulsion of the oocyte. It will be interesting to mention here that there occurs a significant depletion of ovarian catecholamines (norepinephrine) during the pre-ovulatory period in the PMSG-primed prepubertal rat (Bahr et al. 1981). But further studies are needed to determine whether these catecholamines play an essential, modulating or permissive role in ovulation.

Espey and Lipner (1963) first demonstrated rhythmic changes in the antral pressure of mature follicles of rabbits. The average pressure within the follicle was approximately 15 mm Hg. This pressure occasionally underwent rhythmic deviations of 5–10 mm Hg at intervals of 2–5 cycle/min. These rhythmic changes in intrafollicular pressure are believed to be caused by smooth muscle activity (spasms) somewhere in the vicinity of the ovary (possibly within the ovarian hilus). These measurable "contractions" are not a prerequisite for ovulation. These observations were later confirmed by Rondell (1964). The recently described adrenergic mechanisms are also believed to increase both follicle wall tension and pressure (Owman et al. 1980). The sympathetic nerves might also act on the smooth-muscle system in an entirely different way, for example, related to those changes in the vascular bed required to produce ischemia which seems to form a part of the rupture mechanism. It is also possible that nerves do not primarily mediate a motor function, but have a tropic influence on the development of the follicle or its transformation after ovulation.

A distinct role for contractile cells in mammalian ovulation has not yet emerged, because: (1) The presence of smooth-muscle cells in the follicle wall has been challenged as typical smooth-muscle tissue is confined to the hilar and medullary regions of the ovary (Espey 1978 b); (2) the ovarian contractions have not yet been directly correlated to the rupture of follicles (Roca et al. 1976); and (3) intrafollicular pressure does not show much increase prior to follicle rupture as might be expected if contractions were occurring (Blandau and Rumery 1963, Rondell 1964, Virutasmasen et al. 1972, Bronson et al. 1979). But Espey and Lipner (1963) showed pressure fluctuations. However, according to Virutasmasen et al. (1976), the demonstration of increased contractile activity of the ovary about the time of ovulation suggests that ovarian contractions participate in the process of follicular rupture and the extrusion of ova at ovulation. $PGF_{2\alpha}$, norepinephrine and oxytocin are effective in inducing ovarian contractions. Morikawa et al. (1981) have observed that histamine concentration in the human ovary is augmented and the ovarian contractile response to histamine increased at the ovulatory phase. These observations have suggested that apparently histamine is actively involved in ovulation through its effects on the smooth musculature and vascular network. Loeken et al. (1983) have observed stimulatory effect of LH upon relaxin secretion by cultured porcine pre-ovulatory granulosa cells. However, in the porcine pre-ovulatory follicle, the theca is considered to form the principal source of relaxin (Evans et al. 1983 b). These results suggest that relaxin secreted prior to ovulation may have a local ovarian effect, perhaps facilitating ovulation. This possibility needs to be extended and confirmed by carrying out further studies.

Espey (1978 b), after critically reviewing the literature about the ovarian contractility and its relationship to ovulation, has concluded that "ovarian contractility is not necessary for ovulation to take place" and thus the century-old debate about the role of ovarian contractions in ovulation continues. But ovarian contractions are real and appear to play some definable roles in ovarian physiology. Espey (1978 b) has suggested that more attention should be paid to a possible relationship between ovarian contractions and circulatory function. The contraction of ovarian myocytes has been suggested to affect (1) the detachment of the cumulus oophorus, (2) expulsion of the follicular contents after the rupture, or better, opening of the apical wall, (3) the related vascular phenomena causing minor fluctuations in intra-ovarian pressure and (4) the collapse of the follicle and its transformation into corpus luteum (DiDio et al. 1980). Martin and Talbot (1981 a) have recently suggested that the formation of a constriction at the base of the follicle, and possibly the increase in height of pre-ovulatory follicles, and the collapse of the follicle wall after rupture, are probably due to contraction of smooth-muscle cells in the theca externa in the basal hemisphere of the hamster follicle. Talbot and Chacon (1982) have suggested that in-vitro ovulation of hamster oocytes depends on contractions of follicular smooth-muscle.

C. Enzymatic Digestion and Morphological Alterations

Espey (1974, 1978 a) has critically reviewed the previous literature about the ovarian proteolytic enzymes and ovulation, as have Strickland and Beers (1979). Many proteolytic enzymes, such as collagenase, elastase and plasmin, can increase follicle wall distensibility in vitro (Espey 1967, Lipner 1973, Espey 1974, 1978 a, Beers 1975). Lipner (1973), Espey (1967) and Guraya (1974a, 1979c) have also integrated and discussed various observations obtained with the techniques of electron microscopy, histochemistry, and biochemistry for providing a deeper insight into ovulation. The various observations discussed in these reviews have shown that, prior to ovulation, there occurs a progressive dissociation and decomposition of various cellular layers surrounding the apex of the pre-ovulatory follicle (Fig. 50), which is brought about by hydrolytic enzymes especially proteolytic (see also Espey 1978 a, Strickland and Beers 1979, Espey et al. 1981). The basal follicle wall is also subjected to processes similar to those affecting the appearance of the apical wall. This process of decomposition is accompanied primarily by a gradual alteration of the intercellular ground substance of the connective tissue with a corresponding dissociation of the fibrillar and cellular components. The superficial (or surface) epithelium, tunica albuginea, theca externa, theca interna and granulosa cells in areas surrounding the apex of the pre-ovulatory follicle and fibrocytes present in the cortical areas of the follicle wall undergo a progressive alteration and possibly degeneration of some elements prior to ovulation (Espey 1967, 1974, Byskov 1969, Motta et al. 1971, Parr 1974c, Okamura et al. 1980, Yajima 1980, Volkova et al. 1980a, Mori and Uchida 1981, Espey et al. 1981, Martin and Miller-Walkner 1983). All these studies have clearly shown that a variety of proteolytic enzymes can weaken the tensile strength of the follicle wall. Nakajo et al. (1982) have also observed that the proteolytic enzymes, such as protease, collagenase and trypsin, are effective in reducing the tensile strength of quail ovarian follicle wall, while hyaluronidase is ineffective. The results of recent studies using collagenase inhibitors have further extended and confirmed the conclusion that proteolysis by collagenases is indispensable for the ovulatory process (Ichikawa et al. 1983a, b).

The question arises as to why the follicle ruptures at the apex. The wall of the pre-ovulatory follicle is not even in thickness, as the basal wall opposite the apex is much thicker and is supported by the surrounding ovarian tissues, whereas the wall near the surface of the ovary becomes thinner and less supported by the surrounding tissues (Fig. 50a, b). Shortly before rupture the stigma is formed at the apex, which becomes the thinnest region of the follicle wall due to the fact that the internal wall stresses are maximal at this surface (Fig. 50) (Espey 1967, Guraya 1971b, 1979c, Lipner 1973). The apex, which has less tissue support, will then rupture first even though enzymatic degradation of the follicle wall may be generalized.

Various transmission electron microscope observations on ovulation and the mechanism of follicle rupture in the rabbit ovary have confirmed these light microscopic data by revealing the subcellular alterations in more details (Espey 1967, Bjërsing and Cajander 1974a, b, c, d, e, f, 1975a, b, Cajander and Bjërsing 1973, 1975, 1976, Cajander 1976, Volkova et al. 1980, Espey et al. 1981, see other references in Motta and van Blerkom 1980). The dissociation and fragmentation of collagen fibres and cells, and the depolymerization of the intercellular ground substance in the connective tissue of the tunica albuginea and theca externa, are progressively increased by fluid infiltrating the perifollicular zones; this fluid has been observed to accumulate mostly under the surface of the apex of the pre-ovulatory follicle which constitutes the oedematous area (Guraya 1971b, 1979c, Motta et al. 1971, Cherney et al. 1975, Parr 1974c, Bjërsing and Cajander 1974a, b, c, d, e, f, 1975a, b, Cajander 1976, Cajander and Bjërsing 1973, 1975, 1976, Motta and van Blerkom 1975, 1980, Espey et al. 1981). Ultimately, this material penetrates and distends intercellular spaces between superficial cells.

The oedematous zones contain dead or degenerating cells, and the underlaying tunica albuginea shows degenerated fibroblasts and reveals dissociation and fragmentation of the collagen (Espey 1967, Bjërsing and Cajander 1974a, Espey et al. 1981). Prior to ovulation, there is seen increased permeability of perifollicular capillaries of the theca interna, as already discussed. These capillaries become fenestrated and meanwhile there occur changes in the tight junctions between endothelial cells (Anderson 1979, Okuda et al. 1980). The increased permeability of the perifollicular capillaries, just before ovulation, is due to intercellular gaps, as pinocytotic vesicles of perifollicular capillaries of the rabbit ovary do not show significant changes in size, number and distribution through all pre-ovulatory stages (Okuda et al. 1980). During the time of follicular expansion the basement membrane becomes discontinuous, the theca cells form cytoplasmic projections which extend into the granulosa layer (Byskov 1969, Lipner 1973), and degenerative changes become apparent in the theca cells (Moor et al. 1975).

The early leakage of erythrocytes from blood vessels and the common occurence of fibrin in the intercellular space and follicular fluid have shown that an acute inflammatory reaction may be involved in the formation and rupture of the stigma (Parr 1974c, Espey 1980, Espey et al. 1981, 1982), besides the weakening of the follicle wall by hydrolytic enzymes (Strickland and Beers 1979). This suggestion is supported by the fact that the administration of anti-inflammatory drugs before the expected time of ovulation inhibit ovulation (Espey et al. 1982).

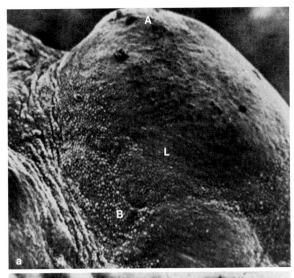

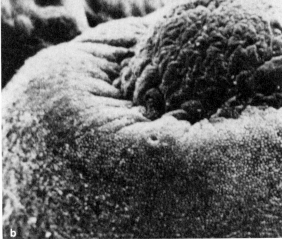

Fig. 51 a, b. Scanning electron micrographs of mature follicles. **a** Some pre-ovulatory follicles from the rat ovary (during the oestrous phase) bulge markedly from the surface of the ovary. Base of follicle *(B)*, lateral surface *(L)* and apex of pre-ovulatory follicle *(A)*; **b** Survey view showing part of a rabbit ovary with a follicle very close to ovulation (12 h after hCG): Freeze-dried (From Bjërsing and Cajander 1974b)

   The results obtained with scanning electron microscopy have further confirmed and extended these observations of progressive decomposition of various cells and tissues at the apex of pre-ovulatory follicles (Nilsson and Munshi 1973, Bjërsing and Cajander 1974b, Motta and van Blerkom 1975, 1980, Cajander and Bjërsing 1973, Cajander 1976, Tsujimoto et al. 1982, Martin and Miller-Walker 1983). The pre-ovulatory follicles, as studied with scanning electron microscopy, appear in the form of blister-like structure which protrude distinctly from the ovarian surface (Fig. 51). (Van Blerkom and Motta 1979, Motta and van Blerkom 1980). At their basal area they are covered with polyhedral cells containing numerous microvilli, whereas on their lateral surfaces superficial cells are elongated and show few microvilli. At the apex of the pre-ovulatory follicle, cells are very flattened and show few microvilli, which are developed only in regions of intercellular contact. The cells in some apical areas are seen in the process of their degeneration. Groups of cells have already "sloughed off", resulting in the exposure of underlying connective tissue of the tunica albuginea (Fig. 52). This particular zone corresponds to the "stigma" of pre-ovulatory follicles studied with light microscope, and due to a translucent appearance at low magnification represents a type of "lucent macula". The translucent appearance of this region of the follicular apex may be the result of reduced thickness of the follicular wall, lack of vascularization and distension by follicular fluid.
   Various scanning electron microscope studies have also shown that a fluid-like material exudes from the intercellular spaces of the superficial epithelium to cover some apical cells (Nilsson and Munshi 1973, Motta and van Blerkom 1975, 1980, van

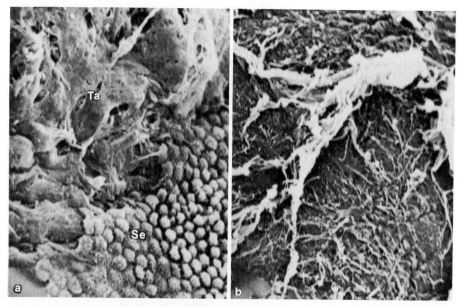

Fig. 52.a Scanning electron micrograph showing the disruption and "sloughing off" of the superficial epithelium *(Se)* in area near the apex of a pre-ovulatory follicle. **b** Its disruption and sloughing off are so great that the connective tissue which composes the tunica albuginea *(Ta)* is revealed (Rabbit ovary, 10 h post coitum) (From Motta and van Blerkom 1975)

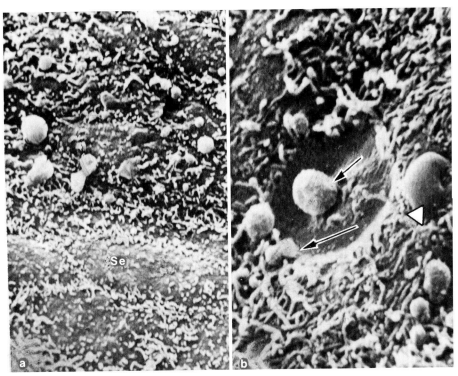

Fig. 53a, b. Scanning electron micrographs of the apex of some pre-ovulatory follicles. **a** Coagulated, fluid-like material is observed free *(arrows)* on the surface of the flattened, superficial epithelium *(Se)* from mouse (in oestrous phase) ovary. The epithelial cells at the apex of the pre-ovulatory follicle are quite flattened and possess few, relatively short microvilli; **b** Showing coagulated material in the form of spherical (or irregular) droplets *(white triangle)* that bulge from the cytoplasm of some cells. The *black arrows* point to the blebs that may be either free on the surface or bulging out from cytoplasm. (Rabbit ovary, 11 h post coitum) (From Motta and van Blerkom 1975)

Blerkom and Motta 1979). The fluid-like material is seen to (1) infiltrate the connective tissue of the tunica albuginea and theca externa, (2) accumulate under the basal lamina and (3) distend intercellular spaces of the superficial epithelium (Fig. 53) (Parr 1974c, Motta and van Blerkom 1975, 1980). The fluid-like material seen between superficial cells apparently mixes with and possibly dissociates collagen fibres. This fluid is also believed to be involved in the labilization of the ground substance of the apical wall, as well as of adjacent cortical areas of the perifollicular stroma. The infiltration of fluid between superficial cells may cause the disruption of intercellular contacts with a consequent change in cell shape and eventual sloughing-off of the superficial cells from the follicular apex. These morphological findings further support the suggestion that an increase in local accumulation of fluids (oedema) may be an important factor that facilitates the distension and final rupture of the weakened pre-ovulatory follicle. The accumulation of fluids (oedema) may depend upon the permeability or fragility of the follicular vessels (Cherney et al. 1975), which is known to increase markedly just prior to ovulation (Bjërsing 1978, Ellinwood et al. 1978, Gillet et al. 1980). The weakening of the follicular wall is more probably the result of lytic activity of enzymes (Espey 1978a). Yajima (1980) has recently observed that in the apical walls of human follicles, enzyme activities for the initial degradation of mature collagen fibres, such as vertebrate collagenase and cathepsin B, are increased towards the time of ovulation. Then these enzyme activities are decreased shortly before ovulation (see also Fukumoto et al. 1981). After collagen fibres are decomposed, the enzyme activities again increase. The collagenolytic peptidase activity increases at the time of ovulation after the above-mentioned decrease of vertebrate collagenase, indicating its lytic role for residual denatured collagen products derived from degradation of native collagens. Fukumoto et al. (1981) have observed that the activities of collagenolytic enzymes in the human follicular apex are slightly higher than those in the base throughout the ovarian cycle. Morales et al. (1983) have studied collagen, collagenase and collagenolytic activity in rat Graafian follicles during follicular growth and ovulation. The collagenolytic enzymes measured apparently are ovulatory enzymes.

Motta and van Blerkom (1975), using scanning electron microscopy, have observed that, just prior to ovulation, large irregular areas of the apex disintegrate and rupture and the oocyte–cumulus cell mass covered with a large amount of fluid oozes out of the gap (Fig. 54a) (van Blerkom and Motta 1979, Motta and van Blerkom 1980, Tsujimoto et al. 1982). At ovulation (Fig. 54b), the oocyte is not completely surrounded with granulosa cells and the zona pellucida is clearly seen. The surface of the zona is quite irregular and contains numerous infoldings, channels and crypts (Motta and van Blerkom 1980). Granulosa cells show polyhedral or star shapes, due to large cytoplasmic evaginations that obliquely penetrate the zone. Both the zona pellucida and corona radiata cells are covered with a fine layer of granular material.

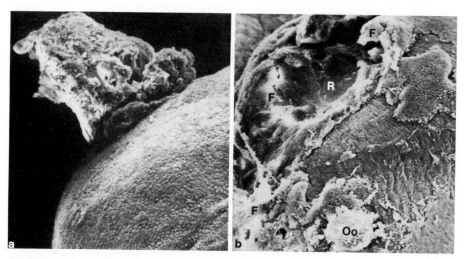

Fig. 54. a Scanning electron micrograph of a rabbit follicle in the process of ovum discharge (From Cajander 1976); **b** Scanning electron micrograph showing the ruptured area of a rabbit follicle just after ovulation *(R)*. Large numbers of follicle cells immersed in fluid *(F)* cover regions adjacent to the ruptured follicle *(R)*. The egg *(Oo)*, which has just escaped from the follicle and is covered by a mass of corona cells, is visible just under the ruptured area (Rabbit ovary, 9 h post coitum) (From Motta and van Blerkom 1975)

The details of alterations that occur in the cumulus oophorus as well as in relationship to the corona radiata cells and oocyte before ovulation are already discussed in Chap. IV.

The so-called abutment nexuses (gap junctions) of granulosa are decreased and continue to do so up to ovulation (Bjersing and Cajander 1974d, Coons and Espey 1977). Coons and Espey (1977) have observed that as a mature follicle approaches rupture, there occurs an appreciable increase in the number of surface nexuses per granulosa cell, as also discussed in Chap. III. The number of interiorized nexuses at this time is slightly decreased. This decrease in both surface nexuses and interiorized nexuses may be a consequence of ovulatory changes during which the rate of granulosa cell division is greater than the rate of formation of new nexuses. Additionally, the disruption of cell-to-cell cohesion during the ovulatory process appears to be independent of the interiorization of surface nexuses. The granulosa cell projections through the basement membrane are formed, and closer to the time of ovulation more and larger granulosa cell protrusions penetrate the partly fragmented basal lamina (Bjersing and Cajander 1974d, Abel et al. 1975). Cran et al. (1979) have also observed that at late oestrus the basal lamina becomes incomplete and thus the boundary of membrana granulosa with the theca interna becomes indistinct. The long basal cytoplasmic processes of granulosa cells expand and develop complex, folded, lateral cell surfaces. The folds interdigitate loosely with similar folds of adjacent cells and thus form broad, tortuous intercellular spaces (Abel et al. 1975). These ultrastructural changes, with more or less open channels into the antrum sometime before ovulation, are believed to permit rapid follicle growth by influx of fluid. At ovulation, the granulosa layer is absent from the stigma in the rat ovary (Espey 1967, Parr 1974c). Clumps of fibrin commonly occur in the extracellular space of the follicle wall and in the follicular fluid, as does debris from degenerating cells. Kobayashi et al. (1983b) have suggested that exclusion of Ca^{2+} and Mg^{2+} from the extracellular environment may be responsible for various structural alterations in the basement membrane of the follicle wall, including breakdown of the intercellular matrix.

The source of proteolytic enzymes necessary for various morphological alterations leading to apex breakdown is still controversial (Espey 1974, Strickland and Beers 1979). The several suggestions in this regard concern the surface epithelium, fibroblasts, theca cells and granulosa cells. By using transmission and scanning electron microscopy, Bjersing and Cajander (174a, b) have shown that at 4 h after injection of hCG the surface epithelial cells are considerably increased in size and many of them lose microvilli and show a progressive development of large, round dense cytoplasmic bodies, especially in the apical region of the pre-ovulatory follicle upto 8 h (Fig. 55). From then on the cytoplasmic bodies show a decrease in amount and, simultaneously, the cells over the apex develop protruding blebs and vesicles. Many of these large dense bodies or granules are believed to form an important source of lysosomal enzymes for causing the weakening of the follicle wall and rupture of the follicle apex in a fairly constant area, the stigma (Cajander and Bjersing 1975, 1976, Cajander 1976). This suggestion is supported by the fact that a few hours (2 h) before the follicle rupture the lysosomal granules decrease and their contents appear to be released towards the underlying tunica albuginea to disintegrate the follicle wall during the hour preceding ovulation (Cajander and Bjersing 1975). This course of events appears appropriate to affect follicle rupture, since the strongest barrier in the apical

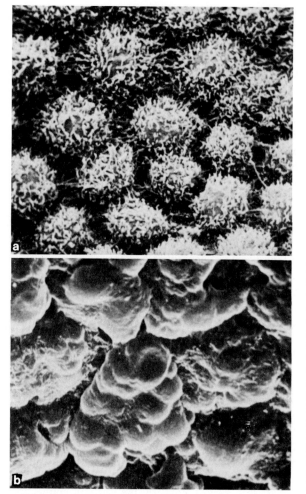

Fig. 55a, b. Scanning electron micrographs of "germinal" or surface epithelium of rabbit ovaries. **a** Normal surface epithelial cells before the ovulatory stimulation. The cells are approximately cubic with a convex surface covered by microvilli; **b** Changed surface epithelial cells on the apex of a follicle 10 h after i.v. injection of 25 i.u. hCG. The cells are considerably larger than earlier and they show only a few microvilli. Several protruding intracellular bodies can be seen bulging against the cell surface (From Bjërsing and Cajander 1975a)

follicle wall is undoubtedly the collagen-rich tunica albuginea. The structural changes of rabbit ovarian follicles, which occur after hCG stimulation, are similar to those occurring after natural ovulatory stimulation (Cajander 1976). From these various studies, Cajander and Bjërsing (1975) have put forward the hypothesis that the surface epithelium contributes lysosomal granules containing proteolytic enzymes which help to disintegrate the follicle apex prior to rupture. These data are not supported by the observations of Rawson and Espey (1977), as both the number and size of electron-dense granules in the surface epithelium covering rabbit ovarian follicles actually increase rather than decrease up to the time of ovulation. Actually, the number of granules does not begin to decrease until after ovulation. Average size of the individual epithelial cells may decrease slightly, rather than increase, at the time of ovulation. The majority of the granules that develop in the surface epithelium are also not lysosomes (see also Espey 1978a). From these data Rawson and Espey (1977) have denied an active role of the surface epithelium in the ovulatory decomposition of the follicle wall.

The pre-ovulatory increase in electron-dense granules in this layer appears to be a response to, rather than the cause of, the traumatic changes which occur in the deeper layers of the follicle wall during ovulation. According to S. S. Guraya (unpublished observations), the large, round dense cytoplasmic bodies of the surface epithelium that develop during ovulation in the rabbit are the highly sudanophilic lipid droplets consisting of triglycerides and phospholipids. These are formed at the apex of the pre-ovulatory follicle as a result of some degenerative changes in its surface epithelium, as such bodies which continue to be seen as large sudanophilic bodies even after ovulation, do not show any appreciable development in the surface epithelium of other regions of the ovary.

Espey (1969, 1971) has correlated the depletion of collagen content of the follicle wall at the apex with the formation and extrusion of many multivesiculated bodies in the theca externa and tunica albuginea, which become maximal about 2 h before ovulation (Fig. 56) (see also Espey et al. 1981). Okamura et al. (1980), studying ovulatory changes in the wall at the apex of the human Graafian follicle with electron microscope, have extended and confirmed these observations. Fibroblasts with little cytoplasm and abundant collagen fibres are present in the theca externa and tunica albuginea at the apex of growing follicles, as also observed by Martin and Miller-Walker (1983), who have developed a technique for the visualization of the three-

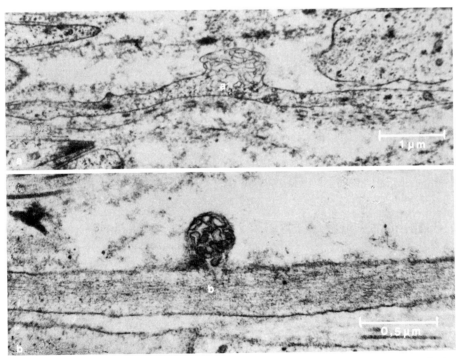

Fig. 56a, b. Electron micrographs of fibroblasts in the rabbit follicle wall showing multivesicular structure which protrude above their plasma membrane. Sometimes these structures appear in the vicinity of free ribosomes *(a)*, whereas other times there are no ribosomes apparent *(b)*. Note that one of the cytoplasmic processes *(b)* contains numerous microfilaments (From Espey 1978a)

dimensional distribution of collagen fibres over pre-ovulatory follicles in the hamster. In the mature human follicles, the fibroblastic cytoplasm is well developed, rich in lysosome-like granules, and shows peripheral multivesicular structures. Intercellular collagen fibres are sparse. The collagen fibres are evidently digested by the content of lysosomal granules and multivesicular bodies, thus aiding follicular rupture. These observations on the human follicles support the data of Espey and his co-workers (Espey et al. 1981) on the rabbit follicle. However, Espey (1967) has been unable to identify any conspicuous, abundant lysosomal granules in thecal fibroblasts. Dimino and Elfont (1980) have also discussed the role of lysosomes in ovulation.

The possibility of participation of granulosa cells and theca interna in the ovulatory process has also to be kept in mind, as they are also greatly altered, as already discussed. A conspicuous feature of follicle walls near the time of ovulation is the rounding up and detachment from the follicle wall of apparently healthy granulosa and theca interna cells (Parr 1974c, Bjërsing and Cajander 1974).

The concentration of β-galactosidase, which is a typical lysosomal hydrolase in rat ovarian bursa fluid at ovulation, has been observed to be seven and a half times greater than its concentration in plasma (Parr 1974a). Its accumulation in bursa fluid has further shown that lysosomal enzymes may be derived from the follicle wall during ovulation and washed into the bursa by the exudation of fluid through the surface of follicles. In a further study Parr (1974b) has shown the absence of neutral proteinase activity in the rat ovarian follicle walls at ovulation. This suggests that the rupture of the ovarian follicle at ovulation is not mediated by a proteolytic enzyme stored in the follicle wall tissue. The proteolytic activity of cathepsin at acid pH can be readily demonstrated and is believed to be the neutral proteinase of polymorphonuclear leukocytes.

The cause of the leukocytic invasion during the pre-ovulatory period is not known, but the consequences may be contributions of histamine, heparin and hyaluronidase to the follicular fluid and follicle wall, as suggested by Lipner (1973). Antihistamine has been shown to block ovulation in the rabbit (Knox et al. 1979), indicating that histamine is necessary for normal ovulation. The interaction between histamine, prostaglandins and other mediators, as well as the temporal sequence of that interactions, are necessary components for ovulation. But according to Espey (pers. commun.) antihistamines do not inhibit ovulation.

It is believed that ovulation requires the LH induction of steroid synthesis in the ovary and that progesterone is the steroid initiating the follicular changes through lytic enzymes (Rondell 1974). Plasmin is a lytic enzyme; therefore its presence in the follicle does not eliminate the possibility that the steroidogenesis is somehow involved in proteolysis. The generation of plasmin from plasminogen in response to LH can be considered an example of the way in which a proteolytic enzyme is produced exclusively in the pre-ovulatory follicle. Plasminogen is present in the follicular fluid of most follicles at levels similar to those in plasma, but plasmin is present only in the tissues of large pre-ovulatory follicles (Beers 1975, Strickland and Beers 1979). The plasminogen activator is believed to be produced by granulosa cells in response to LH and, since only the large nonatretic follicles have granulosa cells with receptors for LH (see Chap. III), these follicles are the only ones with the capacity to produce the activator. Fibroblasts are also well-known sources of this activator (see Espey 1980). The recent experiments reported by Strickland and Beers (1979) have shown that the production of plasminogen activator by granulosa cells is closely correlated with ovulation. FSH controls its levels, as its release by rat granulosa cells is highly specific for FSH activity (Martinat and Combarnous 1983). This hormone initiates the events (or generates the protease plasmin) that weaken the follicle wall. LH is believed to be involved in other functions important at the time of ovulation, such as preparing the cells for luteinizing or initiating the second meiotic division of the oocyte. This idea that FSH controls levels of plasminogen activator (or ovulation) is contrary to the fact that ovulation in mammals occurs as a result of LH surge, as already discussed. Therefore, further work is needed to determine the role of highly purified gonadotrophins in the production of plasminogen activator or other components by the granulosa cells of the follicle, as well as its exact function in ovulation (Espey and Rawson 1979); the source of plasminogen also needs to be determined. Wang (1983) suggests that LH-RH exerts a direct stimulatory effect on plasminogen activator secretion by rat granulosa cells, which may also explain the acute stimulating effects of LH-RH and its agonist on ovulation in hypophysectomized rats. Recent studies of Wang and Leung (1983) have demonstrated that FSH, but not LH, regulates plasminogen activation production by immature granulosa cells from pre-antral follicles. Pre-treatment of the undifferentiated granulosa cells with FSH, choleratoxin, or cAMP induces granulosa cells responsiveness to LH with increased plasminogen activation production. Granulosa cells obtained from pre-ovulatory follicles respond to both FSH and LH with increased plasminogen activation production. These results suggest that, with LH surge at ovulation, plasminogen activation production in follicles is increased and may be important in follicular rupture.

Extracellular plasminogen activator would yield increasing intrafollicular concentration of plasmin (Strickland and Beers 1979), a serine protease that can act on basement membrane and possibly other substrates to weaken the follicle wall (Hirschel and Hunter 1981). Hirschel and Hunter (1981) have shown that the potential protease (plasminogen) activity is significantly higher in bovine cystic follicular fluid than estrual follicular fluid, suggesting that the lytic system is there but is not being activated. The lytic system can be activated in vitro with urokinase. This suggests that plasminogen activator is either deficient or inhibited in the cystic follicle. Cystic follicular fluid contains more protease inhibitor, including an inhibitor for plasminogen activator, than estrual follicular fluid. These results have suggested that bovine cystic follicles lack sufficient activator necessary to convert plasminogen to plasmin. Without this protease, wich may be identical to the "collagenolytic enzyme" of Espey (1978a), ovulation cannot occur. However, Shimada et al. (1983) have suggested that plasminogen activator is not a primary proteolytic enzyme for follicular rupture. They suggest that possibly it may be playing roles in cumulus detachment and/or proliferation of granulosa cells during the ovulatory process.

D. Role of Hormones and Prostaglandins

The results of various studies as discussed here have shown that the collagen-fibroblast matrix of the follicle wall during ovulation is broken down by proteolytic enzymes which are believed to be produced by follicle cells in response to LH, progesterone and prostaglandin (Bjersing and Cajander 1974c, 1975a, Rondell 1974, Espey 1974, 1976).

Despite the variations in progesterone levels before rupture in vivo, a number of in vitro experiments have indicated a role for this steroid hormone in the ovulatory process (Rondell 1974). Baranczuk and Fainstat (1976) observed that in pre-ovulatory hamster ovaries follicle rupture only occurs after progesterone is added to the culture medium. Culture in the presence of progesterone also increases the proportion of rabbit follicles ovulating in vitro (Testart 1983). Inhibition in vitro of the synthesis of steroids suppresses ovulation, although meiotic maturation of the oocyte takes place. In pigs, progesterone enhances the distensibility of the follicle wall in vitro. A similar effect can be produced when the follicle wall is exposed to LH or cAMP. Inhibition of progesterone synthesis with cyanoketone (a 3β-HSDH inhibitor) blocks the action of LH or cAMP on the distensibility of the follicle wall (Lipner 1973, Rondell 1974). The stimulating effects of LH and progesterone on distensibility can be blocked by protein synthesis inhibitors (Rondell 1974). It has, therefore, been suggested that these hormones increase the activity of proteolytic enzymes which bring about the breakdown of the collagenous and mucoploysaccharide substances in follicular tissue, leading to an increased distensibility of the follicle wall, as already discussed (see also Rondell 1974).

Now considerable evidence has been produced to support a role for prostaglandins in the process of ovulation in the rat and rabbit (Fig. 57) (LeMaire and Marsh 1977,

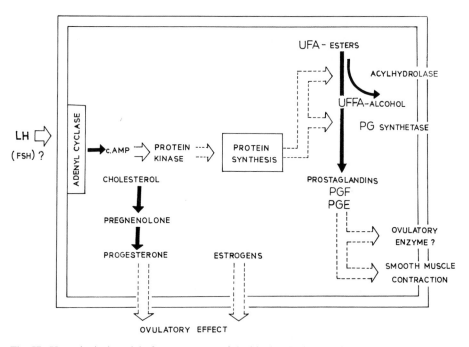

Fig. 57. Hypothetical model of some aspects of the biochemical events in the ovulatory process. *LH* luteinizing hormone; *FSH* follicle stimulating hormone; *cAMP* cyclic AMP; *UFA* unsaturated fatty acids; *UFFA* unsaturated free fatty acids; *alcohols* lysophospholipids cholesterol and diglycerides. *Solid arrows* indicate biochemical conversions. The *open and dashed arrows* indicate stimulation. The *dashed arrows* imply possible, but yet unproven actions (From LeMaire et al. 1979)

LeMaire et al. 1980, Espey 1980, Espey et al. 1981, Espey 1982a, b, Testart 1983, Thebault et al. 1983). $PGF_{2\alpha}$ is increased in the tissues of the pre-ovulatory follicle (Armstrong et al. 1974, LeMaire et al. 1975, 1980, Bauminger and Lindner 1975, see references in Dodson and Watson 1979, Shemesh 1979). Although the role of prostaglandins has not been as extensively studied as in the rat and rabbit, some evidence has been produced to show that prostaglandins are involved in ovulation in many other species (LeMaire et al. 1980). Inhibition of prostaglandin synthesis blocks ovulation in rat, mouse, rabbit and monkey (LeMaire and Marsh 1977, LeMaire et al. 1980, Testart 1983). Espey (1982a, b) has studied PGF production in rabbit ovarian follicles with several anti-inflammatory agents, and correlated their effects with that of indomethacin, a well-known inhibitor of prostaglandin synthesis and ovulation. Nonsteroidal anti-inflammatory agents, fenoprofen calcium and diclofenac sodium, like indomethacin, inhibit ovulation by limiting prostaglandin synthesis. The cytotoxic agents, colchicine and cycloheximide, which have a lesser effect on PGF synthesis, may inhibit ovulation by a mechanism different from the the nonsteroidal anti-inflammatory agents.

Pre-ovulatory increases in prostaglandins are described in swine (Ainsworth et al. 1975). Prostaglandins increase after the pre-ovulatory rise of LH or after exogenous injection of hCG (or LH). The tissue concentrations are highest at ovulation (LeMaire et al. 1973, 1980, Armstrong and Zamecnik 1975, Ainsworth et al. 1975). Espey (1982a) has suggested that prostraglandins need to be produced continuously in the follicle up to the time of actual rupture. Stimulation of ovarian prostaglandin synthesis by LH in vitro has been reported for follicular tissue from bovine (Shemesh and Hansel 1975) and human (Plunkett et al. 1975) origin. Dodson and Watson (1979), using pre-ovulatory pig follicles superfused in vitro with oestradiol, LH or FSH in the presence and absence of indomethacin, have observed that LH and FSH cause an increase in levels of prostaglandins E_2 and $F_{2\alpha}$ (approximately two fold), but the most marked effect is of oestradiol on PGE_2 where a four fold increase in levels above controls is seen. Shemesh (1979) has suggested the presence of a prostaglandin synthetase inibitor as well as a luteinization inhibitor in the bovine follicular fluid from mid-cycle follicles, which disappear in the pre-ovulatory follicle. In follicles that are not expected to ovulate the concentrations of PGs do not change (LeMaire et al. 1973, 1975, 1980). PGs are necessary for inducing the rupture of a follicle. Inhibition of PG synthesis by the systematic or local administration of indomethacin and aspirin, or neutralization of intrafollicular PG activity, blocks rupture (Lau et al. 1974, 1980, Behrman and Caldwell 1974, Hamada et al. 1977, LeMaire et al. 1980, Espey et al. 1981, Espey 1982a, b, Kiesling et al. 1983) bot does not affect LH-induced oocyte maturation and steroidogenesis (Grinwich et al. 1972, Yang et al. 1973, Tsafriri et al. 1973b, Armstrong et al. 1974, Armstrong and Zamecnik 1975, Kiesling et al. 1983). This block is not overcome by LH, but can be reversed by administration of exogenous prostaglandins about the time of ovulation (Armstrong et al. 1974, Bowring et al. 1975; see also Clark et al. 1978a, LeMaire et al. 1980, Wallach et al. 1980). Prostaglandins of the F (PGF) and E (PGE) series are increased in ovulated follicles, but not in follicles which fail to ovulate (see Clark et al. 1976a). These increases are not observed when indomethacin is injected systematically (Yang et al. 1973, Armstrong et al. 1974, LeMaire et al. 1980, Espey 1982a) or intrafollicularly (Armstrong et al. 1974, LeMaire et al. 1980). In addition, intrafollicular injection of a prostaglandin antibody prevents LH-induced ovulation only in the injected follicle (Armstrong et al. 1974). Antiserum is more effective to PGF than to PGE in these experiments in rabbits. The recent data of Espey (1982a, b), based on the effect of anti-inflammatory agents including indomethacin, have suggested the continued need of production of prostaglandins in the follicle up to the time of actual rupture, or else that indomethacin is interfering with some other aspect of the ovulatory process which transpires after the elevation of prostaglandins.

LH stimulation of synthesis of PG is mediated by cAMP and protein synthesis (Marsh et al. 1974, Clark et al. 1976, LeMaire et al. 1980). Graafian follicles isolated from oestrous rabbits and incubated for 5 h with LH (5 µg/ml) produce increased quantities of both PGE and PGF (Marsh et al. 1974). PGF synthesis in rabbit follicles is also stimulated by LH in an organ culture system (Moor er al. 1974). This effect is specific for LH and cannot be elicited in rabbit follicles by FSH or BSA (Marsh et al. 1974). It will be interesting to mention here that if FSH does not induce the essential prostaglandin synthesis then FSH and/or plasminogen activator (as suggested by Strickland and Beers 1979) cannot be important in ovulation.

The addition of cAMP can mimic the action of LH (Clark et al. 1978a, LeMaire et al. 1980). The mediation of LH stimulation of PGE synthesis by cAMP has been further investigated with isolated rat Graafian follicles. It is observed that ATP, ADP 3'-AMP, 5'-AMP, and cGMP do not stimulate PGE synthesis, whereas cAMP and 1-methyl-3-isobutylxanthine cause significant stimulations (Clark et al. 1978b). Derivatives of cAMP, such as dibutyryl (db) cAMP, N^6-monobutyryl (mb)-cAMP, and 8-bromo (Br)-cAMP, are also effective (Clark et al. 1978b). In contrast, it is observed that 8-Br-cGMP stimulates PGE production in rat follicles, although conclusive data on the effect of LH on follicular cGMP are not available (Zor et al. 1977). Cholera toxin is also observed to increase both cAMP and PGE in rat follicles (Clark et al. 1978b).

As the cAMP also stimulates steroidogenesis, it is possible that the action of LH on prostaglandins is mediated by this effect on steroidogenesis (LeMaire and Marsh 1977, LeMaire et al. 1979, 1980). This suggestion is disproven by the observation that the inhibition of steroidogenesis by aminoglutethimide does not inhibit the LH-induced rise in prostaglandins (Bauminger et al. 1975). A time lag of 3 h or more has been observed in vivo (Bauminger and Lindner 1975, Clark et al. 1978a) and in vitro (Clark et al. 1976) between exposure of follicles to LH and detectable increases in prostaglandin synthesis (see also LeMaire et al. 1980). This lag is apparently not due to a requirement of sufficient time for increased steroidogenesis to elapse, as just reported. From the increase in prostaglandins in rabbit follicles in response to cAMP (Marsh et al. 1974) it can be suggested that the time lag is due to a delay in production or action of cAMP by rat follicles (Nilsson et al. 1974). Instead, the delay appears to be in the expression of the action of cAMP, as exogenous cAMP, causes a delay similar to LH in incubated rat follicles (Clark et al. 1978b). A similar lag is also seen in the stimulation of PGE production by 8-Br-cGMP in rat follicles (Zor et al. 1977). The latent period, on the other hand, indicates that there is a need for macromolecular synthesis as a prerequisite to stimulated prostaglandin synthesis (Bauminger and Lindner 1975). Such a need is observed using isolated follicles from PMSG-treated immature rats as a model system. When Graafian follicles obtained from these rats are incubated with LH a marked increase in PGE synthesis is found after 5 h (Clark et al. 1978a). This increase is blocked with simultaneous incubation with 10 µM puromycin and not by 5 μM puromycin aminonucleoside (Clark et al. 1976, LeMaire et al. 1980). Actinomycin D (5 μM) and cycloheximide (10 μM) can also prevent the effect of LH (see Clark et al. 1978a). Clark et al. (1978a) have suggested that the step in LH action needing macromolecular synthesis may be beyond the production of cyclic nucleotides, as the effect of 8-Br-cAMP and 8-Br-cGMP (Zor et al. 1977) can also be blocked by cycloheximide.

The increase in follicular protein and RNA synthesis, which is observed to precede ovulation, appears to be induced by the pre-ovulatory gonadotrophin surge (Mills 1975). Furthermore, the direct intrafollicular administration of inhibitors of RNA and protein synthesis can block ovulation in the rabbit (Pool and Lipner 1966). The nature of the requirements for macromolecular synthesis needs to be determined more precisely (LeMaire et al. 1980).

LeMaire et al. (1979) have summarized the position of prostaglandins in the scheme of events leading to ovulation (Fig. 57). It is not assumed that all the events shown take place in a single cell type, although many of them can occur in isolated granulosa cells. The ovulatory LH surge, to begin with, acts upon a follicular cell by binding to its receptor, activating adenyl cyclase and inducing a rise in intracellular cAMP. In the rat, FSH may also play some role in this process (LeMaire et al. 1979). The steps following the rise in cAMP are not fully understood. They seem to involve protein kinase activation, protein synthesis, increased steroidogenesis and prostaglandin synthesis. Prostaglandin accumulation can be regulated by substrate availability, or by the amount of prostaglandin synthetase, or by alterations in prostaglandin conversion to inactive metabolites. A clear choice between these possibilities cannot be made at this time (LeMaire et al. 1979, 1980). Espey (1980), after discussing ovulation as an inflammatory reaction, has put forward a simple model to explain the basic reactions that occur during the ovulatory process (Fig. 58). This model provides a role for prostaglandin in ovulation by suggesting that this group of inflammatory mediators stimulates the proliferation of thecal fibroblasts and the production of proteolytic enzymes capable of disrupting the follicle wall during ovulation. This hypothesis also provides better assessment of effects of anti-inflammatory agents, along with antihistamines and immunosuppressive drugs, as potential antifertility agents. Kobayashi et al.(1983a) have studied the effect of histamine and histamine blockers on the ovulatory process in the in-vitro perfused rabbit ovary.

For the study of the role of prostaglandin synthesis in ovulation it is necessary to determine which cell type(s) in the follicle produces prostaglandins. The major source of PG in the pre-ovulatory follicle of the rabbit is believed to be the granulosa cells (Challis et al. 1974). The recent data also support the suggestion that granulosa cells are the source of prostaglandins (PGF) (Erickson et al. 1977). PGE is not measured in this study. However, in other studies, it is observed that the majority of the PGE is also produced by the granulosa cells (see Clark et al. 1978a). From these observations it can be stated that the granulosa cells are responsible for most or all of the production of PGE and PGF in the rabbit. Zor et al. (1983) have studied the hormonal regulation of PGE formation in isolated theca and granulosa cells of rat follicles. Exogenous db cAMP or cGMP stimulate PGE production in follicular theca or granulosa cells, but FSH is without any effect on the level of endogenous cGMP in both granulosa and theca cell types. Antibodies to FSH prevent the effect of FSH (but not of LH) on PGE formation by theca and granulosa cells, while antibodies to the β-subunit of LH block

SIMPLE MODEL OF THE OVULATORY PROCESS

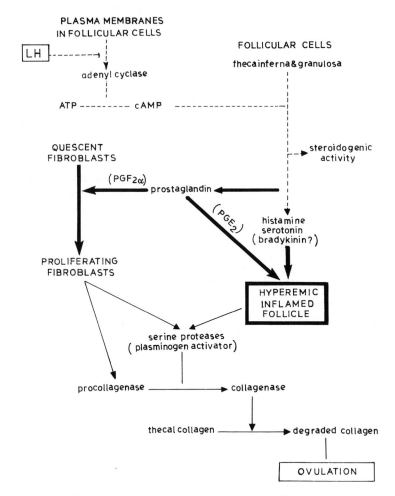

Fig. 58. Simple model of the ovulatory process. *Broken lines* indicate the reactions which take place during the first hours of the pre-ovulatory process; *thick, solid lines* indicate the hypothetical reactions during the intermediate stages of the process; *thin, solid lines* indicate the final reactions which culminate in ovulation (Redrawn from Espey 1980)

the effect of LH but not of FSH. This study has shown that highly purified FSH has a stimulatory effect on PGE formation by the follicular theca.

The mechanisms by which PG induces follicle rupture are not known. It has been suggested that PG mobilizes lysosomes to release enzymes to digest the follicle wall (Bjërsing and Cajander 1975a). Espey et al. (1981), after studying the effect of indomethacin on pre-ovulatory alterations in the ultrastructure of rabbit Graafian follicles, have also suggested that normal prostaglandin synthesis in ovulatory follicles may be important in the tissue decomposition and ultimate thinning of the follicle wall by stimulating the synthesis, release, and/or activation of ovarian proteolytic enzymes by the fibroblasts (see also Fig. 58). The results of ultrastructural study of pre-ovulatory apical development in mouse ovarian follicles are also consistent with the idea that prostaglandins are essential mediators of ovulation and they suggest that these prostaglandins augment apical rupture by mobilizing granulosa cells and stimulating the loss of connective tissue elements (Downs and Longo 1982, 1983). PGs may also act on smooth-muscle cells of the follicle at the time of rupture and thereby enhance ovum expulsion (Wallach et al. 1977, 1980, Amsterdam et al. 1977). But it is not clear whether ovarian smooth-muscle contractility is needed for follicle rupture and ovum expulsion or merely represents a phenomenon associated with ovulation. However, pros-taglandins appear to play a significant role in the process of ovulation. Although the precise function of these PGs in ovulation needs further clarification, they have been observed to affect follicle contractility. In earlier studies, it has been shown that PGE inhibits ovarian contractions and ovulation (Hamada et al. 1977, Lerner et al. 1978). The observations of Martin and Talbot (1981 b) also agree with these results and add to them by showing that PGE inhibits contraction of follicular smooth muscle cells in hamsters. Its mode of action on ovarian smooth muscle cells still needs to be determined. However, Espey et al. (1981) have suggested that prostaglandins mediate the final stages of the ovulatory process perhaps by mobilizing the fibroblasts, by stimulating proteolytic activity in the follicle, or by influencing both of these events. Shimada et al. (1983) have observed that prostaglandins are not involved in the pre-ovulatory synthesis of plasminogen activator induced by hCG in rat ovaries. Evans et al. (1983a) have studied the production of prostaglandins by porcine pre-ovulatory follicular tissues and their roles in intrafollicular function.

Because PGF_2 increases the contractility of follicular strips from the rabbit and human ovary, several workers have suggested that the pre-ovulatory rise in ovarian $PGF_{2\alpha}$ stimulates the contraction of smooth-muscle cells (Martin and Talbot 1981 b). Presumably, the contraction of smooth-muscle cells is affected either by alterations in the proportion of $PGF_{2\alpha}$ to PGE (Yang et al. 1973) or by differences in the distribution of these two prostaglandins within the follicle. Moreover, the contraction of follicular smooth-muscle cells needs an influx of Ca^{2+} and, most likely, specific concentration of PG in the surrounding medium. PG may influence contractile activity of ovarian smooth muscle by regulating intracellular calcium transport.

Evidence obtained using prostaglandin synthesis inhibitors, prostaglandin measure-ments, or prostaglandin antibodies in several species of mammals, as already discussed, support the concept that a gonadotrophin-induced pre-ovulatory increase in follicular prostaglandins is needed for ovulation (see review by Clark et al. 1978a, LeMaire et al. 1980, Wallach et al. 1980). However, several questions in this regard still remain unanswered for the better understanding of this requirement: (1) What is the mechanism by which gonadotrophin can induce prostaglandin synthesis? (2) In which cell type(s) are prostaglandins produced, and upon which do they act? (3) What are the actions of prostaglandins which play a role in ovulation? Some recent advances have partially answered the first two questions, as already discussed. Gonadotrophins seem to stimulate prostaglandin synthesis by increasing the quantities of follicular cAMP (or possibly cGMP). Derivatives of both the nucleotides have been shown to be effective exogenously. The effect of gonadotrophin on prostaglandin synthesis also shows a delay of several hours and needs macromolecule synthesis (LeMaire et al. 1980). In regard to the first portion of question two, the granulosa cells produce the major portion of prostaglandins. But it needs to be determined more precisely which cell(s) respond to the pre-ovulatory rise in follicular prostaglandins and how that response is related to ovulation.

Not all of the available observations, however, support a role for prostaglandins in ovulation. In the chicken, administration of indomethacin (in a dose sufficient to significantly inhibit PGF levels in in pre-ovulatory follicles) fail to block ovulation (Day and Nalbandov 1977). The administration of aspirin to women (Chaudhuri and Elder 1976) and active immunization against PGE_2 and $PGF_{2\alpha}$ in rats (Bauminger 1977) also do not block ovulation, although it is not determined in these studies if in fact follicular prostaglandin increases are inhibited. With these available exceptions, then prostaglandins are observed to play a significant role in the process of ovulation (Espey 1980, 1982a, b, LeMaire et al. 1980, Wallach et al. 1980). However, the function of prostaglandins in ovulation is still obscure and must await further definition of this process of ovulation on a morphological and biochemical basis.

Chapter VI
Luteinization and Steroidogenesis in the Follicle Wall During Pre-ovulatory and Ovulatory Periods

The growing ovarian follicle is the site of many important and complex phenomena during the pre-ovulatory and ovulatory periods, which lead to the processes of luteinization and steroidogenesis, ovum maturation and ovulation. The interaction between various components of the follicle is of great interest for revealing the molecular mechanisms involved in these processes. Here the morphological, histochemical and biochemical features of granulosa and theca interna cells will be discussed in relation to luteinization and steroidogenesis during the pre-ovulatory and ovulatory periods, especially after the pro-oestrous surge of gonadotrophins, as very divergent views have been expressed in this regard.

A. Granulosa Cells

The morphological or ultrastructural modifications typical of steroid-secreting cells (Christensen and Gillim 1969, Guraya 1974 b) do not occur in the granulosa cells prior to ovulation in humans (Guraya 1971 a, Mestwerdt et al. 1979), dog (Abel et al. 1975), sheep (McClellan et al. 1975), rhesus monkey (Amin et al. 1976), rabbit (Bjersing and Cajander 1974 d) and other mammals (Bjersing 1978). Mestwerdt et al. (1979) in their submicroscopic-morphometric studies on the granulosa cells from human pre-ovulatory follicles (LH < 5 ng/ml serum) have demonstrated the presence of large lipid droplets and the predominance of rough endoplasmic reticulum over the smooth reticulum. However, granulosa cells from pre-ovulatory follicles (serum LH levels rising or reaching a peak) show changes in fine structure, which are suggestive of transformation of granulosa cells active in protein synthesis into cells participating in steroid biosynthesis. The most important feature in this regard is the predominance of the smooth endoplasmic reticulum over the rough reticulum in the cytoplasm of granulosa cells. The lipid volume shows a three fold increase but the individual lipid droplets decrease slightly in diameter, a feature pointing to a progressive dispersion of lipid in small droplets. Some other workers have also reported signs of pre-ovulatory granulosa cell luteinization in monkey (Koering 1969), humans (Hertig 1967, Crisp et al. 1970, Delforge et al. 1972, Baird et al. 1975) and other mammalian species (Nicosia 1980 b). According to Dvořák and Tesařík (1980), the occurrence of structures of granular and smooth endoplasmic reticulum (ER), mitochondria with tubular cristae, the Golgi complex and lipid droplets, is typical of granulosa cells in the pre-ovulatory follicles in the human ovary (Fig. 59). This morphological picture together with the increase in the activity of enzymes involved in steroid metabolism (Bjersing 1977, Hoyer 1980) temporally correlates with the pre-ovulatory LH surge (Mestwerdt et al.

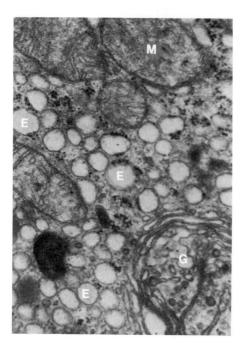

Fig. 59. Electron micrograph of granulosa cells with numerous vesicles of the smooth and granular endoplasmic reticulum *(E)*, the Golgi complex *(G)*, and mitochondria *(M)* (From Dvořák and Tesařík 1980)

1979). Bjërsing (1978) has also discussed the comparative aspects of ultrastructural changes that occur in the granulosa cells during the pre-ovulatory period of different mammalian species. The degree of pre-ovulatory luteinization differs. LH is the hormone which initiates luteinization of the granulosa cells. Abel et al. (1975) observed that after the LH peak, the granulosa cells in the dog enter a period of very active growth and differentiation, and by the end of the first day of oestrus the cell area is by more than a third greater than during the pro-oestrus. Actually, the role of LH, FSH, and other factors in luteinization is complex and still needs to be determined more precisely (see references in Bjërsing 1978). Luteinization is defined by morphological and functional parameters (Rondell 1974). There are two key features to the process: the mitochondria swell up, round off, and form tubulo-vesicular cristae; vesicles, at first granular but later on agranular endoplasmic reticulum, fuse into long anastomosing tubules, and stacked and whorled flattened sacs of agranular ER. Prior to ovulation a continuous change from granular to agranular ER in the granulosa cells generally occurs and whorls of ER develop (Bjërsing 1978). Before the LH peak the chromatin in the nuclei of the granulosa cells is irregular and contains several nucleoli; with the appearance of LH surge the chromatin is more homogeneous and shows a prominent netlike nucleolus. Besides these ultrastructural modifications of organelles, the lipid droplets and several enzyme activities indirectly related to steroidogenesis and 3β-HSDH are also developed at the same time. The functional luteinization concerns the elaboration and secretion of progesterone. Abel et al. (1975) observed that the Golgi apparatus on day 7 of pro-oestrus in the dog becomes larger and more prominent. It is also surrounded by small dense bodies that show acid phosphatase activity. The rise in progesterone levels in the fluid of Graafian follicles is believed to result from synthesis in the granulosa cell layer. Moreover, the oestradiol content in follicular fluid of pre-ovulatory Graafian follicles is increased three-to-four times, compared with the nonovulatory tertiary follicles. Elfont et al. (1977) observed that granulosa cells undergoing luteinization show more acid phosphatase reaction product, suggesting a role of the lysosomal system in steroidogenesis. PMSG-hCG treatment appears to stimulate the development of an extensive lysosomal system within the steroidogenic cells of ovarian follicles.

Prior to ovulation in human, sheep and dog, the ultrastructural and histochemical changes, which occur in the nucleus and cytoplasm of granulosa cells, are mainly indicative of RNA and protein synthesis, as already discussed in Chap. III (Guraya 1971 a, Abel et al. 1975, McClellan et al. 1975, Dvořák and Tesařík 1980). Their synthesis may be more important than steroid synthesis at this stage, especially in the proliferation and growth of mitochondria and granular ER, as the latter will give rise to agranular ER during luteinization of granulosa cells.

Mestwerdt et al. (1979) have made electron microscopic-morphometric studies of freshly ruptured and young corpora lutea in the human ovary. These studies revealed that the serum LH levels are decreased (< 25 ng and 10 ng/ml serum). The process involving conversion of granulosa cells engaged in protein synthesis to cells active in steroid biosynthesis (luteinization) has largely completed. The cellular and nuclear volumes are increased fourfold and twofold, respectively. The surface density of the smooth ER is increased even sixfold over that of granulosa cells from pre-ovulatory follicles during LH stimulation. The lipids are increased in volume, even though the percentage lipid content per cell remains unchanged as compared to that of granulosa cells from pre-ovulatory follicles. Lipid droplets also continue to decrease in diameter. This decrease in lipid pools appears to reflect requirement for the synthesis of the extensively developed membranes of smooth reticulum for progesterone synthesis. The increase in lipids of granulosa cells after LH stimulation is further supported by stimulation of phospholipid synthesis by LH in isolated rat granulosa cells. The selective effects of LH on phospholipid metabolism in granulosa cells may also be related to their steroidogenesis. According to Abel et al. (1975), the Golgi complex is the only organelle which assumes the features characteristic of mature luteal cells. Golgi elements associated with coated vesicles consist of four to seven stacked cisternae. Mestwerdt et al. (1977 b) have recently observed that a structural transformation process from protein synthetic active into steroid biosynthetic active cells occurs in the granulosa cells of the pre-ovulatory follicle of human ovary. Striking changes of paraplasma structures, especially of fat, are seen in various stages in the cytoplasm of its granulosa.

The increasing concentration of progesterone in serum during the pre-ovulatory period demonstrated in several species of mammals (see references in Murdoch and Dunn 1982) has been related to the luteinization of granulosa cells, which begin to develop the morphological features of steroid gland cells, as already discussed. The onset of granulosa cells luteinization also correlates with the increase in progesterone, 17α-hydroxy progesterone and 17β-estradiol levels in the follicular fluid (Edwards et al. 1977, Murdoch and Dunn 1982). This also correlates with the development of receptors for LH in the granulosa, as already discussed in Chap. III. Output of the steroids produced by the granulosa cells into plasma can occur only to a small extent, as the membrana granulosa is nonvascularized and steroids in the follicular fluid are released only slowly into the circulation (Edwards 1974). Several studies have shown that pro-oestrous follicles (mainly granulosa cells) in the rat are the source of pre-ovulatory progesterone increase at 1800 hours on the day of pro-oestrus (Noworyta and Szoltys 1975, Szoltys 1976). Stoklosowa and Szoltys (1978) have studied functional and morphological luteinization of granulosa cells in intact pre-ovulatory follicles and in cultured granulosa cells obtained from the pro-oestrous follicles of the rat. Follicles secrete oestrogens throughout pro-oestrus until 2000 hours. Progestin is increased to maximum levels at 1800 hours. At the same time meiotic divison in the oocytes of pre-ovulatory follicles is seen. At 2400 hours, the granulosa cells of the corona radiata and cumulus morphologically resemble luteal cells. The secretion of large amounts of progesterone before ovulation is preceded by increased Δ^5-3β HSDH activity stimulated by pre-ovulatory gonadotrophin release. This functional luteinization is, however, not expressed by the simultaneous morphological luteinization of granulosa cells, which only appears at 2400 hours. A similar morphological phenomenon has also been reported by Norman and Greenwald (1972) in hamster 2–3 h before ovulation. At that time, hardly any oestrogen is observed in the rat follicle (Stoklosowa and Szoltys 1978).

Klinken and Stevenson (1977) have studied changes in enzyme activities during the artificially gonadotrophin-stimulated transition from follicular to luteal cell types in rat ovary. The changes in enzyme activities, such as cholesterol side-chain cleavage activity, cytoplasmic $NADP^+$-dependent isocitric dehydrogenase, malic enzyme and glucose-6-phosphate dehydrogenase, are correlated with increases in ovarian content of DNA, cellular content of cAMP and the levels of plasma progesterone. The latter follows closely the development of 4-[^{14}C]-cholesterol side-chain cleavage, which is mimicked by the cytoplasmic isocitric dehydrogenase. There is no correlation between cAMP levels and cholesterol side-chain cleavage or progesterone plasma concentrations. Campbell et el. (1980) have demonstrated conspicuous biochemical and structural differences between porcine ovarian follicular and luteal mitochondria. These results have indicated that luteal mitochondria are more active in steroidogenesis and energy production than are follicular mitochondria, as supported by the fact that luteal mitochondria convert five times more 4-[^{14}C]-cholesterol into 4-[^{14}C]-pregnenolone than do follicular mitochondria. The higher levels of activity in luteal mitochondria appear to result, in part, from the presence of additional functional components, as evidenced by increases in mitochondrial size and amounts of cytochrome aa_3 and cytochrome P-450.

Follicular steroidogenesis and luteinization in response to LH is mediated by cAMP (Kolena and Channing 1972b, Marsh 1976, Channing and Tsafriri 1977). The tissue concentrations of cAMP in the pre-ovulatory follicles of the rat increase with the rise in LH concentrations in plasma (Nilsson et al. 1975, 1977a), which is in turn accompanied by an increased ability of the pre-ovulatory follicle to secrete progesterone. cAMP acts as an intracellular mediator of hormone action. But it is still to be determined more precisely whether it also plays an extracellular role in steroidogenesis or follicular rupture (Selstam 1975, Weiss et al. 1976). Marked differences exist between species in the tissue levels of progesterone before rupture. In some species, such as rat, rabbit and hamster, the follicular levels of progesterone peak several hours before and then fall to low levels at the time of rupture, whereas in others, such as monkey, pig and humans, the tissue levels of progesterone increase up to time of rupture. These functional differences also appear to correlate with those of morphological parameters, as the luteinization of granulosa cells in rodents and lagomorphs shows a precocious development, as already discussed. The regulation of inhibition of luteinization in the granulosa cells before pre-ovulatory LH surge is controversial. Actually, it is difficult to state whether luteinization before pre-ovulatory LH increases is blocked (1) by the inhibitory substance which Channing and Ledwitz-Rigby (1975) observed in the follicular fluid of small follicles in the pig, (2) by the inhibitory action of the oocyte itself (El-Fouly et al. 1970), or (3) by the granulosa exerting an inhibitory influence upon oocyte maturation (Foote and Thibault 1969). It is interesting that granulosa cells isolated from rat follicles and cultured separately seem to be affected by the time of their removal from the follicle (Stoklosowa and Szoltys 1978). Rat granulosa cells in vivo and in vitro show similar enzymatic patterns. There is also evidence that LH greatly stimulates 3β-HSDH activity when added to the medium of cultured granulosa cells harvested from follicles at 1100, 1400 and 1600 hours of the pro-oestrous phase (see Stoklosowa and Szoltys 1978). These times coincide with the pre-ovulatory gonadotrophin maximum in vivo.

Nalbandov and his co-workers (El-Fouly et al. 1970, Nekola and Nalbandov 1971, Stoklosowa and Nalbandov 1972) have explained the absence of luteinization in follicles having healthy and viable oocytes. However, oocytes in follicles studied do not show any maturation division. It is possible that at this stage the oocyte can inhibit luteinization, and that when it enters meiotic division it loses its inhibitory capacity. In contrast to this observation, Lieberman et al. (1976) have not observed any influence of ovectomy in rat on progesterone secretion in culture. But Stoklosowa and Szoltys (1978) believe that ovectomy of such small follicles can be injurious and cause atresia instead of luteinization of these follicles. It is difficult to account for this discrepancy with our present knowledge of luteinization. From the experiments of Stoklosowa and Szoltys (1978) it is clear that granulosa cells in pro-oestrus progressively mature hormonally and that LH stimulation causes them to luteinize functionally. Morphological luteinization appears to be the result of functional luteinization. Meanwhile, maturation processes continue in the oocyte whether these two processes are completely independent in vivo as Lieberman et al. (1976) have shown in vitro, needs to be further extended and confirmed.

B. Theca Interna Cells

The theca interna and its blood vessels are greatly altered during ovulation (Bjersing and Cajander 1974f, Abel et al. 1975, McClellan et al. 1975). Mestwerdt et al. (1977b) have observed that the hypertrophied theca interna cells, which form active steroid gland cells in the wall of the growing antral follicle, show a conspicuous size increase of their mitochondria in the pre-ovulatory and freshly ruptured follicles. They have suggested that there is apparently a close relationship between the transformation process in the granulosa and theca of the pre-ovulatory follicle and the increasing concentration of gonadotrophins.

After the injection of hCG, or with the secretion of LH, the theca interna cells in the follicle of rabbit and dog show an increase in the amount of smooth endoplasmic reticulum, lipid droplets and mitochondria with tubular cristae, which are indicative of increasing steroidogenesis (oestrogen) (Young Lai 1972, Bjërsing and Cajander 1974f, Abel et al. 1975). Both abutment and annular nexuses are developed in the theca interna. Its oedema is progressively increased, which is closely accompanied by the development of fenestrations and large gaps or perforations in the endothelium of the blood capillaries. Meanwhile, the surrounding basement membrane is fragmented and partly lost, so that a seemingly free passage from the capillary lumen to the interstitium is eventually established (Byskov 1969, Bjërsing and Cajander 1974f). Some hours before and up to ovulation the pericapillary lumen to the interstitium has also developed broad communication with the cavity of the follicles. Therefore, both pressure and fluid can be passed from the capillaries via the interstitium to the follicle antrum.

C. Shift in Steroid Production in the Ovulatory Follicle

Corresponding to various morphological, histochemical and biochemical changes in the granulosa and theca interna during the pre-ovulatory period, as discussed above and in Chap. III, there occurs a distinct shift in steroid hormone production by the dominant follicle(s) (or large ovulatory follicles) – from a state of oestrogen to progesterone dominance (see Saidapur and Greenwald 1979a, b, c, Terranova 1981a, b, Terranova et al. 1982, Murdoch and Dunn 1982). Blood levels of oestradiol are known to decrease conspicuously in a number of species at the time of or shortly after LH peak (see Murdoch and Dunn 1982), and meanwhile increasing concentrations of progesterone are seen. During the pre-ovulatory phase of the ovine oestrous cycle, and in association with the surge of LH, there is seen a distinct shift from oestradiol to progesterone by the dominant follicle (Murdoch and Dunn 1982). This transition is characterized by an intermediate period of relative steroidogenic quiescence. It will be interesting to mention here that data for venous oestradiol are in accordance with the contention that the pre-ovulatory follicle is the source of the peripheral peak in blood oestrogen which precedes the surge of LH in sheep (Moor et al. 1975). However, the rise in follicular progesterone seen just prior to ovulation is not represented by analogous changes within sera. Therefore, special care must be taken when formulating strict conclusions about the follicular secretion of steroid hormones based solely on the measurement of a blood parameter.

Several in vitro studies have also revealed that there occurs a shift from oestradiol to progesterone during the pre-ovulatory period in response to the LH/FSH surges in large ovulatory follicles of rodents, indicating the onset of follicular progesterone production following LH/FSH stimulation in vivo (see Terranova 1981a, b, Terranova et al. 1982). Both granulosa and theca produce significant amounts of progesterone in vitro. The failure of hamster granulosa cells to produce progesterone prior to the LH surge on pro-oestrus may represent a species-related phenomenon (Terranova et al. 1982), as rat granulosa cells produce progesterone in vitro at that time (Ahrén et al. 1979). However, Makris and Ryan (1977) have shown that hamster granulosa cells produce progesterone after 4–6 h in vitro, suggesting that longer incubations may be needed to observe an increase in granulosa cell progesterone. The increase in progesterone secretion by theca and granulosa cells of hamster follicles after LH/FSH surge coincides with a decrease in androgen production in vitro before or

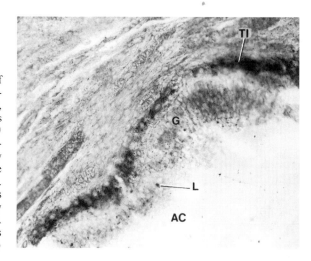

Fig. 60. Photomicrograph of portion of newly ruptured follicle from the human ovary, showing histochemical changes in the membrana granulosa *(G)* and theca interna *(TI)*. Granulosa cells have developed many fine lipid granules besides the large phospholipid bodies *(L)*. Disorganized theca interna cells *(TI)* are filled with highly sudanophilic lipid droplets. A portion of antral cavity *(AC)* is also seen (From Guraya 1968e)

after LH surge. The reason for the decrease in theca androgen production during the pre-ovulatory period may be due to the "inhibitory" action of LH on androgen production during this period (Armstrong and Dorrington 1977). Saidapur and Greenwald (1979c) have demonstrated the acute effects of progesterone and LH on in vivo and in vitro synthesis of oestradiol in the pro-oestrous hamster. The inhibitory effects of LH are most likely mediated through progesterone and/or other factors unknown at present.

In future studies, increasing attention should be paid to the interactions of steroid hormones in which rate-limiting enzymes in steroidogenesis may be directly affected. To cite one possible example: peak pre-ovulatory levels of oestrogen are followed by a drastic decrease as a result of LH surge, as already discussed (see Terranova et al. 1982). It is not known whether this is a direct effect of LH or whether the collapse in oestrogen is mediated by steroids produced as a result of LH activation. However, Guraya (1968e, f) has observed that theca interna cells of newly ruptured follicles are already filled with highly sudanophilic lipid droplets (Fig. 60) consisting mainly of triglycerides (neutral fats), and some cholesterol and/or its esters. The storage of cholesterol-containing lipids in the theca interna cells during ovulation has been attributed to their inactivity in the release of oestrogens, as a result of the disruption of theca interna cells that takes place during ovulation (Guraya 1971a). Thus, the drop in oestrogen production after the pre-ovulatory peak has been attributed to a sudden alteration in the metabolism of cholesterol-containing lipids in the theca interna cells, as they start to function in the storage of hormone precursor rather than in the secretion of oestrogen (Guraya 1971a). Species also seems to be the determining factor as to what the theca and the granulosa do. Therefore, more attention should be paid to species variations in regard to the functions of theca and granulosa. The variation in response among different species is so great that no general conclusions can as yet be formulated about the secretory functions of theca and granulosa of maturing follicle.

Chapter VII
Follicular Atresia

Follicular atresia in the ovaries of mammals is a widespread degenerative process by which the follicles lose their integrity and the major portion of oocytes at variable stages of their development, growth and differentiation, as well as during all stages of the ovarian cycle, is lost (Fig. 3) (Ingram 1962, Turnbull et al. 1977, Byskov 1978, Brand and de Jong 1973, van der Merwe 1979, Peters and McNatty 1980, Ryan 1981). Extensive loss of germ cells as a result of atresia occurs during the development and maturation of mammalian ovary (Guraya 1977a), during puberty, oestrous cycle and pregnancy (Kaur and Guraya 1985). The incidence of follicular atresia is constant within a species but varies between species (Ryan 1981). It occurs more frequently in the advanced stages of follicular growth, even though there are more follicles in the early stages of development in the ovary at any time (Himelstein-Braw et al. 1976, Richards 1980).

The ovarian follicle is a balanced physiological unit whose structure and function depend not only on extrafollicular factors, such as gonadotrophins, but also on a complex system of intrafollicular relationships (see Richards 1980, Nicosia 1980a). A close relationship between the granulosa and the theca layer is essential for normal steroidogenic function of the follicle (see Chap. III, Moor 1977, Richards 1980, Erickson 1982). During atresia, a sequence of degenerative alterations occurs that has a differential effect on the various follicular components of pre-antral and antral follicles (Fig. 61). Over the past 60 years or more numerous morphological investigations have been carried out on atretic follicles in a variety of species (see Ingram 1962, Jones 1970, Guraya 1973a, b, Greenwald 1974, Koering et al. 1982, Kaur and Guraya 1985). Some workers have laid emphasis on the alterations in the follicle wall, others on the fate of the oocyte. However, the majority of them agree that in antral (Graafian) follicles the earliest morphological signs of atresia appear in the granulosa cells, whereas the oocyte appears to be normal (Fig. 61 d, e). Ryan (1981) believes that the presence of pyknotic granulosa cells on the antral surface of the granulosa layer, floating in the fluid, is the most sensitive and reliable criterion (see also Koering et al. 1982). However, Koering et al. (1982) have observed that some pyknosis is normal even in the viable dominant follicle at cycle day 8 or 12 in monkeys. Still, we do not know whether there are one or several processes of atresia. But it is widely accepted that follicles may start to become atretic at any stage in their development (Figs. 3 and 61). However, in very few studies it is possible to determine at exactly what stage during the atretic process degeneration of the oocyte begins. It is also not known whether the atretic processes are the same in the pre-antral and antral follicles, or the prepubertal and adult animal. It is also not yet kwown if atresia starts with changes in the oocyte, granulosa cells or theca cells. The time span between the initiation of the atretic process and disappearance of the follicle

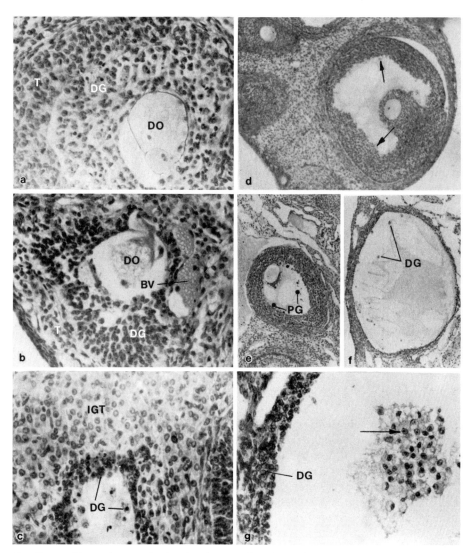

Fig. 61 a–g. Photomicrographs showing degenerative changes in granulosa *(DG)* during successive stages of atresia in pre-antral and antral follicles of the mole rat ovary. **a–c** Degenerative changes in granulosa *(DG)* and oocyte *(DO)* during successive stages of atresia in the large pre-antral follicle. Note the blood vessels *(BV)*. Finally, theca *(T)* forms interstitial gland tissue *(IGT)*; **d–g** Degenrative changes in the granulosa *(DG)* of atretic antral follicles. Atresia begins by the appearance of pyknotic nuclei in the granulosa *(arrows)*. Pyknotic granules *(PG)* are released as particles of varying sizes. As atresia progresses, the granulosa becomes more and more thin, the degenerating granulosa cells *(DG)* swell to approximately twice their original size, assuming the form of signet-ring cells (marked by *arrows*) (P. Kaur and S. S. Guraya, unpublished)

from the ovary still needs to be determined more precisely. This may be different for primordial and for more developed follicles.

The various factors which cause and control follicular atresia are controversial. There may be multiple mechanisms of atresia (Farookhi 1981). Actually, the cause and timing of follicular atresia are still poorly understood. But it is well known that different hormonal treatments influence the rate at which the follicles become atretic (Byskov 1979, Richards 1980, Farookhi 1981, other references in Schwartz and Hunzicker-Dunn 1981). Our meagre knowledge of the physiology of follicular atresia is due to the fact that at the present time it is not possible to recognize the incipient stages of atresia. For example, it is not possible to distinguish morphologically between "healthy-looking" follicles, which are destined to undergo atresia, and "healthy" ones, which will terminate their growth phase by ovulating the egg (Byskov 1979). Morphological (including ultrastructural), histochemical and biochemical alterations of follicular atresia have been worked out extensively during recent years (Guraya 1966b, 1973a, b, Hay and Cran 1978, Byskov 1979, Richards 1980, Farookhi 1981, Schwartz and Hunzicker-Dunn 1981, Kaur and Guraya 1985). The physiological and biochemical events that characterize follicular atresia are still poorly understood and the lack of information makes it difficult to formulate one unifying concept. The recent results obtained with cytological, histochemical and biochemical techniques will be summarized and integrated here in order to gain a deeper insight into the nature and functional significance of follicular atresia. Factors influencing follicular atresia will also be described.

A. Morphology, Histochemistry, Biochemistry and Physiology

Very conspicuous morphological, histochemical and biochemical changes occur in the various components of follicle as atresia progresses. These will be discussed here in relation to primordial, primary, secondary, and tertiary follicles, which have been identified in several previous studies (Mossman and Duke 1973, Peters and McNatty 1980). Since atresia is so common in the mammalian ovary, it is extremely difficult to determine whether structural changes of oocytes and their follicle (or granulosa) cells are normal or degenerative (Adams et al. 1966, Zamboni 1972, Hay et al. 1976, Byskov 1979), thus making comparisons difficult. Green and Zuckerman (1951) stated that the criteria of atresia are difficult to define and vary according to the subjective judgement of the worker at the time of studying the cells. Wilkinson et al. (1979) have described some morphological criteria that are of value in distinguishing healthy from atretic murine ovarian follicles in vivo, as summarized (Table 3) by Ryan (1981) who has also given a summary of histochemical changes associated with atresia (Table 4).

1. Primordial and Primary Follicles

The rate of atresia affecting primordial and primary follicles varies in the ovaries of different species and strains of mammals (Jones and Krohn 1961a, Guraya 1966a, 1967a, 1968a, b, Kolpovskii 1978, Peters and McNatty 1980, Kaur and Guraya 1985). Of the small follicles 50% are lost in the mouse ovary during the first month of life, while only 8% start to develop during this time (Faddy et al. 1976). Very little

Table 3. Histologic criteria of atresia (From Ryan 1981)

1. *Granulosa cells*

    1. Free-floating cells with pyknotic nuclei
    2. Absence of mitosis
    3. Chromatolysis and pyknosis of nuclei
    4. Sudanophilia
    5. Free-floating cumulus
    6. Invasion of zona pellucida
    7. Hyalinization of membrana granulosa
    8. Invasion of connective tissue cells
    9. Hypertrophy and luteinization of theca
    10. Gelatinous and cloudy follicular fluid
    11. Haemorrhage
    12. Cyst formation

2. *Oocyte*

    1. Pyknosis of nucleus
    2. Fragmentation of nucleus
    3. Breakdown of nuclear membrane
    4. Mitochondrial aggregation
    5. Shrinkage and hyalinization of zona
    6. Cytoplasmic crystalloid bodies

Table 4. Histochemical changes associated with atresia (From Ryan 1981)

1. Decreased [^{32}P] uptake
2. Decreased uptake of [^{3}H] thymidine
3. Appearance of Feulgen positive globules
4. Decreased [^{35}S] uptake
5. Increased alkaline phosphatase reaction
6. Appearance of ascorbic acid, lipid droplets and esterase activity
7. Appearance of glycogen in the cumulus oophorus and oocyte
8. Increased activity of acid phosphatase and amino peptidase

information is available about the characteristics of their atresia (Byskov 1979), which can be better studied in frozen gelatin sections showing little shrinkage and distortion. The electron microscope studies have also revealed that the nuclear and ooplasmic components are condensed, causing an appreciable increase in electron density (Odor and Blandau 1972, Nicosia 1980a). The morphological and histochemical characteristics of atretic primordial and primary follicles vary greatly in different species of mammals. The main characteristics by which the atretic primordial oocytes can be distinguished from normal ones are the shrinkage and the crinkled appearance of the nuclear envelope (Ingram 1962, Kaur and Guraya 1985). A secondary and less readily defined criterion is the development of unevenly distributed chromatin material which becomes pyknotic or fragmentary. Recent studies have shown that normal follicular growth as well as atresia of growing antral follicles depend on biochemical and physiological interaction between the theca layer, granulosa and oocyte (Guraya 1971a, Richards 1980, Erickson 1982). But our knowledge is very meagre about the factors which cause the degeneration of small follicles. Their degeneration can be increased greatly by irradiating the animal with small doses of X-ray (Peters and McNatty 1980). A single dose of the carcinogen 9, 10 dimethyl-1, 2 benzanthracene (DMBA) given to mice causes an increase in the number of follicles which degenerate (Krarup 1969, 1970). The rate of degeneration of small follicles can be altered by hypophysectomy in the adult mouse, which significantly retards but does not completely prevent small follicle atresia (Peters and McNatty 1980).

a) Oocyte

The earliest degenerative alteration of the oocyte revealed in vitro is the development of membrane-bound bodies (or lysosomes) containing electron-dense material that varies in structure (Odor and Blandau 1972). Both under in vivo and in vitro conditions, there are mitochondrial changes, and vacuoles and myelin figures are apparent in degenerating small oocytes (Beltermann 1965, Stegner 1967, Vazquez-Nin and Sotelo 1967, Merker and Zimmermann 1970, Zamboni 1972, Odor and Blandau 1972, Nicosia 1980a). According to Byskov and Rasmussen (1973), necrosis of the small oocyte is accompanied by the presence of a dense cytoplasm filled with granular material, free ribosomes, a swollen endoplasmic reticulum and an irregular nucleus. Actually, the various nuclear and ooplasmic contents are greatly changed ultra-structurally during atresia. The number of organelles showing abnormal distribution starts decreasing gradually and meanwhile lipid droplets develop. Histochemical studies have also revealed that the atresia of primordial and primary follicles in the ovaries of rhesus monkey leads to the accumulation of sudanophilic lipids in the oocyte and in granulosa cells. These lipids consist of triglycerides and phospholipids (Guraya 1966a). Such lipids are not present to a comparable degree in atretic follicles of bat, marmoset and cow (Guraya 1967a, 1968a, b).

b) Granulosa Cells

The elimination of small follicles occurs mainly by lysis or phagocytosis of the pyknotic oocyte (Gondos 1969). But the granulosa cells of some atretic primordial and primary follicles continue to persist in the ovaries of some mammals, whereas their oocytes degenerate and disappear completely (Figs. 62, 63a, b) (Guraya 1968b, 1973a, Kaur and Guraya 1985). Odor and Blandau (1972) also observed that the granulosa cells in culture show far fewer degenerative alterations than do oocytes (see also Nicosia 1980a). The persisting granulosa cells of these atretic primordial and primary follicles, which appear to represent the anovular follicles of several previous studies (Brambell 1956), form epithelial cords of variable size (Fig. 62), depending upon the number of granulosa cells present before the start of atresia (Guraya 1968b, 1973b). Epithelial cords are invariably present in the ovarian cortex and may gradually shift towards the corticomedullary parts of the ovary in different mammals such as the bat, rhesus monkey, buffalo, cow, etc. (Guraya 1966c, 1967a, 1968a, 1979a). However, such a shifting of cords has not been observed in the marmoset ovary, as they continue to lie in the broad ovarian cortex or the tunica albuginea (Fig. 62) (Guraya 1973b). As the cords are moved towards the medullary parts of the ovaries in rhesus monkey and cow

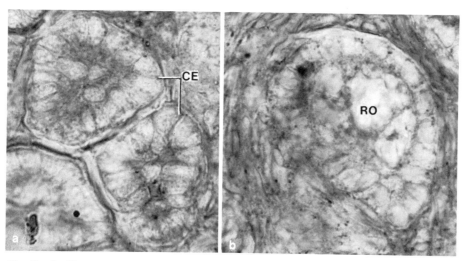

Fig. 62a, b. Photomicrographs showing epithelial cords *(CE)* surrounded by basal lamina, and regressing oocyte *(RO)* in the cortex of marmoset ovary. After the regressing oocyte *(RO)* disappears, the granulosa (or follicle) cells are left behind to form the epithelial cords

(Guraya 1966a, 1968a), they are generally vascularized and meanwhile develop the cytological and histochemical characteristics of steroid-secreting cells. These consist of the presence of diffuse sudanophilic lipoproteins and deeply sudanophilic lipid droplets (Guraya 1968a, b); the diffuse lipoproteins are presumed to derive from the abundantly developed membranes of smooth reticulum of steroid gland cells (Guraya 1971a). These histochemical or chemical changes in the epithelial cells of cords are strongly opposed to the idea of neoformation of oocytes from them in adult mammals (Guraya 1968b). It will be very interesting to determine the details of ultrastructural alterations in the oocyte and its granulosa in the atretic primordial and primary follicle in vivo, as well as in the epithelial cords, for very little work has been carried out along these lines (Byskov 1979, Nicosia 1980a).

2. Secondary and Tertiary Follicles

By employing routine histological techniques, the various morphological changes, which occur in different components of degenerating secondary and tertiary follicles, have been described in detail in several papers and reviews (Figs. 61 and 63) (Brambell 1956, Ingram 1962, Kolpovskii 1978, Wilkinson et al. 1979, Peters and McNatty 1980, Kaur and Guraya 1985). Wilkinson et al. (1979) have determined in vivo identification and structure of atretic murine ovarian follicles. Here more emphasis will be placed on histochemical, biochemical and ultrastructural changes that occur in the granulosa, theca and oocyte of atretic secondary and tertiary follicles (see also Byskov 1979, Nicosia 1980a).

a) Granulosa

Atresia of secondary and tertiary follicles is characterized by necrotic granulosa cells (Figs. 61 and 63) (Kaur and Guraya 1985). Recent SEM studies have also demonstrated extensive dissociation and fragmentation of granulosa cells in more advanced stages of follicular atresia in the rat ovary (Apkarian and Curtis 1981). The healthy follicles are without pyknotic granulosa cells (Ryan 1981). Byskov (1974) has distinguished three stages of atresia in large follicles in the mouse ovary: stage I atresia,

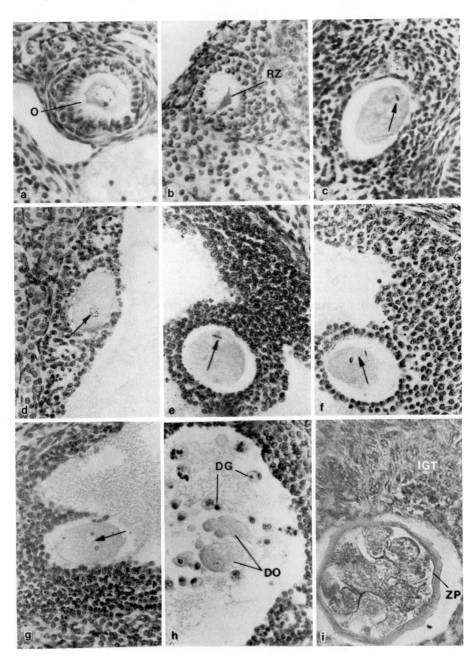

in which mitosis as well as pyknotic nuclei are present within the granulosa layer (see also Adams et al. 1966); stage II atresia, in which few cells are proliferating with many pyknotic granulosa cells; and stage III atresia, which signifies a collapsed nongrowing follicle (see also Peters and McNatty 1980). Similar stages have also been distinguished by Kaur and Guraya (1985) in the ovaries of the Indian mole rat (Figs. 61 and 63).

Nicosia (1980 a) has discussed the morphological alterations that occur during the atresia of follicles during in vitro conditions. Atretic follicles can be recognized by a "dark" central core composed of lipid-laden cells, cell debris and occasionally by peculiar calcific aggregates. Wilkinson et al. (1979) have observed that follicles in the healthy "class" appear lucent and spherical in outline, which is in contrast to follicles in the atretic class which are more opaque, pink in colour, and often more oval.

Peluso et al. (1977a) have studied surface ultrastructural changes in granulosa cells of atretic follicles. In nonatretic antral follicles of the rat, adjacent granulosa cells are in close contact, spherical in shape, and possess irregular cytoplasmic evaginations. With the start of atresia, the intercellular space between the granulosa cells is increased and these cells become flattened and lose their cytoplasmic evaginations, as also reported for the mouse by Byskov (1979). According to Wilkinson et al. (1979), the granulosa cells of atretic follicles are degenerating and characteristically have pyknotic nuclei and dense-staining ribosome-like particles filling the cytoplasm. Additionally, these degenerating granulosa cells have spherical intracellular structures that appear to be remnants of or are derived from rough endoplasmic reticulum. These spherical bodies are frequently located close to the plasma membrane, often appearing to have fused with the plasmalemma and to have released their contents into the intracellullar space. The granulosa cells in healthy follicles are isomorphous and show none of the alterations associated with degenerating granulosa cells. Uilenbroek et al. (1981) have observed that on day 3, after two injections of sodium pentobarbital, one at pro-oestrus and one at oestrus, pyknotic cells lining the antral cavity of rat follicle are seen, while one day later the thickness of the granulosa wall is decreased and oocytes of all atretic follicles show resumption of meiosis. This has suggested that blockage of ovulation with sodium pentobarbital results in atresia of pre-ovulatory follicles, which follows a constant pattern of morphological changes.

Van der Merwe (1979) studied the growth and atresia of follicles in the Natal clinging bat *(Miniopterus schreibersi natalensis)*. Basically two types of regression occur; generally destruction of the stratum granulosum takes place before the oocyte is

---

Fig. 63a–i. Photomicrographs showing changes in oocyte nucleus during successive stages of atresia in pre-antral and antral follicles of the mole rat. **a** Single-layered atretic follicle with shrunken oocyte *(O)*; **b** Swollen nucleus in the end-stage of atresia in small follicle having healthy granulosa cells and remains of regressing zona pellucida *(RZ)*; **c** Metaphase of first meiotic division *(arrow)* in atretic large pre-antral follicle; **d–f** Note resumption of first meiotic division *(arrows)* in oocyte nucleus of atretic antral follicles; **g** Formation of three pseudonuclei *(arrow)*, each with one distinct nucleolus; **h** Fragments of the degenerated oocyte *(DO)* are clearly visible among the degenerating granulosa cells *(DG)* in the antral cavity (P. Kaur and S. S. Guraya, unpublished); **i** Portion of atretic follicle (late stage in the small Indian mongoose ovary) showing many fragments of the degenerated oocyte, surrounded by zona pellucida *(ZP)*, which form a morula-like structure. Relatively undifferentiated interstitial gland tissue *(IGT)* of thecal origin is also seen

affected. The reverse condition is also seen, but is very uncommon. Corpora lutea atretica resulting from the luteinization of medium-sized Graafian follicles are commonly seen. A correlative histological, histochemical and ultrastructural study of the structural changes that characterize primary, secondary and tertiary atresia in sheep Graafian follicles has revealed that in primary atresia vacuoles representing swollen endoplasmic reticulum are conspicuously developed along the antral border together with disorganized granulosa cells containing pyknotic nuclei, which form DNA-positive masses (Hay et al. 1976, Hay and Cran 1978). Even in follicles identified as nonatretic, a few antral vacuoles and occasional pyknotic nuclei are seen. During secondary atresia, pyknotic nuclei are fused to form the characteristic Feulgen-positive atretic bodies of variable size. A second area of degeneration is frequently seen in the membrana granulosa, two or three cell layers from the basal lamina, and it is at this level that exfoliation of granulosa cells occurs in tertiary atresia, leading to a greater disorganization of membrana granulosa. The basal lamina remains intact. In addition, a few granulosa cells remain which, as judged by their ultrastructural appearance, are viable. Such cells may be seen in close apposition to areas of gross degeneration. This suggests that all cells in the granulosa are not equally affected by the atretic process. If such follicles are cultured, luteinization of granulosa appears to start within 24–36 h (Hay and Moor 1978). It seems clear that the source of the granulosa cells that luteinize in the atretic follicles in culture are those granulosa cells that escape the damaging effects of the atretic process.

Individual granulosa cells in the basal layer of sheep follicle often show numerous, large, osmiophilic membrane-bound bodies (Hay and Cran 1978). Such cells are especially frequent in the cumulus, although they may occasionally be seen in the granulosa. The presence of such bodies is apparently related to follicle and cumulus size. They are seen throughout the cycle in the larger follicles, most of which are destined to become atretic, as well as in the pre-ovulatory follicle. The nature of these organelles is not known. Histochemical tests do not demonstrate any acid phosphatase or aryl sulphatase activity, and thus there is no positive evidence to reveal that they have any clearly defined lysosomal role. However, the frequency with which they are seen does suggest that they may have an important function. Other workers have also shown that degenerative changes associated with early atresia of ovine follicles is limited to the membrana granulosa directly adjacent to the antral cavity, whereas widespread degeneration of granulosa cells together with disappearance of the cumulus-oocyte unit characterizes secondary atresia, and decreased thecal integrity is observed in tertiary atresia (Brand and de Jong 1973, Turnbull et al. 1977). These cellular alterations are accompanied by a change in blood-flow within the thecal microcirculation (Hay et al. 1976) and a decrease in the steroidogenic capacity of the follicle, especially its ability to synthesize oestrogen (Moor et al. 1978). Moor et al. (1978) have postulated that limited oestrogen synthesis is a predisposing factor in atresia (see Richards 1980).

Electron microscope studies have revealed that granulosa cell junctions, both of "gap" and intermediate type (described in Chap. III) of cultured follicles are moderately decreased in number after 48–72 h, and markedly after 120–144 h (see Nicosia 1980a). Disappearance of gap junction has also been seen in vivo; and this is believed to occur through their internalization and disposal within autophagic vacuoles. Since follicular growth is very likely dependent on metabolic and electric coordination of granulosa cell function, as already discussed in Chap. III, a decrease in the number or presence of gap junctions may accelerate follicular atresia.

In contrast to the membrana granulosa, there are during secondary atresia only slight indications of degenerations in the cumulus oophorus of cow (Guraya 1968a), sheep (Hay et al. 1976, Hay and Cran 1978), mouse (Byskov 1979), and Indian mole rat (Kaur and Guraya 1985). Even during tertiary atresia, the cumulus in some follicles shows no sign of necrosis (Hay and Cran 1978), in others there are degenerate cells with pyknotic nuclei within the cumulus mass. As atresia advances, the cumulus ultimately breaks up and the oocyte with a few cumulus cells attached to it is set free within the antral cavity. Hay et al. (1976) observed that, despite some disintegration of the cumulus, the integrity of the oocyte is maintained in the sheep and its nucleus remains vesicular (see also Hay and Cran 1978), as also observed for the cow (S. S. Guraya, unpublished observations).

With the advancement of atresia, the antrum is first filled with a fine web of connective tissue, then it collapses and the follicle contracts (Figs. 61 and 63). Such changes have also been observed during the atresia of follicles occurring in vitro (Nicosia 1980a). Dekel et al. (1977) observed that 79% of 298 recovered oocytes during late human pregnancy and 78% of their cumuli are degenerative. Degeneration is correlated to the appearance of phagocytes in the follicular fluid. During atresia of Graafian follicles, the physico-chemical characteristics of the follicular fluid are greatly altered both qualitatively and quantitatively (see Ryan 1981). Familiari et al. (1981) studied follicular fluid in healthy and atretic mouse ovarian follicles by using electron microscope localization of lanthanum nitrate and polycation ruthenium red. Fol-

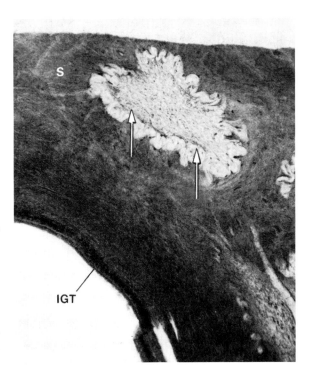

Fig. 64. Photomicrograph of histochemical preparation of an ovary from a nonpregnant woman, showing stroma and relatively undifferentiated theca-type interstitial gland tissue *(IGT)* surrounding the follicular cavity of an atretic follicle. Clear scars *(arrows)* left after the degeneration of follicles are seen (From Guraya 1978)

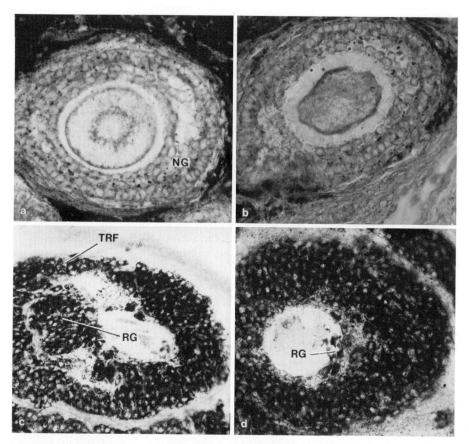

Fig. 65 a–d. Photomicrographs of normal and atretic pre-antral follicles from the rat ovary showing lipid changes in granulosa and theca during degeneration. **a** Normal pre-antral follicle of large size showing heterogeneous lipid bodies in the normal granulosa *(NG)*; **b** Large pre-antral follicle showing the start of atresia. Besides the heterogeneous lipid bodies, some small lipid granules have started appearing in the granulosa; **c** Atretic pre-antral follicle showing an accumulation of sudanophilic lipid droplets in the regressing granulosa *(RG)* and hypertrophied theca *(TRF)*; **d** Late stage of atresia in the pre-antral follicle, showing some remnants of regressing granulosa *(RG)* adjacent to the degenerating oocyte. Whole patch is constituted of hypertrophied theca (or interstitial gland tissue) filled with highly sudanophilic lipids (From Guraya and Greenwald 1964 b)

licular fluid shows a dense globular aspect in growing follicles but becomes more dispersed and globular-fibrillar in atretic follicles. It is formed by rounded negatively charged particles (20 nm in diameter) that are closely packed in growing follicles, but spaced and united by filaments in atretic follicles. These alterations may be related to a changed role of granulosa cells and also to an altered passage of nutritive substances from the plasma into the follicle.

In some follicles the granulosa cells disappear but the follicular fluid-filled cavity continues to be present and this is called cystic atresia (Watzka 1957). It is not known how long these fluid-filled bodies persist in the ovary. In late atresia, after most

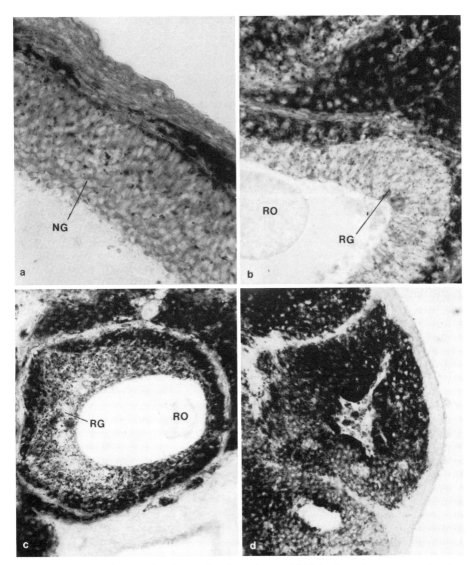

Fig. 66 a–d. Photomicrographs of normal and atretic antral follicles from the rat ovary, showing lipid changes in the granulosa and theca during degeneration. **a** Normal antral follicle showing heterogeneous lipid bodies in the normal granulosa *(NG)*; **b** Atretic antral follicle showing the appearance of some sudanophilic lipid droplets in the regressing granulosa *(RG)*; regressing oocyte *(RO)* is also visible. Note the disappearance of heterogeneous lipid bodies which were present in normal granulosa; **c** Atretic antral follicle showing further increase in the amount of sudanophilic lipid droplets in the regressing granulosa *(RG)*. Theca, which also contains lipid droplets, forms a distinct band around the regressing granulosa. Regressing oocyte *(RO)* showing no lipids is also seen; **d** Very late stage of atresia in the antral follicle. Regressing granulosa has nearly disappeared and its place has been occupied by the hypertrophied theca (or interstitial gland tissue) filled with sudanophilic lipid droplets (From Guraya and Greenwald 1964 b)

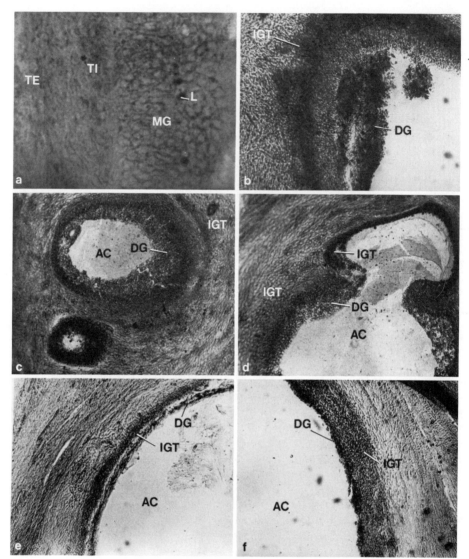

Fig. 67a–f. Photomicrographs showing morphological and sudanophilic lipid changes in the membrana granulosa *(MG)*, theca interna *(TI)* and theca externa *(TE)* during degeneration of antral follicles from the nonpregnant buffalo ovary. **a** Appearance of some diffusely distributed sudanophilic lipids in the membrana granulosa *(MG)* and thecal layer *(TI and TE)* with the initiation of atresia in antral follicle; mambrana granulosa shows heterogeneous lipid bodies *(L)* usually seen in normal follicles. Morphologically, membrana granulosa and thecal layers are intact; **b** Portion of antral follicle in its late stage of atresia illustrating accumulation of sudanophilic lipids in the relatively undifferentiated interstitial gland tissue *(IGT)* of thecal origin, and degenerating granulosa cells *(DG)*; **c** Accumulation of sudanophilic lipids in the degenerating granulosa *(DG)* and relatively undifferentiated theca-type interstitial gland tissue *(IGT)* of small atretic antral follicle with cavity *(AC)*, as well as in the granulosa of pre-antral atretic follicle; **d–f** Portions of atretic Graafian follicles with antral cavities *(AC)* showing variable amounts of sudanophilic lipids in the relatively undifferentiated interstitial gland tissue *(IGT)* of thecal origin, and disintegrated and degenerated granulosa *(DG)*.

components of the follicle have disappeared, there may be left behind a cord of connective tissue (glassy membrane), which is seen in the area of the inner surface of the theca; its folded, wavy body indicates the zone in which the antral follicle once developed (Fig. 64).

Conspicuous changes occur in the amount and nature of sudanophilic lipids in the granulosa of atretic follicles in different species of mammals (Figs. 65, 66 and 67) (Guraya and Greenwald 1964a, b, 1965, Guraya 1966a, b, 1967a, b, 1978a, c, d, Byskov 1979, Parshad, V. R., and Guraya 1983, 1984). Some electron microscope studies have also demonstrated the accumulation of lipid droplets in the degenerating granulosa cells (Beltermann 1965, Stegner 1967, Byskov 1979). Degenerative changes (mostly autophagocytosis and accumulation of lipid droplets) have also been observed by electron microscopic examination of granulosa cells of cultured follicles (Nicosia 1980a). The lipid droplets in the granulosa of atretic pre-antral and antral follicles in mammals (Figs. 65, 66 and 67) are composed of triglycerides, cholesterol and/or its esters and some phospholipids. Actually, atresia leads to excessive storage of neutral fats (triglycerides) and to a decrease in the amount of phospholipids. However, several species variations in regard to the amount of lipids stored in the degenrating granulosae are noticed. The amount of lipids stored also appear to vary greatly with the stage of follicular development as well as in different parts of the same follicle (Guraya 1979a, Parshad, V. R., and Guraya 1983, 1984). The accumulation of lipid droplets in the incipient stages of atresia in the mouse ovary are first seen in the granulosa cells close to the basal lamina (Byskov 1979). The granulosa cells lying below the basal lamina in pre-ovulatory Graafian follicles of rats treated with pentobarbitone also develop relatively more cholesterol-containing lipids which are lacking in the cumulus oophorus (Parshad, V. R., and Guraya 1984). But staining uniformly for diffusely distributed phospholipids and lipoproteins is increased throughout the granulosa. In healthy large follicles very little lipid is present (Parshad, V. R., and Guraya 1983, 1984). Byskov (1979) demonstrated some alterations in the lipid droplets of atretic follicles after treatment of the mice with PMSG. The granulosa cells of atretic secondary and small tertiary follicles store relatively more sudanophilic lipid droplets than the large tertiary follicles in the hamster, rat, etc. (Guraya 1971a, 1973a, b). It is interesting that the granulosa cells of atretic follicles in the nonpregnant civet cat *(Paradoxirus hermaphroditus)* gradually regress and disappear without storing sudanophilic lipid droplets (Fig. 68) (Guraya 1981).

There also occurs a variety of other histochemical alterations associated with atresia (see Table 4). A decrease of radioactive phosphorus – [^{32}P] – uptake has been observed during follicular atresia (Jacoby 1962, Ingram 1962).

The decreased uptake of [^{32}P] and [^{3}H] thymidine appears to be related to the cessation of mitosis in atretic follicles, whereas the development of Feulgen-positive globules suggests lysis of nuclei (Ryan 1981). These observations indicate that cessation of DNA synthesis and granulosa cell division may be early events in the atretic process. The decreased [^{35}S] uptake may be indicative of decreased synthesis of sulphated mucopolysaccharides or increased sulphatase activity resulting in a faster [^{35}S]O$_4$ turnover. Other metabolic activities also occur with the initiation of atresia, as evidenced by a lower incorporation of [^{3}H]-proline in atretic follicles than in healthy follicles (Byskov 1979). In autoradiograms of mouse ovaries obtained 1 h after injection of [^{3}H]-proline, the grain number over granulosa cells of healthy follicles is

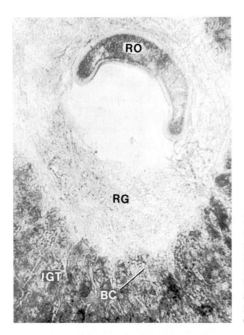

Fig. 68. Photomicrograph showing very little accumulation of sudanophilic lipids in the regressing granulosa *(RG)* of degenerated follicle in the nonpregnant civet cat ovary. Regressing oocyte *(RO)* is also seen. Interstitial gland tissue *(IGT)* of thecal origin having abundant blood vascularity *(BC)* shows diffusely distributed sudanophilic lipids (From Guraya 1981)

three times higher than that over granulosa cells from atretic stage I follicles. The acid phosphatase-positive cells limited to the junction of granulosa and thecal layers in guinea-pig ovary are believed to be the macrophages (Adams et al. 1966).

In parallel with the accumulation of lipids and decrease of [^{32}P], [^{3}H] thymidine and [^{35}S] uptake in the atretic follicles, granulosa cells of rat ovary develop a marked activity of two hydrolytic enzymes [acid phosphatase and amino peptidase (L-leucyl naphthalamidase)] which form the characteristic components of lysosomal enzymes (Deane 1952, Lobel et al. 1961, Parshad, V. R., and Guraya 1984). These enzyme activities often precede the appearance of morphologic manifestations of atresia (Ryan 1981). Elfont et al. (1977) also observed that the major site of the acid phosphatase reaction in the ovaries of untreated immature rats is in autophagic vacuoles associated with numerous atretic follicles. Little reaction product is seen in cells of intact preantral follicles. In older cultures of mouse ovaries, the majority of granulosa cells develop lysosomal-like bodies and a variable number of lipid droplets (Odor and Blandau 1972, Nicosia 1980a). Acid phosphatase and amino peptidase are lyosomal enzymes which are not seen in the granulosa of normal follicles (Parshad, V. R., and Guraya 1984). Ryan (1981) has assigned an early and perhaps causative role to increased lysosomal activity. However, hydrolytic enzymes indicative of lysosomal activity do not develop in atretic granulosa cells of the guinea-pig ovary, which also do not store many lipids (Adams et al. 1966, Guraya 1968c). Similarly, β-glucuronidase activity (of lysosomal origin) in the bovine follicle does not appear to be associated with atresia (Lobel and Levy 1968). The granulosa cells of atretic follicles in the guinea-pig ovary continue to show their nucleoside mucophosphatase (adenosine monophosphatase) activity until they are shed. But it does disappear in the sloughed granulosa

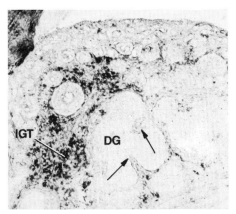

Fig. 69. Photomicrograph of portion of 5-day-old, postnatal guinea-pig ovary showing the invasion of degenerating granulosa *(DG)* of atretic follicle by the stromal tissue *(arrows)*. Interstitial gland tissue *(IGT)* is also seen. The ovarian cortex contains numerous primordial follicles and small pre-antral follicle

cells. Schmidtler (1980) has stated that atresia of follicles can be detected by enzyme histochemistry in the ginea-pig ovary. The enzymes studied by him include β-glucuronidase, acid phosphatase and nonspecific esterase. The latter is also increased in the granulosa of atretic Graafian follicles of pentobarbitone-treated rats (Parshad, V. R., and Guraya 1984).

Adams et al. (1966) observed an invasion of the granulosa and cumulus of apparently normal medium-sized follicles in the guinea-pig ovary by one or two long strands of connective tissue and a small blood-vessel from the theca interna. These invading cells, which can be easily distinguished due to their nucleoside polyphosphatase activity, are not the alkaline phosphatase-positive theca interna cells. It is not known whether this invasion is a cause of or a response to a subtle initiation of atresia. Such invasion also occurs in the 5-day postnatal ovary of the guinea-pig (Fig. 69) (Guraya 1977b) and adult buffalo ovary. In regard to the distribution of alkaline phosphatase activity and its change during atresia, there are some discrepancies, the reason for which is not known (see Ryan 1981).

According to Adams et al. (1966), additional evidence of atresia is provided by the presence of a faint generalized nucleoside phosphatase activity in the cytoplasm of the basal granulosa layers of follicles, which is absent in the corresponding site of normal follicles. NAD diaphorase also becomes intensely active in the granulosa of atretic follicles (Baillie et al. 1966). In the developing mouse ovary (Hart et al. 1966) between birth and 3 weeks, 3α-, 3β- and 16β-HSDH activities, which are indicative of steroid biosynthesis or metabolism, develop in virtually all follicles, and the large number of histochemically active follicles appears to be related to the high rate of follicular atresia. After 3 weeks, the atretic follicles develop extensive formazan distributed throughout the membrana granulosa with activity distributed around the periphery of the cells. Rubin et al. (1963) also claimed that 3β-HSDH activity is developed in the granulosa cells of atretic follicles of rat and mouse ovaries. The pentobarbitone treatment of rats also causes increase of 3β-HSDH and appearance of 17α-HSDH activities in the granulosa cells of atretic pre-ovulatory Graafian follicles in which ovulation is blocked (Parshad, V. R., and Guraya 1984). 3β-HSDH also develops in the atretic granulosa of the human ovary (Deane et al. 1962). Miyamoto et al. (1980)

observed a weak or very weak G-6-PDH activity in atretic follicles of the goat ovary, suggesting that they are not active in steroidogenesis. Succinate dehydrogenase (SDH) activity also becomes weaker in atretic follicles than in normal follicles. Atresia results in increased amounts of sudanophilic lipids in the granulosa cells and decreased activities of SDH and LDH, and NADH- and NADPH-tetrazolium reductases during later stages of degeneration in the ovaries of Indian gerbil, suggesting that atresia affects the mitochondrial functions and reoxidation of NADH and NADPH coenzymes (Parshad, V. R., and Guraya 1983). Weak activities of 3β- and 17α-HSDH seen in the granulosa of Graafian follicles also disappear during later stages of their atresia.

The development of cholesterol-containing lipid droplets and of some hydroxysteroid dehydrogenase activities in the granulosa cells of atretic tertiary follicles suggests that their granulosa cells undergo abortive luteinization in response to some luteinizing factors, which are present during atresia. This luteinization appears to be nonfunctional, as the granulosa cells of atretic follicles do not develop any appreciable blood vascularity for the transport of secretory products; furthermore, their lipids are not mobilized after strong gonadotrophin (PMSG) stimulation (Guraya 1972a). But the lipid droplets of their thecae (or interstitial gland tissue) are mobilized. Byskov (1979), however, observed that a characteristic effect of PMSG treatment 24 h previously on atretic follicles is a decrease in the cytoplasmic content of lipid droplets compared to controls. Lipid droplets appear to disperse within the cytoplasm of granulosa cells in stage III atresia. Recent studies have demonstrated that during the atretic process, granulosa cells are unable to produce significant quantities of steroid hormones (Terranova et al. 1982), thus supporting the suggestion that granulosa cells of atretic follicles develop abortive luteinization. Their capacity to bind [^{125}I]-labelled FSH, hCG and prolactin is greatly altered during atresia of follicles in the hamster ovary (Oxberry and Greenwald 1982). FSH, hCG and prolactin binding is always notably reduced or absent in granulosa of atretic follicles (see also Richards 1980, Shaha and Greenwald 1982). This shows that there occurs a decrease in gonadotrophin receptors with advancing atresia (Uilenbroek et al. 1980). In the hamster, cyclic nucleotides decrease with time in atretic follicles and this may be related to changes in the numbers or properties of gonadotrophin receptors (Hubbard and Greenwald 1981b). Collectively, various observations suggest that morphological, histochemical and biochemical changes are probably the major factors responsible for the rapid development of refractoriness to gonadotrophins. Granulosa cells in atretic follicles do not suddenly cease to function but instead gradually lose their functional synthetic capacity (Hubbard and Greenwald 1983). With progressive development of degenerative changes, the atretic follicles reaches a point where it can no longer respond at all and thus becomes functionally defunct.

Various other studies have revealed that probably one of the first events to occur in the degenerating follicle is the loss of its steroid and protein hormone receptors (Richards and Midgley 1976, Richards et al. 1978, 1979, Richards 1980). Receptors are essential to bind hormones to their target tissues (see Chap. III) and, therefore, the loss of receptors may interfere with progressive development of the follicle and its steroidogenesis, thus leading to atresia.

Glycogen is generally absent from normal granulosa cells but develops as an early manifestation of atresia in the cumulus oophorus of the atretic follicles in the rat

(Deane 1951, 1952). Ascorbic acid, lipid droplets and esterase activity also increase at the same time (Deane and Andrews 1953).

The granulosa cells of secondary follicles appear more resistant to events causing atresia than those of tertiary follicles, because in the latter case they show a more rapid regression and disintegration (Figs. 61 and 63) (Guraya and Greenwald 1964b, 1965, Byskov 1979), suggesting some differences at the molecular (or metabolic) level. Granulosa cells of atretic follicles usually undergo pycnosis or karyolysis and are simultaneously detached from the membrana granulosa (Figs. 61 and 63), as already discussed. They are removed by lysis and phagocytosis. In this way cellular debris accumulates in the follicular fluid. Invading leucocytes and macrophages participate in the removal of granulosa cell debris. The granulosa cells and their lipid accumulations and enzyme systems in atretic follicles of all sizes degenerate and disappear earlier than their oocytes. Other workers have also observed that there is no synchrony in the development of degenerative changes in the oocyte and granulosa cells (Beltermann 1965, Stegner 1967, Vazquez-Nin and Sotelo 1967, Odor and Blandau 1972). At later stages of in-vitro atresia, obliteration of the antrum is accomplished by shrinkage, disruption of the follicle's basement lamina and by penetrating connective fibres and theca cells (Nicosia 1980a). The similarity between in vivo and in vitro atretic follicles has also been documented at the ultrastructural level (Nicosia 1980a).

From the discussion of recent morphological, histochemical and biochemical data it can be concluded that granulosa cells of atretic secondary and tertiary follicles undergo conspicuous morphological (including ultrastructural), histochemical and biochemical alterations that vary with the stage of follicle development and mammalian species.

b) Theca

The degenerative changes in the granulosa cells of atretic follicles in the mammalian ovary are closely accompanied by specific modifications in their theca interna and surrounding stroma, which persist and gradually hypertrophy to form interstitial gland cells (Figs. 61, 65–68) (Guraya and Greenwald 1964a, b, 1965, Guraya 1966a, b, 1967b, 1968a, c, d, 1973c, 1978, Anderson and Meurling 1977, Byskov 1979, Hillier and Ross 1979, Richards 1980, Schwall and Erickson 1981). Indirect evidence for structural and functional preservation of theca cells during atresia is also provided by incorporation studies with the radio labelled DNA precursor, thymidine (Nicosia 1980a) by binding of [^{125}I]-labelled hCG (Shaha and Greenwald 1982). These studies have revealed that, in contrast to granulosa cells in which DNA synthesis decreases dramatically, DNA synthesis remains unaltered in theca cells during atresia. The functional preservation of theca interna cells during atresia of follicles is also supported by the fact that their various enzymes such as SDH, LDH, NADH- and NADPH-tetrazolium reductases and 3β- and 17α-HSDH are not affected significantly by atresia (Parshad, V. R., and Guraya 1983). Except for the appearance of weak activities of acid phosphatase and leucine aminopeptidase in some follicles of the rat ovary, the thecal layers also do not show marked histochemical change in the activities of 3β- and 17α-HSDH during atresia of the Graafian pre-ovulatory follicles (Parshad, V. R., and Guraya 1984); their atresia was induced after administration of pentobarbitone sodium which blocks ovulation in the rat. Various structural alterations suggest that granulosa cells of large healthy follicles may exert an inhibitory influence on theca cells, preventing hypertrophy. It is also possible that something released from the pyknotic granulosa cells may stimulate theca cell hypertrophy (Richards 1980). But these suggestions need to be confirmed by biochemical data. However, hypertrophy of theca cells is not observed consistently during atresia of small follicles, indicating that functional differences in theca may be involved.

The degree of hypertrophy of theca-type interstitial gland cells greatly varies in different mammalian species (Figs. 65–68). The hypertrophied thecae of atretic follicles ultimately constitute conspicuous masses of interstitial gland tissue in the ovarian stroma (Fig. 70), which meanwhile accumulate abundant diffuse lipoproteins, lipid droplets of variable size and chemical composition, intense alkaline phosphatase activity and various enzyme activities (Fig. 35c) related to steroid biosynthesis

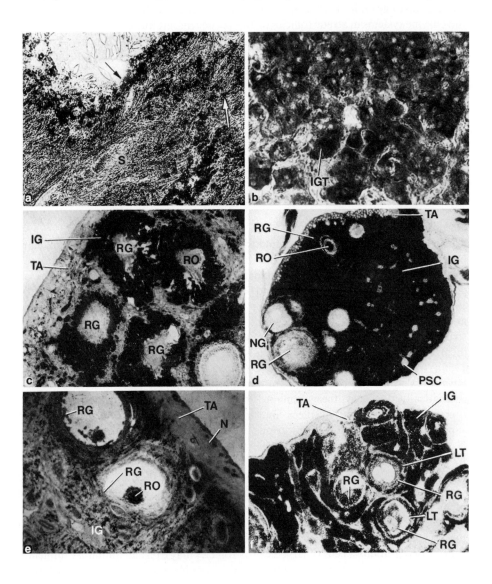

(Guraya 1971a, 1973a). Electron microscopic studies of rat atretic follicles have revealed the development of lipid droplets, agranular endoplasmic reticulum and mitochondria with tubular cristae during the transformation of theca cells into interstitial gland cells (Fig. 71) (Schwall and Erickson 1981). Theca cells, in contrast to granulosa cells of in vitro follicles , do not undergo regressive changes, except focally during late stages of culture (Nicosia 1980a). Instead, the theca layer of atretic follicles inceases in thickness due to cell hypertrophy and, very likely, to cell proliferation. In addition, theca cells develop focal cytologic changes suggestive of steroidogenesis (i.e. mitochondria and with tubulo-vesicular cristae).

The amount and nature of lipid droplets in the theca-type interstitial gland cells show conspicuous variations not only in different mammalian species but also with the ovarian cycle (Figs. 3, 71), suggesting variable metabolism in relation to steroid biosynthesis. Adams et al. (1966) have reported the presence of faint acid phosphatase activity in the perinuclear distribution during the transformation of thecal cells into interstitial cells. The thecae internae of large atretic tertiary follicles in the human ovary during nonpregnancy contribute relatively little interstitial gland tissue (Fig. 72) (Mossman et al. 1964, Guraya 1966b).

The role of the theca and of its vascularity and nerve supply in the initiation of follicular atresia is not known, though it has been suggested that an insufficient development of the theca leads to follicular degeneration. Volkova et al. (1980a) have demonstrated a definite correlation between the size of the follicles and their vascularization in the course of growth and atresia. A primary role for the microcirculatory bed in initiation of atresia is suggested. Hay et al. (1976) observed that changes in the thecal microcirculation appear to play a key role in atresia of sheep follicles: adjacent to the basal lamina of nonatretic follicles there is seen a well-

---

Fig. 70 a–f. Photomicrographs showing variations in the development and distribution of interstitial gland tissue of ovaries not only in different species of mammals but also under different physiological situations in the human. **a** Showing relatively very little development of lipid-containing interstitial gland tissue *(arrows)* of thecal origin in the stroma *(S)* of nonpregnant human ovary; **b** Portion of human ovary obtained during 39th week of pregnancy showing diffusely distributed lipids and some discrete lipid bodies in the greatly developed and hypertrophied interstitial gland tissue *(IGT)* of thecal origin; **c** Portion of oestrous ovary of cat showing patches of interstitial gland tissue *(IG)* derived by the hypertrophy of theca and possibly surrounding stromal tissue of atretic pre-antral follicles in which regressing granulosa *(RG)* and oocyte *(RO)* are still seen. Besides the patches, the sudanophilic interstitial tissue also occurs as singly scattered cells in the general stroma. Tunica albuginea *(TA)* is well developed (From Guraya and Greenwald 1964b); **d** Bat ovary showing most highly developed interstitial gland tissue *(IG)* filled with sudanophilic lipids, narrow tunica albuginea *(TA)* with numerous primordial follicles, regressing granulosa *(RG)* and oocyte *(RO)* of atretic pre-antral and antral follicles, and normal granulosa *(NG)* of pre-antral follicle. Some sparsely scattered patches of small epithelial cells *(PSC)*, which do not contain lipid droplets of the interstitial type, are present (From Guraya and Greenwald 1964b); **e** Portion of ovary of pseudopregnant dog, showing the formation of sparsely scattered interstitial tissue *(IG)* from the theca and surrounding stromal tissue of atretic pre-antral and antral follicles *(RG* regressing granulosa; *RO* regressing oocyte). Some cords or nodules *(N)* of epithelial cells are seen arising by the invagination of germinal epithelium (From Guraya and Greenwald 1964b); **f** Portion of rat ovary showing scattered patches of interstitial tissue *(IG)*, theca of regressing follicles *(TRF)*, and regressing granulosa of atretic pre-antral and antral follicles. Tunica albuginea *(TA)* forms a narrow zone. A portion of corpus luteum is also seen (From Guraya and Greenwald 1964b)

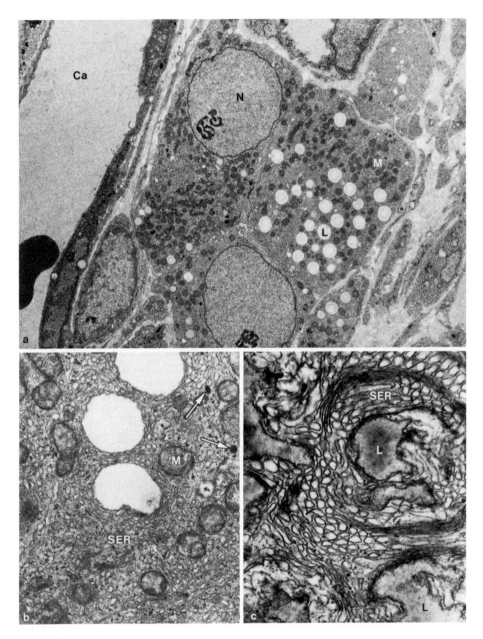

Fig. 71 a–c. Electron micrographs of interstitial gland cells of rabbit ovary. **a** Low-power view of fully developed interstitial cells obtained post coitum. Perivascular spaces are seen. Interstitial gland cells show spherical nucleus *(N)* with nucleolus and abundant cytoplasm which contains lipid droplets *(L)* and mitochondria *(M)*. *Ca* blood capillary; **b** Portion of fully developed interstitial gland cell, showing abundant smooth endoplasmic reticulum *(SER)*, mitochondria *(M)*, lipid droplets *(L)* and microperoxisomes *(arrowed)*; **c** Higher-power view of portion of fully developed interstitial cell, showing morphological association of lipid droplets *(L)* with smooth endoplasmic reticulum *(SER)* (From Guraya and Motta 1980)

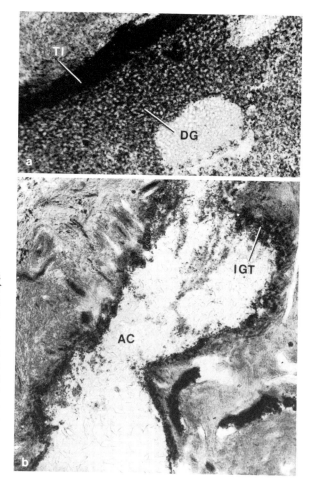

Fig. 72 a, b. Photomicrographs of histochemical preparations of portions of degenerating antral follicles with cavity *(AC)* from the nonpregnant human ovary. **a** Accumulation of abundant sudanophilic lipids in the theca interna *(TI)* and moderate amount of lipid droplets in the degenerating granulosa *(DG);* **b** Disappearance of degenerated granulosa cells, leaving behind relatively undifferentiated interstial gland tissue *(IGT)* filled with sudanophilic lipid droplets around the antral cavity *(AC)*

developed capillary network which is significantly decreased as atresia progresses (Hay and Cran 1978). This reduction in the size of the capillary bed adjacent to the membrana granulosa has been suggested to restrict the availability of nutrients to the granulosa cells, leading to atresia of the follicle. Byskov (1979) observed that as luteinization and degradation of the granulosa layer progress beyond stage III atresia, luteinized theca cells and blood-vessels of the theca layer mix with remnants of the luteinized granulosa layer.

Gimbo et al. (1980) studied the innervation of large atretic follicles (or atretic corpora lutea) in the camel ovary. In the early atretic corpora lutea, numerous neurofibrils from the theca folliculi penetrate the degenerating granulosa cells. These fibrils are associated with fibroblastic projections and are directed towards the central part where they form a network in the loose connective tissue. In the more developed atretic corpora lutea, the fibrils distribute themselves along the basement lamina in the recesses of the theca lutein cells (or theca-type interstitial gland cells) always forming a central mediastinal dense network. The areas occupied by the interstitial cells present a rich innervation. It is recommended that future studies should be directed to the better understanding of development and alterations in surrounding stroma including theca and its vascularity and innervations during follicular growth and atresia, which may help us to determine causative factors of atresia more precisely. The study of role of extra-ovarian and intra-ovarian factors in this regard will be very rewarding.

c) Oocyte and Zona Pellucida

The oocyte generally develops degenerative changes later than granulosa in secondary and tertiary follicles (Figs. 61 and 63). Brietenecker et al. (1978) observed that in both younger and older women one out of three oocytes is degenerated. Degenerated oocytes are seen in some nonovulatory follicles as well as in the follicles with pre-ovulatory changes in their walls. Degenerative changes in the oocytes of atretic follicles vary and depend on the stage of growth reached before the follicles degenerate. Sanyal (1979) classified human follicular oocytes in four different types, which were collected from different patients without ovarian pathology. These oocyte types include (1) oocytes with nuclei containing diffused or filamentous chromatin (type 1), (2) mature oocytes with metaphase chromosomes or first polar body (type 2), (3) oocytes with clumped chromatin within the nuclei (type 3), and (4) degenerated oocytes (type 4). More than 50% of the oocytes (types 3 and 4) formed the abnormal population of follicular oocytes. During primary and secondary atresia in the sheep (Hay and Cran 1978), the oocytes remain in the germinal vesicle stage and are in all respects structurally similar to oocytes in nonatretic follicles. When pyknotic nuclei first develop in the cumulus layer during tertiary atresia, the oocyte is generally in the germinal vesicle stage and processes from the surrounding cumulus cells clearly penetrate the zona (Hay et al. 1976). Oocytes, which are still in the growth phase, may show several kinds of degeneration: cells may penetrate the zona pellucida. In more advanced stages of degeneration many cumulus macrophages are seen within the zona pellucida (Himelstein-Braw et al. 1976).

   During atresia, the oocyte may shrink and become distorted (Fig. 63a, b). The nucleus may show necrotic changes; occasionally cleavage of the oocyte is observed. The reasons for these changes are not known. Byskov (1979) observed that in some follicles of the mouse ovary atresia appears first in the oocyte, which may shrink, become necrotic, or lose its connections with the cumulus cells. These events often occur simultaneously with or are followed by maturation division, pseudocleavage or fragmentation of the oocyte (Fig. 63c–i). Subsequently, phagocytosing cells penetrate the zona pellucida and remove the oocyte (Peters and McNatty 1980). Within a 24–48 h culture, oocytes begin to show clustering of mitochondria, organelle-free cortical areas, storage of lipid droplets, diffuse loss of junctional contacts with granulosa cells, interruption of the nuclear membrane, and occasional pronuclear formation and fragmentation (Nicosia 1980a). Actually, atresia which occurs in cultured follicles, occurs in oocytes prior to granulosa cells, as these last cells maintain as intact morphology upto days 3–4 of culture. According to Adams et al. (1966) the first sign of start of atresia in the oocyte of guinea-pig is the development of precocious eccentricity of the nucleus; the latter also shows a variety of bizarre morphological changes described in detail by Brambell (1956) and Ingram (1962).

   In the sheep ovary, as degeneration of the cumulus progresses, germinal vesicle breakdown has occurred in oocytes, some of which are in metaphase I or II (Hay and Cran 1978). In such oocytes processes no longer penetrate the zona. These either have been withdrawn or degenerated. Fragmentation of the folicular oocyte is an extremely rare event in sheep follicles of more than 2 mm in diameter. Moor and Trounson (1977) studied the viability of oocytes in atretic follicles during secondary and tertiary stages of atresia and observed that 50% of these oocytes have the capacity for normal embryonic development, as compared with 46% of oocytes from large nonatretic follicles. During atresia, the complex system of gap junctions within the granulosa and cumulus complex (see Chap. III) appears interrupted (see Nicosia 1980a), but apparently without serious deleterious effect upon the oocyte. However, germinal vesicle breakdown does occur when atresia progresses to a stage in which the close spatial relationship between healthy cumulus cells and the oocyte is lost (Fig. 63). The formation of a spindle or other events associated with the maturation division (Fig. 63c–f) do not occur during the atresia of small follicles (Himelstein-Braw et al. 1976, Byskov 1979). Apkarian and Curtis (1981), using SEM, observed the disappearance of oocyte microvilli and granulosa cell projections from the zona pellucida of rat oocytes in early atretic follicles. In more advanced stage of atresia oocyte fragmentation is also seen.

Peluso (1979) determined the rate of germinal vesicle breakdown in oocytes collected from follicles in various stages of atresia. These observations suggest that once an atretic oocyte is removed from the influence of the follicle, whether by being placed in culture or by degeneration of its follicles, it spontaneously fragments. By 72 h after PMSG treatment, the follicle begins to degenerate, as revealed by the presence of acid phosphatase activity within the granulosa cell layer (Peluso et al. 1977a). But degenerative changes within the oocyte occur 96 h after PSMG injection (Peluso et al. 1979). Since the granulosa cells form the nutrient source for the oocyte, as already discussed in Chap. III, any degenerative changes within granulosa cells could be detrimental to the oocyte. The temporal relationship between degenerative changes within the follicle and oocyte suggests that the follicle deteriorates prior to the oocyte.

Oocytes, which have attained their full size, generally degenerate in different ways (Fig. 63). Many resume meiosis and are blocked in metaphase I, the so-called pseudomaturation division (Peters and McNatty 1980); the later stages of follicular atresia in the rat ovary are also accompanied by meiosis-like changes in the oocytes (Magnusson et al. 1983). Another type of degeneration is the fragmentation of the egg, which is similar to the segmentation of the oocyte (Fig. 63g–i). Fragmentation exhibits a certain pattern beginning with the clumping of chromosomes and followed by the development of vacuoles around the chromosome clumps, thus simultating the appearance of separate nuclei. Even a single chromosome can form such a nucleus-like body. Their DNA content is related to the amount of chromosomal material included in them (Burkl 1962); truly segmented oocytes in atretic follicles are only rarely seen. Kaur and Guraya (1985) have observed that, as atresia in the ovary of the Indian mole rat progresses, the oocyte in large pre-antral and antral atretic follicles contains two–three pseudonuclei or a single nucleus with three–five nucleoli; or it may be invaded by granulosa cells (Fig. 63g, h) Resumption of meiosis in oocyte nucleus is observed in atretic antral follicles. In later stages, shape of the oocyte is distorted. Sometimes two oocytes are seen in the antral cavity.

Burkl and Schiechl (1976) made a light and electron microscope study of the pseudonuclei of fragmented egg cells in the atretic follicles of the rat ovary. Each pseudonucleus shows one or more Feulgen-positive particles and is surrounded by a membrane on its surface. The pseudonuclei are formed in the course of degeneration of the oocyte from contracted chromosomes during meiotic metaphases I and II, probably with the involvement of the mixoplasma. Zenzes and Engel (1976) observed that in the rat ovary oocyte degeneration by fragmentation starts on the 22nd day of life. The fragmented oocytes derived from mature oocytes appear for the first time in 21-day-old animals. The highest number of fragmented oocytes is seen in the 27-day-old rat and it decreases at about the time ovulation starts. Chromatin is scattered over the fragments. Fragmentation is a sequential process. It starts with two fragments and reaches a highly fragmented stage 72 h later. In early fragmental stages, the fragments are identical in size and shape. Since they resemble early cleavage stages with respect to the distribution of the cytoplasm, both early fragmentation and cleavage may be induced by the same factors. These cleavage factors may be present in the oocyte cytoplasm and become active about 24 h after the oocyte has reached metaphase II of meiosis. Recent studies of numerous fragmented oocytes in the rat ovary have revealed an asynchronous relationship between nuclear and cell divisions (Shinohara and Matsuda 1982).

Von Weymarn et al. (1980b) made a cytogenetic study of development and degeneration of oocytes from juvenile mice. Five different types of degenerating oocytes are distinguished: (1) Atretic oocytes present in all age groups studied with pycnotic or disintegrating nuclei. (2) Precociously matured oocytes first observed in mice aged 15 days containing normal diakinesis, and metaphase II chromosomes. (3) Vacuolated oocytes first isolated on day 18, and having metaphase II chromosomes scattered within the cytoplasm. (4) Fragmented oocytes also first observed 18 days post-partum containing chromatin dissociated into micronuclei of different sizes within the cytoplasmic fragments. (5) Ghost oocytes present in all age groups, characterized by cytoplasmic disintegration so that only the surrounding zona pellucida can be clearly seen.

Wassarman et al. (1977) observed that mouse oocytes are induced by cytochalasin B to undergo pseudocleavage in vitro in two compartments, only one of which shows microvilli. This particular response to cytochalasin B appears to be related to oocyte size and possibly to the acquisition of meiotic competence by the oocyte during its growth phase. The morphological events, which characterize pseudocleavage, include an initial withdrawal of microvilli from the oocyte surface, subsequent formation of a pseudocleavage furrow and contractile ring, and the development of microvilli and associated microfilaments in one of the two resulting oocyte compartments. These alterations in surface architecture are reflected in the distribution of fluorescein-conjugated lectins bound to the oocyte surface during pseudocleavage. Cleavage of the oocyte, as observed in various studies, may be suggestive of removal of a meiotic inhibitory activity present in the follicular fluid (see Chaps. III and IV).

Zenzes and Engel (1977) compared LDH activities in fragmented oocytes and ova with those already known in early rat cleavage stages. Atretic oocytes are totally devoid of LDH activity. But fragmented oocytes and ova show nearly identical LDH activities to early cleavage stages. Fragmentation and cleavage appear to be homologous processes, induced by the same maternally transmitted factors in the oocyte cytoplasm.

It has been suggested in the earlier histochemical studies that the first alterations that develop with the start of atresia are modifications in the distribution of argyrophilic granules and the shrinkage of oocyte (Ingram 1962). But various electron microscopic studies have indicated that the earliest signs of follicular atresia in vivo and in vitro are the retraction of the granulosa cell processes followed by withdrawal of oocytic microvilli (Beltermann 1965, Stegner 1967, Vazquez-Nin and Sotelo 1967, Odor and Blandau 1972). These changes at the subcellular level suggest that shrinkage of oocyte occurs. This is not an artefact, because it has been observed by the present author even in frozen gelatin sections of ovaries.

With the start of atresia, vacuoles which contain some cell organelles and lipid inclusions in their interior also develop in the outer ooplasm (Fig. 73) (Guraya 1977b). With electron microscopy, they are seen to form digestive or autophagic vacuoles or

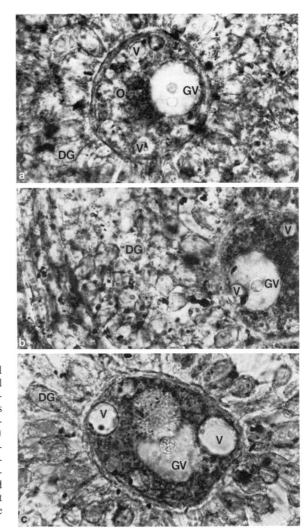

Fig. 73a–c. Photomicrographs of portions of large pre-antral follicles from the postnatal day 2 guinea-pig ovary, showing the development of vacuoles *(V)* of variable size in the peripheral portions of oocytes *(O)* with the start of atresia. Sudanophilic lipids begin to accumulate in the degenerating granulosa *(DG)*, thecal layer and oocyte *(O)*. Such lipids are not seen in the germinal vesicle *(GV)*

cytolysosomes, which have been reported in atretic oocytes in culture (Odor and Blandau 1972, Nicosia 1980a). The various organelles, such as mitochondria, Golgi complex, endoplasmic reticulum, etc. are greatly changed in their distribution and fine structure (Beltermann 1965, Stegner 1967, Vazquez-Nin and Sotelo 1967, Odor and Blandau 1972, Peluso et al. 1979, Nicosia 1980a). Meanwhile the organelles also start losing from their amount and lipid droplets are accumulated. As a result of these degenerative changes myelin figures are also developed.

Some lysosomes and bundles of cytoplasmic filaments are formed in the oocytes of atretic follicles. It is still controversial whether these filaments form normal or abnormal ooplasmic components, since they have been observed at different stages of oocyte growth in situ in hamster (Weakley 1966, 1967), rat (Szöllösi 1972) and mouse (Szöllösi 1972, Zamboni 1972). Beltermann (1965) and Vazquez-Nin and Sotelo (1967) believe that they form abnormal ooplasmic elements in the eggs of mice and rats respectively. Merker and Zimmermann (1970) and Odor and Blandau (1972) believe that the cytoplasmic fibrous lamellae of rodent oocytes in vitro are indicative of premature cytoplasmic maturation, which is not associated with the normal nuclear changes that lead to the formation of polar bodies; however, Odor and Blandau (1972) observed no evidence of polar body formation.

Degenerating oocytes of all sizes in the ovaries of rat, hamster, bat, cat, marmoset, and man are gradually resorbed and thus finally disappear without showing the accumulation of lipids which, however, are stored in the granulosa and theca interna (Guraya 1973a, b). The atretic oocytes of all sizes in the opossum ovary, and large oocytes of dog, buffalo, cow, guinea-pig, and rhesus monkey ovaries store variable amounts of sudanophilic lipid droplets which consist of triglycerides and very few phospholipids (Fig. 74). These lipid droplets generally increase in size or coalesce with each other in contrast to the normal distribution. Adams et al. (1966) also observed the formation of heavy deposits of lipids in the large atretic oocytes of the guinea-pig ovary, which occasionally develop acid phosphatase activity in paranuclear clumps rather than in dispersed granules; acid phosphatase-positive granules with increasing granulosal shedding are strikingly decreased in number and subsequently disappear. The atretic oocytes of small follicles may develop a dense ring of nucleoside triphosphatase (adenosine triphosphatase) activity at the oocyte surface, which has been interpreted as enlargement or even fusion of the terminals of the projection of the granulosa cells. This contrasts sharply with the normal situation.

Hedberg (1953) observed that as the volume of the atretic oocyte decreases, its total dry weight/unit volume increases in comparison to the normal, showing shrinkage of oocyte. Atresia has also been associated with an increase in the amount of protein unit/volume of tissue, a change which appears to be related to the shrinkage of the cell. Lipid accumulations associated with the zonae pellucidae of atretic oocytes are commomly seen in the rhesus ovary (Guraya 1966a). Their accumulation suggests the strong possibility that with the start of atresia their transport into the oocyte stops (as demonstrated in normal oocytes, see Chap. III). They consist mainly of triglycerides and very few phospholipids. Such lipid accumulations do not develop in association with the regressing zonae pellucidae in other mammals (Guraya 1973a, b). The degenerating and shrunken oocytes and their associated crenated zonae pellucidae continue to persist even after the granulosa cells have completely degenerated and disappeared (Guraya 1973a, b, see also Byskov 1979).

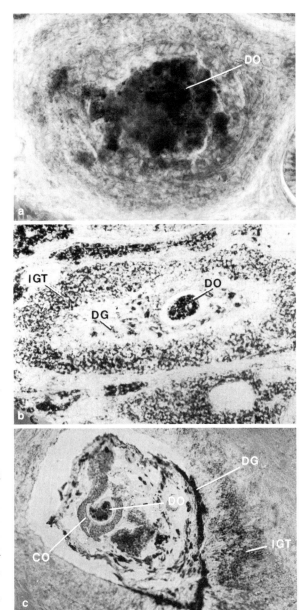

Fig. 74 a–c. Photomicrographs showing accumulation of sudanophilic lipid droplets in the degenerating oocytes of different mammals. **a** Degenerating oocyte *(DO)* from the atretic follicle of the American opossum ovary; **b** Degenerating oocyte *(DO)*, some remnants of degenerated granulosa *(DG)* and interstitial gland tissue *(IGT)* of guinea-pig ovary; **c** Degenerating oocyte *(DO)* and granulosa *(DG)*, and interstitial gland tissue *(IGT)* in the wall of atretic follicle of the cattle ovary. Note the normal appearance of cumulus oophorus *(CO)* (Guraya 1968 a)

Familiari et al. (1981) studied by electron microscope the changes occurring in the fine structure of zona pellucida of growing and atretic follicles in mice, using lanthanum nitrate and ruthenium red. The zona pellucida colours intensely only in atretic follicles, where it consists of a dense globular fibrous matrix, but not in growing follicles, where it shows a weak reaction with ruthenium red. The globular negatively charged particles (20 nm in diameter), probably due to sialic acid moieties, are rarely seen in the zona pellucida of growing follicles, but they become very numerous and connected by microfilamentous structures in atretic follicles. The changed electrical charge of polyanions may reflect a different chemical permeability of the zona pellucida. The increase of negative-charged particles in the zona may have a role in a process of sequestration of the oocyte during atresia.

Some electron microscopic studies have indicated that the macrophages gradually replace some or all of the granulosa cells bordering the zonae pellucidae of atretic oocytes (Odor and Blandau 1972). They are believed to be involved in the removal of all traces of the degenerated oocyte cytoplasm. However, the cells of cumulus oophorus surrounding the oocyte of degenerated antral follicle in the cow remain normal and intact after the other granulosa cells have completely disappeared (Fig. 74c) (Guraya 1968a). The zonae pellucidae, surrounded usually by macrophages, are the last components to regress and disappear, as the crenated zonae pellucidae continue to be seen in abundance in the ovarian stroma for a longer time (Kolpovskii 1978, Byskov 1979, Guraya 1982b).

B. Factors Influencing Follicular Atresia

A remarkable feature of folliculogenesis in mammals is that only a few follicles survive to complete the process, as already discussed (see also Richards 1980). The mechanisms which initiate and regulate atresia of follicles are poorly understood. The various factors influencing follicular atresia have been suggested in the previous literature (Ingram 1962, Richards 1980, Nicosia 1980a, Ryan 1981, Farookhi 1981). These include age, stage of the reproductive cycle, pregnancy, lactation, hypophysectomy, unilateral ovariectomy, hormones of extra-ovarian or intra-ovarian sources, chemical messenger, genetics, nutrition, ischaemia, grafting, X-irradiation. Some insight has been gained recently into the intra-ovarian and extra-ovarian factors which determine the ultimate fate, ovulation and atresia of follicles. It has been suggested that atresia is the inevitable outcome of follicular development and that all follicles would age and become atretic unless rescued at a critical development stage by appropriate trophic stimuli (Richards 1980, Ryan 1981).

1. Hormonal and Nonhormonal Chemical Factors

It is now well established that the intensity of follicular atresia varies greatly with different phases of the ovarian cycle in mammals (Ingram 1962, Guraya 1966a, b, 1969a, b, 1973a, Marion et al. 1968, Dekel et al. 1977a, Richards 1980, Peters and McNatty 1980, Kaur and Guraya 1985). Hormones from extra-ovarian or intra-ovarian sources are believed to regulate follicular growth during various reproductive phases (Richards 1980) as well as in in vitro systems (Nicosia 1980a). They are believed to influence cell viability, receptor formation, granulosa cell proliferation, steroidogenesis or the vascular network (Richards and Bogorich 1980, Nicosia 1980a, Richards 1980, Farookhi 1981). In other words, the interaction of both gonadotrophins and steroid hormones is known to play an important role in the regulation of follicular development. The products derived from a degenerating follicle (e.g. hormones and enzymes) are also believed to influence the rate of other follicles in the vicinity (Peters and McNatty 1980). Actually, the presence within the follicle of two cell types – the granulosa and theca – each with differing hormonal sensitivities and steroidogenic capabilities, has significant implications with respect to the selection process (or atresia) of follicles (Richards and Bogovich 1980, Richards 1980, Nicosia 1980a, Erickson 1982). In order to become a pre-ovulatory follicle, theca and granulosa cells must complete all steps in the selection process or in their differentiation. Interruption of the maturation sequence at any step in the process would prevent the production of pre-ovulatory follicles and result in atresia. Very contradictory views exist in regard to hormonal and nonhormonal factors which may interrupt the maturation sequence (Richards 1980, Nicosia 1980a, Schwartz and Hunzicker-Dunn 1981, Farookhi 1981).

Several lines of evidence have indicated that the process of atresia in the mammalian ovary is either due to the lack of proper gonadotrophic stimulation or to imperfect balance of various hormones, including steroids. The steroid and protein hormones appear to have a decisive influence on atresia in the large follicle. The balance between oestrogen and androgen plays an especially important role in determining whether a follicle becomes atretic or not (Richards and Bogovich 1980, Richards 1980). This suggestion is supported by the recent findings of Clark et al. (1981), which have indicated that elevated concentrations of oestradiol or progesterone during the development of the rhesus monkey follicle, otherwise destined to ovulate, disrupt the pre-ovulatory development and results in atresia. Based on declining amounts of oestrogen in the peripheral circulation, this effect may be exerted within the first day of treatment. Furthermore, the consistent association between increased FSH and continuous follicular development suggests, but does not prove, a cause and effect relationship between these events. Louvet et al. (1975) presented data that implicated intra-ovarian androgen in the control of pre-antral follicular maturation and possibly atresia. The precise controlling mechanisms are still to be determined. But it is established that abnormalities in follicular development can occur in association with an increase in the prodution of ovarian androgens (Inaba et al. 1976). In other words, the ability of a follicle to maintain a high capacity for oestrogen production appears central to pre-ovulation follicular growth (McNatty et al. 1979). Thus the understanding of control mechanisms involved in follicular oestrogen production appears basic to the understanding of follicle selection (or atresia) (Richards 1980, Farookhi 1981, Schwartz and Hunzicker-Dunn 1981, Erickson 1982).

The growth and atresia of Graafian follicles in the adult mammalian ovary occur in a cyclical fashion (see references in Kaur and Guraya 1983). Follicular growth is mainly initiated and maintained by FSH secreted by the pituitary gland (Richards and Bogovich 1980, Richards 1980). The in vitro observations also support the concept that FSH and/or FSH-stimulated oestrogen secretion is essential for follicular growth and antrum formation (Nicosia 1980a). The best preservation of follicular morphology is observed when FSH is used in combination with LH in vitro system, showing synergistic effect in maintaining follicular antrum and size in vitro, as also demonstrated in in vivo studies (see also Farookhi 1981). A group of follicles start their growth but only a small proportion ovulate, depending upon the species (usually one in man). It is possible that the peak of LH secretion that occurs shortly before ovulation induces atresia in those follicles which will not ovulate (second rank follicles; Sturgis 1961). The other effect of LH appears to be to induce the resumption of meiosis in the egg and the formation of the ovulatory stigma on the follicle wall (Schuetz 1969, Neal and Baker 1973), as already discussed in Chaps. IV and VI. In hamster ovaries treated with PMSG on each of the days of the oestrous cycle, numerous large pre-antral follicles, having four to seven layers of granulosa cells, are stimulated to ovulate, resulting in superovulation (Guraya and Greenwald 1965). Normally, such follicles become atretic, showing that a lack of sufficient amounts of endogenous pituitary gonadotrophins, especially FSH and some LH, prevents these follicles from maturing. The premature induction of the pre-ovulatory LH surge in women is known to cause rapid regression of the follicle which was destined to ovulate (Cargille et al. 1973, Friedrich et al. 1975). Neutralization of endogenous gonadotrophins also leads to the degeneration of large follicles (Uilenbroek et al. 1976). In support of these observations Peters et al. (1975) have also observed that injection of gonadotrophins may lead to a decrease in the number of atretic follicles and an increase in the number of healthy ones, thus altering the balance between healthy and atretic follicles with gonadotrophins. Hypophysectomy is well known to cause the follicular atresia (Braw et al. 1981).

In the light of studies on hypophysectomized immature rats, the function of gonadotrophins has also been assessed. Treatment of these animals with gonadotrophins as well as with oestrogens decreases the number of atretic follicles (see Peters and McNatty 1980). However, the results of combined treatment with anti-oestrogens have revealed that the anti-atretic effects of gonadotrophins most likely is mediated by a local stimulation of oestrogen synthesis (Louvet et al. 1975). Exogenous oestrogen alone can also modify the number of follicles that degenerate; the greater the dose of oestrogen injected, the lower the percentage of degenerating follicles (Harman et al. 1975). All these observations support the suggestion that gonadotrophins (FSH and LH) and oestrogen in proper proportions are needed for the normal development of the follicle.

Clark et al. (1979) observed that high concentrations of exogenous oestrogen induce rapid regression of the follicle destined to ovulate in the ovaries of rhesus monkeys. It should not be determined whether the exogenous oestrogen acts directly on the ovary or at the hypothalamo-hypophyseal level. However, Williams (1956) observed that injection of stilbestrol into hypophysectomized immature rats retards atresia affecting growing and Graafian follicles, whereas treatment with PMSG accelerates follicular atresia. Williams also observed that the effects of PMSG are altered with increasing dose from a follicle-stimulating effect to atresia, and then to luteinization. Dekel et al. (1977) suggested that endocrine status of women in late pregnancy apparently does not prevent early follicular growth but induces premature atresia. These observations further support the concept that imperfect balance of hormones leads to atresia of follicles. Ledwitz-Rigby (1979) suggested the presence of a luteinization inhibitor in fluid from small follicles, which is gradually lost as the follicle matures. Follicles that do not experience such a change in fluid composition may be unable to respond to LH stimulation and may be more susceptible to other factors that initiate atresia.

The well-established mitogenic effect of oestrogens, in addition to their anti-atretic effect (Goldenberg et al. 1972) and the increased incidence of atresia after injection of testosterone (Payne and Runser 1958, Richards 1980), further indicate the important role of these steroid hormones in the determination of the fate of follicles. Administration of exogenous androgen to oestrogen-treated hypophysectomized rats increases the incidence of pyknosis and enhances the morphological disintegration of the membrana granulosa (Richards 1980). But intrafollicular, biochemical events responsible for atresia need to be determined more precisely, in spite of the fact that receptors for androgens have been reported in the rat granulosa cells. Some studies have suggested that the high androgen concentration may be interfering with oestradiol or FSH-mediated events (Farookhi 1981). Androgen decreased FSH stimulation of LH receptor (J. Richards, pers. commun.). Androgens come predominantly from the theca (Chap. III), implicating indirectly the theca as the initiator of atresia. Androgens produced by the theca can act directly on the granulosa cells which possess receptors for androgens (Richards 1980, Richards and Bogovich 1980). Androstenedione is produced only by the theca of the hamster antral follicle and then transported into the granulosa cells for conversion into testosterone and oestrogens (Makris and Ryan 1975, Hubbard and Greenwald 1982). Hence, androstenedione appears to be the pivotal steroid in the regulation of further production of C-18 and C-19 steroids in the hamster.

Harman et al. (1975) observed that oestrogens plus hCG adminstered to hypophysectomized rats increase follicular atresia. This effect could be inhibited by treatment with anti-androgens or antisera against testosterone (Louvet et al. 1975, Richards 1980). Louvet et al. (1975) also believe that the increased growth rate is follicles of hypophysectomized rats observed after treatment with gonadotrophins in brought about by the initiation of local oestrogen secretion. Thus, the anti-atretic effect seen 24 h after the administration of PMSG into intact 21-day-old mice (Peters et al. 1975) may be due to local stimulation of oestrogen production. However, in-vitro studies have indicated that oestradiol 17β (0.001–1.0 µg/ml) exerts only a sporadic protective effect on follicular size and antral morphology (Nicosia 1980a). This lack of an anti-atretogenic effect of oestradiol has been interpreted on the basis of loss of oestrogen receptors in cultured follicles. It is also possible that follicular cells are sensitive only to FSH-stimulated and endogenously produced oestrogen. The influence of oestradiol on antrum maintenance may also be mediated by its well-known effect on vascular permeability (see Nicosia 1980a). Alternatively, FSH may exert its anti-atretogenic (Peters 1969) effect without oestrogen mediation. Similarly, the mitogenic effect of FSH (Peluso and Steger 1978) may be due to FSH-induced mitogens other than oestradiol (van der Haegen 1978, Gospodarowitz and Bialecki 1979).

Support for an anti-atretogenic role of endogenously produced oestrogens is provided by the in vitro observations with androstenedione, which is apparently converted to oestradiol (Nicosia 1980a). This suggestion is further supported by the fact that the anti-atretogenic effect of androstenedione is not duplicated by dihydrotestosterone (DHT), a nonaromatizable androgen (Richards 1980). The non-aromatizable androgen DHT prevents hormone induction of LH receptor in developing antral follicles of immature intact rats (Richards 1980, Farookhi 1981). The ability of DHT to block this oestradiol- and FSH-dependent process is not associated with an apparent alteration in the number of FSH receptors in granulosa cells or with a reduction in endogenous ovarian follicular production of oestradiol (Farookhi 1981). Rather, DHT apparently changes the ability of follicles to respond to oestradiol. Although DHT may change the intracellular content or distribution of oestradiol receptors in granulosa cells, androgen may also block an oestradiol-dependent process (see Richards 1980). Further work is needed to determine the effects of androgen on follicular function more precisely. Beyer et al. (1974) also observed that only aromatizable androgens are capable of stimulating follicular growth.

Both atretogenic and anti-atretogenic effects of progesterone on follicular development have been observed (Nicosia 1980a, Richards and Bogovich 1980). The primary
effect of progesterone in vivo is to suppress LH secretion and inhibit the LH surge. Its
very high intrafollicular concentrations, stimulated due to LH surge, may exert local
inhibitory effects. Progesterone does not act to decrease oestradiol receptors in
granulosa cells but it may inhibit aromatase activity (see Richards and Bogovich 1980).

The atretic follicles in the PMSG-treated ovaries are not affected, indicating that
once atresia is advanced the follicles cannot revert to their normal pattern of growth
(Guraya 1972a). Byskov (1979) observed that PMSG-treated ovaries of mice show
some follicles in stage I atresia, but very few are in stage II. The frequency of stage III
appears to be comparable to the untreated mice. These findings have indicated that
"healthy" large follicles, which are in some stage of early atresia, are "rescued" after
treatment with PMSG (see also Peters et al. 1975). They appear to become rescued by a
self-elimination of dying cells and by a recovered growth ability of the granulosa cells.
In "rescued atretic" follicles, the lipid droplets appear to be dissolved and/or become
exocytosed. Wilkinson et al. (1979) observed that the volume of PMSG-treated
follicles is increased about seven times. Many granulosa cells are in mitotic division,
cells are widely spaced, and the interphase cells are highly pleomorphic (as evidenced
by the presence of numerous microvillar and pseudopodial processes). Their
endoplasmic reticulum is greatly distended and often is located in attenuated surface
projections. Often single or small groups of mitochondria are seen in the interstitial
spaces, suggesting that some granulosa cells are lysed due to their exposure to PMSG.
A peculiar feature of cells in the treated follicles is the consistent presence of "foamy"-
looking membranous structures at the surface of granulosa cells in PMSG-treated
animals, which appear to derive from pseudopodial processes. It is not known whether
this phenomenon is related to the processes described in degenerating granulosa cells
(spherical bodies). Peluso et al. (1980) have observed that after PMSG treatment of
immature rats, pre-ovulatory follicles develop, but subsequently degenerate. Before
the appearance of pycnotic nuclei (stage I of atresia), degenerative alterations are seen
in granulosa cell-layer focal areas as evidenced by the presence of cytoplasm blebbing
and changes in granulosa cells shape. The appearance of these degenerative alterations
coincides with a decrease in ovarian concentrations of oestradiol and testosterone.
Since both these steroids maintain the follicle (Richards 1980, Richards and Bogovich
1980), a decrease in their amounts could be responsible for the further degenerative
changes that cause the complete deterioration of the pre-ovulatory follicle. In stage I
atretic follicles, lysosome-derived autophagic vacuoles, develop and macrophages
invade both the thecal and granulosa cell layers. The combined actions of the
autophagic vacuoles and macrophages could destroy both the granulosa cell and
thecal layers and thereby transform the pre-ovulatory follicle into an ovarian cyst.

In different species of mammals it has been established that following unilateral
ovariectomy the remaining ovary undergoes compensatory hypertrophy by developing
follicles which usually occur in two normal ovaries (Arai 1920a, b, Mandl and
Zuckerman 1951, Jones and Krohn 1961a, b, Greenwald 1962, Chatterjee and
Greewald 1971, Baker et al. 1980). This hypertrophy only affects those follicles which
are found in the ovary at the time of surgery; oocytes are not produced from the
germinal epithelium or from any other source (Mandl and Zuckerman 1951). The
ovulation rate of the single remaining ovary becomes double so that initially litter size

remains the same as in the intact female, but the total number of offspring produced during the reproductive life span is halved (Biggers et al. 1962, McLaren 1966). The results of these studies suggested that the removal of one ovary may influence the rate of atresia in the other as a result of availability of more gonadotrophins to the remaining ovary. The slower rate of loss of pre-antral follicles following hypophysectomy may be due to a lowered rate of initiation of primary follicle development or to the lack of a pituitary factor that hastens or initiates atresia. Baker et al. (1980) observed that on a per-ovary basis follicular pool size is reduced and the number of growing follicles is increased after unilateral ovariectomy, these changes occurring within the first 4 weeks after surgery. Thereafter oocyte depletion occurs at the control rates. There is no change in the atresia rate after surgery. McLaren (1966) observed that the level of gonadotrophins in these unilaterally ovariectomized females remains fairly constant and promotes a response in one ovary which would normally occur in two.

The results of recent in vivo and in vitro studies have shown that one of the first steps leading to atresia is the loss of receptors for gonadotrophins (Richards 1980, Nicosia 1980a). Lack of development or loss of these receptors may interfere with follicular responsiveness to gonadotrophins, with follicular steroidogenesis (Hubbard and Greenwald 1983) and with autoregulation of hormone receptors, and may lead ultimately to atresia. The relationship between binding and atresia is not clear, although atresia is known to be associated with loss of LH receptor in rat granulosa cells (Peluso et al. 1977b, Richards 1980). The stage of atresia should be kept in mind when studies of gonadotrophin binding are being carried out on follicular components of ovaries (Richards and Bogovich 1980). Richards (1978) proposed that the LH surge may act on some pre-antral or small antral follicles to cause a loss of gonadotrophic hormone (FSH and/or LH) receptors, to decrease responsiveness of these follicles to hormonal stimulation (or inhibition of aromatization of androgen into oestrogen), and possibly to initiate follicular atresia. The recent studies of Carson et al. (1979) have demonstrated a decrease in FSH binding to granulosa from ovine atretic follicles, as also demonstrated for the granulosa of atretic follicles in hamster (Oxberry and Greenwald 1982). FSH is needed to maintain aromatase activities in granulosa cells, as already discussed in Chap. III. The loss of FSH binding in atretic follicles might explain the loss of aromatizing capacity of these follicles (see Richards et al. 1978, 1979, Richards 1980). This is not to suggest that loss of binding sites is the cause of atresia. Actually, many of the morphological signs of atresia become apparent before any significant alterations in binding are seen. The decrease in FSH binding associated with atresia may also explain the decrease in FSH binding to large follicles, since a greater proportion of large follicles are atretic (classes III–V) (Carson et al. 1979). Since oestrogen is necessary for maintenance of granulosa cells and induction of follicular LH receptors (Richards and Midgley 1976; Richards und Bogovich 1980, Richards 1980), a lack of oestrogen could lead to atresia. However, Hirshfield (1982) concluded that follicles of rats treated with aromatase inhibitor grew faster than follicles of control rats, suggesting that suppression of endogenous oestradiol production appears to enhance, rather than inhibit, the growth of large antral follicles. The correlations of the biological response of follicles and gonadotrophin binding capacity of granulosa cell suspensions to the stage of follicular atresia should be determined in future studies.

Through the synergistic action of oestrogen, FSH, and increasing LH concentrations, LH receptors are induced in the granulosa cells of antral follicles, as already discussed in Chap. III (see also Richards 1980, Richards and Bogovich 1980). Pre-ovulatory follicles show, at least, two important characteristics: (1) the induction of LH receptor in granulosa and (2) the increase in oestrogen synthesis capability. In atretic follicles, one or both of these characteristics are absent or not maintained (Bill and Greenwald 1979, 1981, Peluso and Steger 1978, Carson et al. 1979, Hubbard and Greenewald 1983). Since FSH and LH play such significant functions in follicular growth and maturation, it is reasonable to suggest that inappropriate alterations in these gonadotrophins at critical points in follicular growth may cause the loss of follicular function and thus the atresia (Richards 1980, Farookhi 1981, Erickson 1982). For example, a decrease in or loss of FSH activity would decrease oestrogen synthesis since, as discussed in Chap. III, induction of aromatase and androgen substrate production are dependent on the action of FSH and LH, respectively. Alternatively, increased androgen production could arise as a consequence of the LH surge and this too could change the function of the follicle. In order to examine these possibilities, Farookhi (1981) has studied the effects of (1) decreased FSH on all small antral follicles, (2) decreased LH on large antral follicles and (3) androgens on follicular development. These results support the suggestion that inappropriate changes in FSH and LH at critical points affect the functions of follicle. The site of action of androgen could not be determined in this study, as suggested by the absence of any demonstrable effects on oestrogen synthesis or FSH responsiveness (Farookhi 1980).

In the human, clinical data have suggested suppressive effects of prolactin on follicular function (Franks et al. 1975, Yen 1978). Recent studies have revealed direct actions of prolactin on ovarian cells. The presence of specific receptors for prolactin in the nonluteal ovarian tissue (e.g. follicle and interstitial gland tissue) of several species (see references in Oxberry and Greenwald 1983) suggests that prolactin may play an important role in regulating follicular growth, maturation and steroidogenesis. In human and porcine granulosa cells maintained in monolayer culture, prolactin exerts direct inhibitory effects on progesterone synthesis (see references in Veldhuis et al. 1981). Some earlier studies did indicate that increased levels of prolactin in hyperprolactinaemia and lactation are often associated with suppression of follicle maturation and reduced ovarian oestrogen secretion (Frantz et al. 1972). Taya and Greenwald (1982a, b) have suggested that high levels of progesterone and prolactin during the first half of lactation along with sucking stimuli may be involved in the suppression of follicular development, presumably by lowering basal levels of serum LH. Magoffin and Erickson (1981) suggested that prolactin inhibition of androgen production by the theca cells may adversely affect follicle development and cause atresia (Fig. 42) (see also Erickson 1982). Prolactin-mediated inhibition of aromatase activity and thus of oestrogen secretion from the granulosa cells may also result in the termination of follicular growth and the initiation of atresia in some follicles (Wang et al. 1981). These findings have important implications in follicle selection (see Erickson 1982). Schreiber et al. (1981) suggested that progesterone inihibits follicular oestrogen production, a factor necessary for mitotic activity of the granulosa cell, which is needed for follicular growth. A receptor-like protein mediate inhibitory effect of progestins. The direct demonstration in culture of significant interactions between oestradiol and prolactin in both immature and mature granulosa cells has suggested a critical role for oestrogens in the regulation of both suppressive and facilitative effects of prolactin in the porcine ovary (Schreiber et al. 1981). Oxberry and Greenwald (1983) have suggested that LH/FSH surge might initiate a down-regulation of prolactin receptors in the hamster ovary. The physiological significance of this down-regulation of ovarian prolactin receptors is not known. Such a down-regulation of prolactin receptors may serve as a mechanism by which the ability of prolactin to affect peri-ovulatory events (follicular maturation, oocyte maturation, ovulation, luteinization, etc.) in the ovary might be regulated.

In addition to prolactin, glucocorticoids and GnRH have also been shown to inhibit the oestrogenic responses of FSH and LH on the two follicle-cell populations (Fig. 42) (Richards and Bogovich 1980, Erickson 1982), indicating that many hormones influence follicular oestrogen biosynthesis and thus influence the process of follicle maturation and atresia. From the correlation of results obtained in different studies, Erickson (1982) has proposed that disturbances in the responsiveness of the granulosa and theca cells to the gonadotrophin signal, which enhance and maintain maximal rates of oestrogen and androgen biosynthesis, result in cessation of follicle growth and the initiation of atresia (see also Baird 1983). But this proposal does not provide any explanation for the atresia of pre-antral follicles which do not possess the differentiated theca. Terranova (1981a) has also suggested that atresia of Graafian follicles by ovulatory delay in the hamster may be due to an LH-like (luteinization-like) action on the follicle.

Previous studies have suggested that androgen may have an important role in the induction of atresia (For review see Armstrong and Dorrington 1977). Anti-androgens prevent hCG from depressing ovarian weights and atresia in hypophysectomized oestrogen-treated rats (Ross et al. 1979). This is in agreement with early reports that neonatal androgen administration enhances follicular atresia (Payne et al. 1956). In addition, FSH prevents LH-induced atresia in hypophysectomized rats (Richards and Midgley 1976). Armstrong and Dorrington (1977) have suggested that anti-androgen serum ans FSH prevent atresia by a common mechanism, i.e. prevention of the atretic action of dihydrotestosterone (DHT) on the follicle. DHT is a potent inhibitor of aromatase (Schwarzel et al. 1973), indicating that a lack of oestradiol 17β induces atresia. Possibly, anti-androgen serum binds testosterone, preventing the formation of DHT (Armstrong and Dorrington 1977); whereas FSH increases aromatase activity thereby depleting androgen for DHT formation (Armstrong and Dorrington 1977). Jarrell and Robaire (1981) have suggested that the follicles that do become atretic in the rat are those in which FSH-dependent aromatase enzyme activity is not sufficiently stimulated to produce adequate amounts of oestradiol. As a result, DHT levels may increase in the granulosa cell, resulting in follicular atresia. These data are consistent with the hypothesis that the relative production of testosterone and oestradiol in the follicle regulates 5α-hydroxysteroid dehydrogenase activity and, in so doing, determines the ultimate fate of that follicle: atresia and growth and ovulation. Ware (1982) also suggested that the process through which follicles are selected for ovulation is extremely sensitive to the androgenic environment and that the developmental pathway leading to ovulation or pre-ovulatory follicular atresia are closely linked.

Besides the gonadotrophins and steroid hormones, other factors, such as inhibin (De Jong and Sharpe 1976, Schwartz and Channing 1977, Channing 1979b, Channing et al. 1981), ovarian GnRH (Hsueh et al. 1980, Erickson 1982), gonadocrinins (Ying and Guillemin 1980), "atretogenic factor" (Channing et al. 1981), etc. may be involved in follicular atresia. In addition to prolactin, other hormones, such as glucocorticoids and GnRH, inhibit oestrogenic responses of granulosa cells to FSH and LH, resulting in cessation of follicle growth and initiation of atresia (Fig. 42) (Erickson 1982). Recent data of Channing et al. (1981) are suggestive of factor(s) being present in atretic, but absent in viable, mare follicular fluid, which are capable of causing degeneration of granulosa cells in vitro as well as reducing their production of progesterone. But it needs to be determined whether it is a causative factor in producing atresia or is a product of that process. It is probably a local follicular product, since viable and atretic follicles can reside adjacent to each other in the same ovary. Whether or not the follicular fluid atretogenic factor is specific for granulosa cells remains to be tested. The presence of gonadotrophin-binding inhibitors in the follicular fluid has been shown (Chap. III) (Ledwitz-Rigby et al. 1973, Darga and Reichert 1978). The results of various in vivo and in vitro studies, as discussed here, suggest that interplay of intra-ovarian and extra-ovarian factors on follicular development and atresia still cannot be defined more precisely. But from the recent biochemical data a hypothesis has emerged, which indicates that in the presence of an FSH-driven aromatase system the follicle environment is oestrogenic and conducive to growth, while in its absence androgens may accumulate and follicles may undergo atresia. According to this hypothesis, the synergism between FSH and LH and between androgens and oestrogens is suggestive of a similar synergism between theca and granulosa cells, as discussed in Chap. III. Since FSH and LH play such significant roles in follicular development, it is reasonable to suggest that inappropriate changes in these gonadotrophins at critical points in follicular development may lead to loss of follicular function and atresia (Baird 1983).

Before a specific role is assigned to androgens for causing follicular atresia, it is essential to define the ability of theca cells of follicles at various stages of maturation to produce androgens. In response to LH, androgen production appears to increase only in developing pre-ovulatory follicles (Bogovich et al. 1981, Carson et al. 1981). Therefore, androgen enhancement of atresia may occur only in a specific class of large follicles that possibly for lack of FSH may not have sufficient aromatase activity needed for conversion of androgens into oestradiol, or are unable to respond appropriately to oestradiol. Prior to and at the time of the LH surge, increased androgen production may increase the rate of atresia of large follicles not capable of ovulating. This suggestion does not provide an explanation for the extensive atresia of follicles at earlier stages of growth. Therefore, we have to look for other factors influencing atresia of such follicles. Various lines of evidence have indicated that in the absence of an appropriate sustained increment of LH, FSH, or both, the follicles in earlier stages of growth are unable to enter the pre-ovulatory stage and thus degenrate. Basal levels of gonadotrophins are either too low to stimulate theca cell differentiation and increased androgen production (Carson et al. 1981) or too low to enhance granulosa cell differentiation and aromatase activity. All these studies clearly indicate that both LH and FSH are needed to sustain pre-ovulatory follicular growth, as FSH deprivation leads to the rapid loss of granulosa cell receptors for FSH, LH and oestradiol, and an inability of these large follicles to ovulate or to luteinize (Richards 1980). Carson et al. (1981) have concluded that functional differentiation of theca interna cells, which is the key to the maturation of the Graafian follicle, is dependent primarily upon modest increases in basal LH activity and can occur in the presence of elevated levels of serum progesterone in pregnant rats.

Farookhi (1981) has proposed that follicular atresia can be viewed as an inflammatory response (Espey 1980) and that the development and selection of the ovulatory follicle is regulated not only by the reproductive system, i.e. hormones, but also by the immune response. But further work is needed to determine more precisely the role of the immune system. The results of several other studies have also suggested that granulosa cell development can be inhibited by a number of regulatory agents: epidermal growth factor (Mondschein et al. 1981), follicular fluid (Anderson and Channing 1978, Ledwitz-Rigby et al. 1977), serum (Erickson et al. 1979), and gonadotrophin-releasing hormone (Hsueh and Erickson 1979, Hsueh and Ling 1979). Presumably then, positive effectors, such as FSH and insulin (Anderson and Channing 1978, May et al. 1980), must act to overcome negative influences in order that more highly differentiated states can be attained. The balance between positive and negative effectors may determine whether a follicle differentiates to the ovulatory stage or undergoes atresia.

2. Aging Factor

With aging, the stock of primordial follicles is depleted. No ovulation occurs in post-menopausal women due to the absence of oocytes. It is interesting that some Graafian follicles are often seen in the ovaries of post-menopausal women (Guraya 1976c). Similar follicles have also been described in the ovaries of newly born and young children (Potter 1963). These follicles apparently do not ovulate due to the lack of proper gonadotrophic stimulation and often degenerate. Both the theca interna and the granulosa cells of these atretic Graafian follicles are highly luteinized during the early post-menopausal period and develop some resemblance to corpora lutea, inasmuch as the cytological and histochemical features of steroid-secreting luteal cells appear (Guraya 1976c). These histochemical characteristics consist of the presence of abundant diffuse sudanophilic lipoproteins and some deeply sudanophilic lipid droplets in the cytoplasm of luteinized thecal and granulosa cells (Fig. 75). The amount of lipid droplets is relatively higher in the granulosa lutein cells. Such luteinized follicles containing hyalinized eggs have also been described in the ovaries of young children and are usually called corpora lutea atretica, although there is no biochemical and physiological evidence that they have the same secretory pattern as the normal glands. These resemblances in the morphological and histochemical features of atretic follicles during the prepubertal and post-menopausal periods indicate that there are some similar physiological or gonadotrophic stimulations to which the human ovary is being exposed. The absence of ovulation, which leads to follicular atresia, may be due to the lack of proper ovulatory hormone stimulation during these periods.

3. Ischaemia and Dietary Factors

The follicles, which have started their growth, are very sensitive to ischaemia, possibly due to deprivation of oxygen and gonadotrophins (Ingram 1962). Dietary and other factors apparently influence follicular atresia via their effect on the pituitary, as there is no evidence to suggest that nutrition has any direct effect on atresia. Printz and

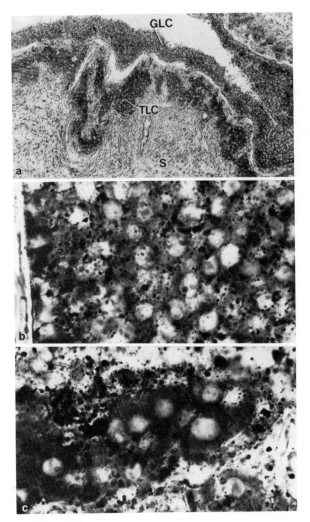

Fig. 75a–c. Photomicrographs of histochemical preparations of portions of corpus luteum atreticum (or atretic Graafian follicle) from the post-menopausal ovary of women of 45 years. **a** Low-power view of granulosa lutein cells *(GLC)* and theca lutein cells *(TLC)* showing sudanophilic lipids which are not seen in the surrounding stroma *(S)*; **b** Higher-power view of granulosa lutein cells showing diffusely distributed sudanophilic lipids and highly sudanophilic lipid droplets which are not seen in their nuclei; **c** Higher-power view of theca lutein cells showing diffusely distributed lipids and lipid droplets of discrete nature (From Guraya 1976b)

Greenwald (1970) also concluded that the alterations in the reproductive cycle caused by starvation are affected by inhibition of FSH release and synthesis, and by an increased LH release, rather than by a direct effect on the ovary.

4. Intra-ovarian Factors

The extensive atresia or degeneration of germ cells in the developing mammalian ovary (see Baker 1972, Guraya 1977a, b) also suggests the presence of some intra-ovarian factors influencing atresia, about which our knowledge is still very meagre. Genetic, environmental, and metabolic factors have been suggested in this regard (Baker 1972, Ryan 1981). The environmental factors may well be related to the proximity of a blood supply. Evidence for an important role of vascularity in follicular function is recently supplied by Zeleznik et al. (1981). These workers observed that in monkeys the dominant follicle is surrounded by a dense network of branching vessels that penetrate up to the basal lamina. Other large, antral follicles, however, were much less richly vascularized. The authors proposed that this difference could be important in selection of the dominant follicle. Farookhi (1981) has proposed that atresia is due to an alteration in follicular function of which there may be several causes, which leads to inappropriate changes in basement membrane permeability. This change in basement membrane permeability leads to the inappropriate entry of serum proteins into the follicle following a localized, premature inflammatory response. It is this mechanism that is recognized as morphological atresia. It is suggested that the surrounding stroma including the thecal layer and its blood and nerve supply should be studied in depth during follicular growth and atresia in order to determine the causative factors of atresia. The understanding of interaction between the extra-ovarian and intra-ovarian factors in relation to differentiation and function of theca will help us to determine the mechanisms that regulate follicular growth and atresia. It has been shown that the integrity of the perifollicular vascular network, which is obviously disrupted by in vitro experiments, is essential for normal follicular function (Nicosia 1980a). Recent studies have demonstrated an extensive nerve supply in the thecal layer, as discussed in Chap. III. But the details of regulation of differentiation and function of the thecal layer by the blood and nerve supply needs to be determined, as the granulosa cell layer is without such a supply. Secretory products, such as epinephrine, and norepinephrine of nerves distributed in the thecal layer, may be regulating/modulating its various biochemical activities which are essential for the normal growth, maturation and ovulation of the follicle.

5. Irradiation

X-irradiation directly causes follicular atresia, which greatly varies with the stage of oocyte development as well as with the animal species used (Baker 1971, 1972). The dose of radiation needed to destroy a given population of oocytes varies greatly between species and also depends upon age (see Lacassagne et al. 1962, Mandl 1964, Baker 1971, 1972). Primordial oocytes in rats and mice are highly sensitive (Mandl 1959, Peters 1969), while those in guinea-pigs, rhesus monkeys, and possibly the human female, are fairly resistant to X-irradiation and hypophysectomy. In rats, the dose of X-rays which destroyed all the oocytes in intact animals (ca. 400 r) showed a lesser effect in those from which the pituitary had been removed. Treatment of these animals with PMSG, however, evoked a similar response to that of the intact animals. It is generally agreed that the degenerative changes following irradiation are not strictly analogous to those occurring spontaneously during atresia (Zuckerman 1960, Ingram 1962), and that no conclusion about the lifespan of the atretic oocyte may be drawn from irradiation experiments.

C. Significance of Follicular Atresia

1. Formation of Interstitial Gland Tissue and its Role in Steroidogenesis

Follicular atresia in the mammalian ovary appears to be related to the formation of ovarian interstitial gland cells from the theca interna and surrounding stroma of atretic follicles (Figs. 65–68) (see Guraya 1978, Nicosia 1980a, Schwall and Erickson 1981). These interstitial gland cells form the most characteristic and important glandular-looking tissue whose presence has been clearly demonstrated in the ovaries of every mammalian species investigated so far (Fig. 70) (Guraya 1971a, 1973a, b, c, 1978, 1980a, Anderson and Meurling 1977, Guraya and Motta 1980); its presence has also been shown convincingly even in the ovaries of nonpregnant women, rhesus monkey, cow and buffalo (Mossman et al. 1964, Guraya 1966b, 1967b, 1968a, 1973a, b, 1978, 1980a, Guraya and Motta 1980), in which it was denied by earlier workers using routine histochemical techniques. Interstitial gland cells of thecal origin have been demonstrated in the buffalo ovary with histochemical techniques (Fig. 67) (Guraya 1973b, 1979a). The granulosa cells of atretic follicles in the civet cat *(P. hermaphroditus)* ovary gradually regress and disappear without showing the accumulation of lipid droplets (Guraya 1981). But their thecae are left behind to constitute large masses of interstitial gland tissue which occupy the whole ovary and are highly vascularized during the phase of follicular development (Fig. 68).

Although the interstitial gland cells greatly vary in their amounts, distribution, and differentiation in different mammalian species as well as with the ovarian cycle of the same species, they show the cytological, histochemical and biochemical features of well-established steroid-secreting cells (testicular Leydig cells and luteal cells) (see reviews by Guraya 1971a, 1972a, 1973a, b, 1978, 1980a, 1981, Guraya and Motta 1980, Schwall and Erickson 1981). The most striking features are: (1) abundant diffuse lipids (lipoproteins) in the cytoplasm, which have been presumed to derive from the abundant ultrastructural membranes of agranular endoplasmic reticulum; (2) well-developed cell organelles, especially the pleomorphic mitochondria with a complex system of internal cristae that are predominantly tubular; (3) the development of diffuse lipoproteins (or membranes of smooth reticulum), closely accompanied by the appearance of enzyme activities related to the biosynthesis of steroid hormones; (4) under certain physiological situations (with low levels of gonadotrophins) the accumulation of steroidal lipid droplets (hormone precursor material) in the cytoplasm, which consist of triglycerides, cholesterol and cholesterol esters, and phoshpholipids; (5) receptors for LH/hCG and prolactin (Oxberry and Greenwald 1982, Shaha and Greenwald 1982); (6) a high capacity to form a variety of steroids in vivo and in vitro. The highly vascularized glandular-looking interstitial gland tissue of the civet cat *(P. hermaphroditus)* ovary shows the presence of abundant diffuse lipoproteins and a few lipid granules composed of phospholipids (Fig. 68) (Guraya 1981); cholesterol and/or its esters are absent, indicating that these interstitial gland cells are active in steroid hormone synthesis.

The results of all these ultrastructural, histochemical and biochemical studies correlate well with each other and are compatible with the endocrine function of ovarian interstitial gland cells (or stromal cells) which appear to secrete various steroid hormones, such as androgens, oestrogens and progestogens, depending upon the mammalian species and possibly the physiological state (Guraya 1971a, 1972a, 1973a, 1978, 1980a, Guraya and Motta 1980, Nicosia 1980a, Koninckx et al. 1983). Schwall and Erickson (1981) have clearly shown that the interstitial cells in the rat ovary make large amounts of progestins and a very small amount of androgens. Atresia of hamster follicles is also associated with a transformation of theca cells into interstitial gland cells (Figs. 65 and 86) (Guraya and Greenwald 1965) that secrete large amounts of progesterone and small amounts of androstenedione (Taya et al. 1980, Silavin and Greewald 1981). Collagen-dispersed interstitial cells from 5-day hypophysectomized hamsters produce progesterone and responds to ovine LH stimulation in vitro with a dose-dependent increase in progesterone (Silavin and Greenwald 1982). FSH and prolactin have no effect. Phenobarbitol blockade of ovulation for 3 days in the pro-oestrous hamster induces atresia of pre-ovulatory follicles. Thecae (or interstitial gland cells) of these experimentally induced atretic follicles also form the source of progesterone (Terranova et al. 1982, Hubbard and Greenwald 1983). Follicles, which undergo in vitro atresia, show a limited capacity for steroidogenesis and secrete progesterone in relatively greater amounts than oestrogen (Nicosia 1980a). Based on ultrastructural observations, theca cells (or theca-type interstitial gland cells), and not granulosa cells, of such atretic follicles appear responsible for steroidogenesis.

The pre-ovulatory rise in the secretion of progestins studied biochemically for some mammalian species has also been related to the interstitial gland cells from which the cholesterol-positive lipid droplets are simultaneously depleted in response to high levels of ovulatory hormone (LH) (Fig. 76) (Guraya 1973a, 1978, 1980a, Guraya and Motta 1980).

Recent in vitro experiments have increased our knowledge about the alterations or shifts that occur in the secretion of steroid hormones with the initiation of atresia in Graafian follicles (see references in Hubbard and Greenwald 1983). Bill and Greenwald (1979, 1981) have developed an animal (hamster) model which provides a large synchronous population of atretic antral follicles. By using this model, they have observed an immediate decrease in oestradiol 17β and testosterone in the serum, medium and tissue. Serum progesterone remains unchanged, whereas progesterone in the medium and tissue rises gradually upto 24 h (see also Hubbard and Greenwald 1981a, b). Recent studies of Hubbard and Greenwald (1983) have also shown that atretic follicles rapidly lose their ability to synthesize oestradiol and testosterone, while production of 17α-hydroxyprogesterone, progesterone and cAMP declines more slowly as atresia advances. The source of this progesterone appears to be interstitial gland tissue of thecal origin in the atretic follicles, the wall of which is conspicuously developed in the hamster ovary (Figs. 65 and 66) (Guraya and Greenwald 1965). Brunning et al. (1979) also observed decreases in oestradiol 17β and androgens, while progesterone levels rose during the observed periods. Inhibition of aromatase activity is believed to be the possible cause for these changes. However, the simultaneous drop in oestradiol 17β and testosterone contradicts this theory. Hubbard and Greenwald (1981a, b, 1983) have further reported correlative changes in steroids (oestradiol, testosterone and progesterone) and nucleotides (cAMP and cGMP) in isolated atretic antral follicles of hamster. Both cAMP and cGMP per follicle show increases at 2 and 4 h, respectively, before gradually dropping to their lowest levels at 72 h. It will also be interesting to mention here that both the nucleotides are elevated at different time periods. These biochemical activities have been related to the theca (or newly formed interstitial tissue) of atretic follicles, as their granulosa compartment has undergone histologic degeneration (Figs. 65 and 66).

Continued blockade of ovulation with sodium pentobarbital in the cyclic rat and hamster has been observed to induce atresia of pre-ovulatory follicles (Uilenbroek et al. 1981, Terranova 1980, 1981a, b, Mizuno et al. 1983). These atretic antral follicles also lose the capacity to synthesize androgen and oestrogen in vitro in response to LH, retain androgen aromatizing capability and produce large amounts of progesterone; production of 20α-hydroxyprogesterone is increased as compared to pro-oestrous follicles. The synthesis of progesterone has been related to the differentiation of theca-type interstitial gland tissue in the wall of these atretic follicles (Figs. 65 and 66) (Terranova et al. 1982), which continues to show activities of steroid dehydrogenases, such as 3β-HSDH and 17α-HSDH (Parshad, V. R., and Guraya 1984). This change of oestrogen, androgen, and progesterone synthesis during delayed ovulation of follicles resembles the oestrogen-progesterone shift during the post-ovulatory period (Armstrong and Dorrington 1977, Saidapur and Greenwald 1979a, b, c, Terranova et al. 1982), which has been attributed to the inhibitory action of LH on follicular oestrogen production (see Terranova 1981a, Hubbard and Greenwald 1983). The inhibition of the 17α hydroxylase-C_{17-20} lyase system is believed to decrease intrafollicular androgen and oestrogen and possibly increase the accumulation of progesterone. Uilenbroek et al. (1981) have also concluded that total steroidogenic activity is not decreased, but that the steroidogenic pathway is altered during atresia in follicles. A steroidogenic block occurs beyond progesterone formation in the nonovulated follicles. Based on these various in vitro studies, it can be suggested that the early breakdown of the granulosa may cause the initial fall in oestradiol and testosterone, while the theca-type interstitial gland cells in the wall of atretic follicles continue to produce 17α-hydroxyprogesterone, progesterone and cAMP until advanced stages of atresia. Stepwise decline in steroids and cAMP is observed. High dose (200 ng/ml) of LH leads to excessive stimulation of progesterone by hamster atretic Graafian follicles in vitro, depending upon the degree of their atresia (Hubbard and Greenwald 1983). The site of action of LH is the theca interna (or interstitial gland cells) of atretic follicles, as LH receptors persist on the theca interna of atretic follicles (Shaha and Greenwald 1982). Hubbard and Greenwald (1983) have also suggested that a high dose of LH directly or indirectly blocks steroidogenesis somewhere beyond progesterone production, causing low levels of testosterone and oestradiol in the hamster atretic follicle and medium. Previous studies using human follicles (both normal and atretic) have reported a differential response in steroid production to either high or low doses of LH (McNatty et al. 1980). Low doses of LH (1–10 ng/ml) produced large increase in thecal output of androstenedione in both normal and atretic follicles. This was also the case for oestradiol and progesterone in normal follicles. High doses of LH (50 ng/ml) depressed overall steroid output when compared with lower levels of LH (10 ng/ml). It is also suggested that prolactin may play a role in the decrease of oestradiol biosynthesis of ovulatory follicles. Terranova (1981a) believes that atresia induced by ovulatory delay is due to an LH-like (luteinization-like) action on the follicles. Braw et al. (1981) have observed that atresia of rat Graafian follicles is induced by hypophysectomy on the morning of the day of pro-oestrus. These atretic follicles are characterized by increased progesterone and decreased androgen and oestradiol production. Addition of LH to the culture medium stimulates progesterone accumulation in follicles from rats upto 24 h after hypophysectomy but not 48 h after the operation, indicating that follicles gradually lose their responsiveness to LH during the atretic process, as also observed by Hubbard and Greenwald (1983) for the hamster. Cran (1983) has shown that the membrana granulosa is lost from the majority of the follicular cysts produced after the administration of PMSG to sheep. However, their theca interna is luteinized to develop a structure of steroid-secreting cells. The major hormone secreted by such luteinized follicles is progesterone which derives from the luteinized theca. Terranova and Ascanio (1982) have studied alterations of ovarian steroidogenesis induced by ovulatory delay in the immature rats treated with PMSG. This study has indicated that (1) the PMSG-treated immature rat is a useful model for the atresia, (2) ovulatory delay inhibits ovarian androgen secretion (and oestradiol, but this depends on the type of PMSG), (3) prolactin may be casually related to the alteration in androgen synthesis, and (4) pre-ovulatory follicles are sensitive to gonadotrophin deprivation during ovulatory delay caused by injection of phenobarbital.

The interstitial gland cells of thecal origin show several variations in regard to their permanency, which appear to be regulated by various factors, such as the nature and amount of gonadotrophins, blood and nerve supply, genetic and metabolic factors, etc. (see Guraya 1973c, 1978). They are very transient in nature in some nonpregnant mammalian species (Figs. 67, 70a and 72) (women, rhesus monkey, cow and buffalo) in which, after persisting for some time in the wall of atretic follicles, they are generally reverted to the relatively embryonic, compressed stromal tissue from which they originated during follicular growth (Guraya 1973a, b, c, 1978, 1979a, 1980a, Guraya and Motta 1980). The interstitial gland cells actually do not show much permanency or accumulation in those ovaries in which there is some sort of balance between the formation of new interstitial cells from the theca interna during follicular atresia and their reversion or de-differentiation to the original stromal tissue, thus causing no accumulation in the ovary. This short duration of the existence of interstitial gland cells led most of the earlier workers to deny their presence in the ovaries of nonpregnant women, rhesus monkey, cow, and buffalo. In the ovaries of mammals showing a greater development and differentiation of interstitial gland cells, there is seen accumulation rather than their reversion back to the original stromal tissue. Thus they form the major portion of the ovarian stroma (Figs. 68, 70, and 76). All these alterations in ovarian interstitial gland tissue appear to be regulated by the qualitiy and quantity of gonadotrophins secreted in cyclic fashion. Besides reverting back to the original stromal tissue, some interstitial gland cells undergo degeneration (Guraya 1967c, 1978, Davies and Broadus 1968), become refractory to gonadotrophic stimulation, retain their lipid droplets (Guraya 1978, Guraya and Motta 1980), and exhibit altered forms of cytoplasmic organelles. Besides the gonadotrophins, catecholamines derived from adrenergic innervation of the interstitium appear to influence the functional life of ovarian interstitial cells. This suggestion is supported by the fact that catecholamines stimulate progesterone production in interstitial cells of hypophysectomized hamsters and propranolol blocks this effect without LH-induced progesterone production (Silavin and Greenwald 1982). Adrenergic receptors are also present in the interstitium.

Since interstitial gland cells formed from the theca interna of atretic follicles are seen in young animals as well as in aging animals, and show cycles of abundance and

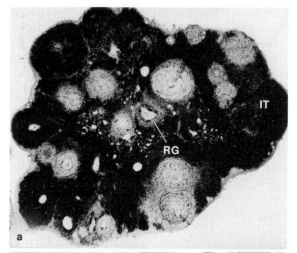

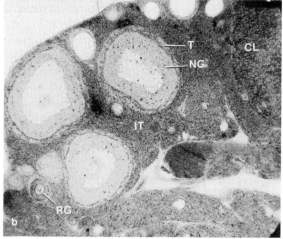

Fig. 76a, b. Photomicrographs of hamster ovary. **a** Untreated day 2 morning ovary showing interstitial gland tissue *(IT)* filled with sudanophilic lipid droplets, and many pre-antral follicles. Regressing granulosa *(RG)* of atretic follicles show accumulation of sudanophilic lipids; **b** Portion of day 2 evening ovary treated with 60 i.u. PMSG on day 1, showing the depletion of most of the sudanophilic lipid droplets from the interstitial gland tissue (compare it with **a**). Many pre-antral follicles seen in **a** have been stimulated to grow by the PMSG. Follicles having regressing granulosa *(RG)* at the time of injections are not affected. The lipid accumulations formed by the degeneration of granulosa and corpora lutea *(CL)* remain intact. Follicles with normal granulosa *(NG)* and theca *(T)* are also seen (From Guraya and Greenwald 1965)

differentiation correlated with reproductive age and cycles, they appear to be an important ovarian gland (Guraya 1978, 1980a, Guraya and Motta 1980). Anderson and Meurling (1977) observed that the accumulation of interstitial gland tissue during the maturation of the ovary in wild rabbits *(Oryctolagus cuniculus)* in South Sweden is closely related to follicular atresia and it appears to be essential for attaining full reproductive capacity, as is also observed for the civet cat (Guraya 1981). The steroid secretions of ovarian interstitial gland cells (or theca interna of atretic follicles), which are now being identified in in-vitro systems, as already discussed, are likely to play some significant role in the initiation of puberty and cyclic ovarian activity in the females. Silavin and Greenwald (1982) have suggested that the ability of interstitial cells to produce steroid after gonadotrophin deprivation may be necessary for "priming" follicular growth in the ovary at the onset of the breeding season. Sturgis (1961) suggested that the accumulation of interstitial gland tissue through follicular atresia is needed to trigger an adequate response by the pituitary gland. Thus, significant amounts of gonadotrophins would be secreted by the pituitary to enable some follicles to achieve "full maturation at or after puberty" (Sturgis 1961). Gachechiladze and Labadze (1977) believe that the cell elements of the atretic follicles are important for the regulation of rat ovaries during post-natal development. This view is based upon the dynamic of the atretic follicles and upon their structural and ultrastructural characteristics.

2. Limiting the Number of Offspring

Atresia of large follicles after puberty prevents excess ovulation and thus limits the number of off-spring to that characteristic of the species (Guraya 1973b). This is related to internal fertilization and implantation. There can be little doubt that this is regulated by the pituitary hormones, in view of recent occurrence of multiple births in women who have been treated with gonadotrophins (Gemzell 1963, Aiken 1969).

3. Decrease in Fertility

The only function that has been assigned to atresia of primordial follicles is that it may serve as a selective process in which only the "fittest" oocytes survive (Ingram 1962, Guraya 1973a, b). The result of a high incidence of follicular atresia in the mammalian ovary is that old animals have a greatly reduced "stock" of oocytes which decreases fertility with age: other factors also play their part in senile sterility.

D. General Discussion and Conclusions

The correlation of recent morphological, histochemical, biochemical and physiological data shows that at present our knowledge about the various aspects of follicular atresia is still very meagre. It is still not known how long it takes for an atretic follicle to be eliminated from the ovary. One of the most challenging and fundamental problems in reproductive biology is why one follicle should be selected to undergo development, meiosis and ovulation, while its immediate neighbours should undergo degeneration and be eliminated from the ovary. Intra-ovarian and extra-ovarian factors regulating the development of receptors for hormones (see Chap. III) must be involved in this type of selection process. Recently, the specific actions of the gonadotrophic and steroid hormones in the gross development of follicles and at the cellular level have been identified. The ability of a developing follicle to release high concentrations of oestrogens that stimulate growth and cell differentiation is central to the selection of a given follicle for maturation and ovulation. Interruption of oestrogen production at any step results in atresia of Graafian follicles. However, this concept developed recently as a result of biochemical studies does not provide an explanation for atresia of primary and secondary follicles, i.e. pre-antral follicles, which lack some component of the systems needed for oestrogen synthesis. Thus, we still need to know what initiates the first changes in the follicular apparatus that lead to its growth and development.

The basic biochemical or endocrinological aspects of the onset of follicular atresia are still unknown for any mammal in spite of the fact that our knowledge about the biochemical actions of gonadotrophins and steroid hormones in the developing follicle has greatly increased in recent years. A great variety of endocrine insults have been shown to cause atresia at all levels of follicular development. Factors causing follicular atresia appear to be variable and dependent upon the stage of follicular growth. However, it is becoming increasingly clear that the process of atresia is regulated by the interaction of gonadotrophins and steroids. The imperfect balance or lack of these hormones may be of great significance in this regard, at least in the mammalian ovary.

Ryan (1981) has recently speculated on the mechanisms for production of atresia. He has two possible alternatives in this regard. First, the granulosa cells are genetically programmed to die. Second, their death is mediated by a chemical massenger. Ryan has discussed the merits and demerits of these two concepts, and suggests that atresia appears to be mediated by a chemical messenger, latent within the follicle. Such a component might be the complement system which exists in the serum and in follicular fluid. This system has the capacity to kill cells, normally occurs in an inactive state and needs a positive event for activation. This speculation needs to be confirmed by experimental data.

With the initiation of atresia, various components of the follicle start undergoing conspicuous morphological, histochemical and biochemical alterations. It still remains to be determined whether these modifications are the cause of or response to a subtle beginning of atresia. But they are apparently related to digestion and resorption of different structures of atretic follicles. Lysosomes are well known to be involved in their regression (Dimino and Elfont 1980). The most important criterion for follicular atresia at all stages including primordial follicles is the shrinkage of oocyte and the granulosa cell layer; however, there is little, if any, reduction in follicular size during cystic follicular atresia. Morphological and physiological relationships among oocytic microvilli, the zona pellucida, and follicular cell processes are disrupted and the basement membrane disappears. Enzyme activities required for transport of substances into the oocyte, and radioactive precursor substances into granulosa cells, disappear. The shrinkage of the oocyte clearly indicates that some drastic changes take place in the structure and permeability properties of cellular membranes of follicle. Menezo and Gerard (1977) have studied in vitro macromolecular transfer between calf and macaque Graafian follicles and culture media. Macromolecules traverse the follicle membrane only if follicles remain healthy during culture. When granulosa cells are pycnotic at the end of culture, transport completely disappears. In healthy follicles, macromolecule transport appears to be dependent on the molecular weight of the substances. Atretic follicles in vivo do not seem to play any role in intra- or extra-ovarian regulation of folliculogenesis. The release of enzymes into the medium may furnish an indication of the metabolic changes caused by gonadotrophins.

The changes due to atresia cause abnormal structure, composition, distribution, and function of organelles in degenerating oocytes and granulosa cells; the mitochondrial alterations form the most significant signs of early degeneration. As a result of various modifications in organelles, the living machinery of the follicle is gradually disrupted, and new cellular components, namely lysosomes, cytolysosomes (or vacuoles), and myelin figures containing hydrolytic enzymes, are formed to digest the various structures of atretic follicles; macrophages also play significant role in their digestion.

The receptors for hormones disappear in the granulosa cells of atretic follicles. However, with the start of atresia, lipid droplets consisting mainly of triglycerides, and some phospholipids and glycogen, develop in the granulosa cells and oocyte. Species variations in regard to the amount and possibly nature of the substances occur, but their significance is still to be determined. The development of cholesterol-positive lipid droplets, diffuse lipoproteins and some enzyme activities related to steroidogenesis in the granulosae of atretic secondary and tertiary follicles, indicates that they apparently undergo an incipient, nonfunctional luteinization in response to some luteinizing factors which begin to become available with the initiation of atresia; the degree of their luteinization varies in follicles of different sizes under different physiological situations which are apparently determined by their previous exposure to hormones before the start of atresia.

Granulosa cells of primordial, primary and small secondary follicles are relatively more resistant towards atresia. In some mammals they continue to persist even after the complete disappearance of oocytes, and form epithelial cords which may develop the histochemical features of steroid-secreting cells. These differences in the behaviour of granulosa cells in follicles of different sizes during atresia in vivo, as well as in vitro, clearly indicate that they undergo some basic, metabolic changes during follicular growth, apparently in response to gonadotrophins, steroids and some other environmental factors. Channing (1969, 1970), working on granulosa cells from follicles of variable size in culture under various physiological conditions, has also demonstrated great differences in their behaviour in relation to luteinization and secretion of progesterone. Further studies using a variety of techniques are needed to determine the nature of these differences in vivo, especially in relation to the regulation of development of receptors for hormones.

Some changes, such as the disruption of the normal relationship among oocytic microvilli, the zona pellucida, and corona radiata, and development of ooplasmic lamellae during follicular atresia, are very similar to those of pre-ovulatory maturation of the ovum (see Chap. IV); these alterations during atresia may also be accompanied by premature nuclear changes of bizarre appearance. There apparently is a definite interrelationship between the oocyte, the corona and the cumulus cells, as discussed in Chaps. III and IV. Increasing evidence also indicates that if the egg is not completely surrounded by granulosa cells, meiosis proceeds to diakinesis and follicular atresia results.

The granulosa cells and their lipid accumulations in atretic secondary and tertiary follicles degenerate and disappear, leaving behind their hypertrophied thecae internae which finally form conspicuous masses of interstitial gland cells in the ovarian stroma. In their centre are seen, sometimes, the remnants of degenerating oocytes and zonae pellucidae; the latter are the last to disappear. The interstitial gland cells of thecal origin develop abundant diffuse lipoproteins, lipid droplets, intense alkaline phosphatase activity, and various emzyme activities indicative of steroidogenesis (see Guraya 1971a, 1972b, 1973a, 1978, 1980a, Guraya and Motta 1980).

The most important physiological function of follicular atresia is the removal of oocytes. It is also related to the accumulation of interstitial gland tissue in the ovary, and consequently of the steroid hormones secreted by the ovary (Guraya 1978, 1980a, Guraya and Motta 1980). Since interstitial gland cells derived from the theca interna of atretic follicles are seen from young to old age and show cycles of abundance and differentiation correlated with the reproductive age and cycles, they may be an important ovarian gland cell. Their steroid secretions may be of great physiological significance in the initiation of puberty and cyclic ovarian activity in the female.

References

Abel JH Jr, Verhage HG, McClellan MC, Niswender GN (1975) Cell Tiss Res 160:155
Adams EC, Hertig AT (1964) J Cell Biol 21:397
Adams EC, Hertig AT, Foster S (1966) Am J Anat 119:303
Ahrén K, Dekel N, Hamberger L, Hillensjö T, Hultborn R, Magnusson C, Tsafriri A (1978) Ann Biol Anim Biochem Biophys 18:409
Ahrén K, Hamberger L, Hillensjö T, Nilsson L, Nordenstrom K (1979) J Steroid Biochem 11:791
Ahuja KK, Bolwell GP (1983) J Reprod Fertil 69:49
Ahuja KK, Tzartos SJ (1981) J Reprod Fertil 61:257
Aiken RA (1969) J Obstet Gynaecol Br Commonw 76:684
Ainsworth L, Baker RD, Armstrong DT (1975) Prostaglandins 9:915
Aitken RJ, Richardson DW (1981) J. Exp. Zool 216:149
Aitken RJ, Rudgk EA, Richardson DW, Dor J, Djahanbahkch O, Templeton AA (1981) J Reprod Fertil 62:597
Albertini DF (1980) In:Motta PM, Hafez ESE (eds) Biology of the ovary. Nijhoff, The Hague, pp 138–149
Albertini DF, Anderson EA (1974) J Cell Biol 63:234
Albertini DF, Anderson EA (1975) Anat Rec 181:171
Albertini DF, Anderson EA (1977) J Cell Biol 73:111
Albertini DF, Kravit NG (1981) J. Biol Chem 256:2484
Albertini DF, Fawcett DW, Olds PJ (1975) Tiss Cell 7:389
Amenta F, Cavallotti C (1980) Acta Histochem Cytochem 13:619
Amenta F, Allen DJ, Didio LJA, Motta P (1979) J Submicrosc Cytol 11:39
Amin H, Richart RM, Brinson AO (1976) Obstet Gynaecol 47:562
Amsterdam A, Lindner HR (1979) In: Midgley AR Jr, Sadler WA (eds) Ovarian follicular development and function. Raven Press, New York, pp 137–138
Amsterdam A, Josephs R, Liebermann E, Lindner HR (1974a) J Cell Biol 63 (Pt 2):8a
Amsterdam A, Josephs R, Liebermann E, Lindner HR (1974b) Isr J Med Sci 10:1578 (Abstr)
Amsterdam A, Koch Y, Liebermann E, Lindner HR (1975) J Cell Biol 67:894
Amsterdam A, Josephs R, Liebermann E, Lindner HR (1976) J Cell Sci 21:93
Amsterdam A, Lindner HR, Groschel-Stewart U (1977) Anat Rec 37:187
Amsterdam A, Kohen F, Nimrod A, Lindner HR (1979a) In: Channing CP, Marshand J, Sadler WA (eds) Ovarian follicular and *Corpus Luteum* function. Plenum Press, New York, pp 69–75
Amsterdam A, Shemesh M, Salomon Y (1979b) In: Channing CP, Marsh J, Sadler A (eds) Ovarian follicular and *Corpus Luteum* function. Plenum Press, New York, pp 401–406
Andersen MM, Meurling P (1977) Acta Zool (Stockholm) 58:95
Andersen MM, Kroll J, Byskov AG, Faber M (1976) J Reprod Fertil 48:109
Anderson E (1967) J Cell Biol 35:160
Anderson E (1971) Anat Rec 169:473
Anderson E (1972) In: Biggers JD, Schuetz AW (eds) Oogenesis. Univ Park Press, Baltimore, pp 87–117
Anderson E (1979) In: Midgley AR Jr, Sadler WA (eds) Ovarian follicular development and function. Raven Press, New York, pp 91–105
Anderson E, Albertini DF (1976) J Cell Biol 71:680
Anderson E, Wilkisson RF, Lee G, Meller S (1978) J Morphol 156:339

Anderson LD, Channing CP (1978) Fed Proc 37:283
Anderson LD, Hoover DJ (1982) In: Channing CP, Segal SJ (eds) Intraovarian control mechanisms. Plenum Press, New York, pp 53–78
Anderson LD, Schaerf FW, Channing CP (1979) In: Channing CP Jr, Marsh J, Sadler WA (eds) Ovarian follicular and *Corpus Luteum* function. Plenum Press, New York, pp 187–193
Aonuma S, Okabe M, Kawai Y, Kawaguchi M (1978) Chem Pharm Bull (Tokyo) 26:405
Apkarian R, Curtis JC(1981) Scan Electr Microsc 1981:165
Arai H (1920a) Am J Anat 27:405
Arai H (1920b) Am J Anat 28:59
Armstrong DT (1980) In: Tozzini RI, Reeves G, Pineda RL (eds) Endocrine physiopathology of the ovary. Elsevier/North-Holland, Biomedical Press, Amsterdam, pp 165–178
Armstrong DT, Dorrington JH (1977) In: Thomas JA, Singhal RH (eds) Regulatory mechanisms affecting gonadal hormone action. Univ Park Press, Baltimore, pp 217–258
Armstrong DT, Zamecnik J (1975) Mol Cell Endocrinol 2:125
Armstrong DT, Grinwich DL, Moon YS, Zamecnik J (1974) Life Sci 14:129
Armstrong DT, Goff AK, Dorrington JH (1979) In: Midgley AR Jr, Sadler WA (eds) Ovarian follicular development and function. Raven Press, New York, pp 169–182
Arnaud R St, Walker P, Kelley PA, Labrie F (1983) In: Greenwald GS, Terranova PF (eds) Factors regulating ovarian function. Raven Press, New York, pp 209–214
Auerbach S, Brinster RL (1967) Exp Cell Res 46:89
Austin CR (1956) Exp Cell Res 10:533
Austin CR (1961) The mammalian egg. Blackwell, Oxford
Austin CR, Braden AWH (1954) Aust J Biol Sci 7:195
Austin CR, Braden AWH (1956) J Exp Biol 33:358
Austin CR, Lovelock J (1958) Exp Cell Res 15:267
Ax RL, Ryan AJ (1979) Biol Reprod 20:1123
Ax RL, LaBarbera AR, Ryan RJ (1979) In: Channing CP, Marsh J, Sadler WA (eds) Ovarian follicular and *Corpus Luteum* function. Plenum Press, New York, pp 77–82
Ayalon D, Tsafriri A, Lindner HR, Cordovat T, Harella A (1972) J Reprod Fertil 31:51
Baaken AH (1976) J Cell Biol 70:144A
Baca M, Zamboni L (1967) J Ultrastruct Res 19:354
Bachvarova R (1974) Dev Biol 40:52
Bachvarova R (1981) Dev Biol 86:384
Bachvarova R, De Leon V (1977) Dev Biol 58:248
Bachvarova R, Baran MM, Tejblum A (1980) J Exp Zool 211:159
Bachvarova R, De Leon V, Spiegelman I (1981) J Embryol Exp Morphol 62:153
Bae I-H, Foote RH (1975) J Reprod Fertil 42:357
Bae I-H, Foote RH (1980) J Reprod Fertil 59:11
Baeckelan E, Heinen E (1980) Bull Assoc Anat 64:173
Bahr J, Kao L, Nalbandov AV (1974) Biol Reprod 10:273
Bahr J, Shahabi N, Gardner R, Critchlow L (1979) In: Channing CP, Marsh J, Sadler WA (eds) Ovarian follicular and *Corpus Luteum* function. Plenum Press, New York, pp 219–224
Bahr JM, Arbogast LA, Wang S-C, Dial OK, Ben-Jonathan N (1981) In: Schwartz NB, Hunzicker-Dunn M (eds) Dynamics of ovarian function. Raven Press, New York, pp 129–133
Baillie AH, Ferguson MM, Hart DM (1966) Developments in steroid histochemistry. Academic Press, London New York
Baird DT (1977a) In: James VHT (ed) Endocrinology, vol I Excerpta Medica. Amsterdam, p 330
Baird DT (1977b) J Reprod Fertil 50:183
Baird DT (1983) J Reprod Fertil 68:343
Baird DT, Baker TG, McNatty KP, Neal P (1975) J Reprod Fertil 45:611
Baker TG (1969) In: Siker MR, Mahlum DD (eds) Radiation biology of the fetal and juvenile mammal. US Atom Energ Comm CoNF-690501, p 955
Baker TG (1970) Adv Biosci 6:7
Baker TG (1971) Mut Res 11:9
Baker TG (1972) In: Balin H, Glasser S (eds) Reproductive biology. Excerpta Medica, Amsterdam, pp 398–437

Baker TG (1979) In: Midgley AR Jr, Sadler WA (eds) Ovarian follicular development and
    function. Raven Press, New York, pp 353–364
Baker TG, Franchi LL (1967) Chromosoma 22:358
Baker TG, Franchi LL (1969) Z Zellforsch 93:54
Baker TG, Hunter RHF (1978) Ann Biol Anim Biochem Biophys 18:419
Baker TG, Neal P (1972) In: Biggers JD, Schuetz AW (eds) Oogenesis. Univ Park Press,
    Baltimore, pp 377–396
Baker TG, Beaumont HM, Franchi LL (1969) J Cell Sci 4:655
Baker TG, Hunter RHF, Biggs JSG (1977) In: James VHT (ed) Endocrinology, vol I. Int Congr
    Ser No 402. Excerpta Medica, Amsterdam, pp 351–355
Baker TG, Challoner S, Burgoyne PS (1980) J Reprod Fertil 60:449
Balabanov YuV, Danilovskii MA (1976) Tsitologiya 18:985
Balachandran PK, Moodbidri SB, Nandedkar TD (1983) Experientia 39:792
Balakier H (1978) Exp Cell Res 112:137
Balakier H, Dzolowska R (1977) Exp Cell Res 110:466
Balboni GC (1976) In: James VHT, Serio M, Giusti G (eds) The endocrine function of the human
    ovary. Academic Press, London New York, pp 125–134
Bar-Ami S, Tsafriri A (1981) Gamete Res 4:463
Bar-Ami S, Nimrod A, Brodie AMH, Tsafriri A (1983) J Steroid Biochem 19:965
Baranczuk RJ, Fainstat T (1976) Am J Obstet Gynaecol 124:517
Baranska W, Konwinski M, Kujawa M (1975) J Exp Zool 192:193
Barton BR, Hertig AT (1972) Biol Reprod 6:98
Barton BR, Hertig AT (1975) J Anat 120:227
Batten BE, Anderson E (1981) Am J Anat 161:101
Batten BE, Anderson E (1983) Am J Anat 167:119
Bauminger S (1977) 59th Annu Meet Endocr Soc, New York, Abstr 81
Bauminger S, Lindner HR (1975) Prostaglandins 9:737
Bauminger S, Lieberman ME, Lindner HR (1975) Prostaglandins 9:753
Bedford JM (1974) In: Coutinho EM, Fuchs F (eds) Physiology and genetics of reproduction.
    Plenum Press, New York, pp 55–68
Bedford JM (1977) Anat Rec 188:477
Beers WH (1975) Cell 6:379
Beers WH, Dekel N (1981) In: Schwartz NB, Hunzicker-Dunn M (eds) Dynamics of ovarian
    function. Raven Press, New York, pp 95–104
Beers WH, Strickland S, Reich E (1975) Cell 6:387
Behrman, HR, Caldwell BV (1974) In: Greep RO (ed) Reproductive Physiology, vol VIII. Univ
    Park Press, Baltimore, pp 63–94
Beltermann R (1965) Arch Gynaekol 200:601
Bernard J, Psychoyos A (1977) J Reprod Fertil 49:355
Beyer C, Cruz ML, Gay VL, Jaffe RB (1974) Endocrinology 95:722
Bier, HMH (1968) Biochim Biophys Acta 160:289
Biggers JD (1972) In: Biggers JD, Schuetz AW (eds) Oogenesis. Univ Park Press, Baltimore, pp
    241–251
Biggers JD, Powers RD (1979) In: Midgley AR Jr, Sadler WA (eds) Ovarian follicular
    development and function. Raven Press, New York, pp 365–373
Biggers JD, Schuetz AW (eds) (1972) Oogenesis, Univ Park Press, Baltimore
Biggers JD, Finn CA, McLaren A (1962) J Reprod Fertil 3:303
Biggers JD, Whittingham DG, Donahue RP (1967) Proc Natl Acad Sci USA 58:560
Bill CH, Greenwald GS (1979) Biol Reprod 20 Suppl 1, Abstr 46a
Bill CH, Greenwald GS (1981) Biol Reprod 24:913
Binov Z, Wolf DP (1979) J Reprod Fertil 56:309
Bjärsing L (1977) In: Zuckerman S, Weir BJ (eds) The ovary. Academic Press, London New York,
    pp 303–391
Bjärsing L (1978) In: Jones RE (ed) The vertebrate ovary. Plenum Press, New York, pp 181–214
Bjärsing L (1982) In: Channing CP, Segal SJ (eds) Intraovarian control mechanisms. Plenum
    Press, New York, pp 1–14
Bjärsing L, Cajander S (1974a) Cell Tiss Res 149:287

Bjërsing L, Cajander S (1974b) Cell Tiss Res 149:301
Bjërsing L, Cajander S (1974c) Cell Tiss Res 149:313
Bjërsing L, Cajander S (1974d) Cell Tiss Res 153:1
Bjërsing L, Cajander S (1974e) Cell Tiss Res 153:15
Bjërsing L, Cajander S (1974f) Cell Tiss Res 153:31
Bjërsing L, Cajander S (1975a) Experientia 31:605
Bjërsing L, Cajander S (1975b) In: Adv Abstr Pap. 10th Acta Endocrinol Congr, Amsterdam, August 1975, Suppl 199, p 105
Björkman N (1962) Acta Anat 51:125
Blandau RJ (1966) In: Greenblatt RB (ed) Ovulation: stimulation, suppression and detection. Lippincott, Philadelphia, p 3
Blandau RJ, Rumery RE (1963) Fertil Steril 14:330
Blank M, Soo L, Britten JS (1976) J Membr Biol 29:410
Bleil JD, Wassarman PM (1978) J Cell Biol 79:173a
Bleil JD, Wassarman PM (1980a) Proc Natl Acad Sci USA 77:1029
Bleil JD, Wassarman PM (1980b) Dev Biol 7:185
Bleil JD, Wassarman PM (1980c) Cell 20:873
Bleil JD, Beall CF, Wassarman PM (1981) Dev Biol 86:189
Blerkom van J (1977) In: Johnson MH, Edidin M (eds) Immunobiology of the gametes. Univ Press, Cambridge, pp 187–206
Blerkom van J (1979) Dev Biol 72:188
Blerkom van J (1980) In: Motta PM, Hafez ESE (eds) Biology of the ovary. Nijhoff, The Hague, pp 179–190
Blerkom van J (1981a) In: Glasser SR, Bullock DW (eds) Molecular and cellular aspects of implantation. Plenum Press, New York, pp 155–176
Blerkom van J (1981b) Proc Natl Acad Sci USA 78:7629
Blerkom van J (1983) In: Metz CB, Monroy A (eds) Biology of fertilization. Academic Press, London New York (in press)
Blerkom van J (1984) In: Venezick C (ed) Control of cell growth and proliferation. Norstrand, in press
Blerkom van J, Manes C (1977) In: Sherman MI (ed) Concepts in mammalian embryogenesis. MIT Press, Cambridge, Mass, pp 37–94
Blerkom van J, McGaughey RW (1978a) Dev Biol 63:139
Blerkom van J, McGaughey RW (1978b) Dev Biol 63:151
Blerkom van J, Motta P (1979) The cellular basis of mammalian reproduction. Urban & Schwarzenberg, Munich
Blerkom van J, Runner MN (1983) (submitted)
Bloom AM, Mukherjee BB (1972) Exp Cell Res 74:577
Bogovich K, Richards JS, Reichert Jr LE (1981) Endocrinology 109:860
Bohr J, Gardner R, Schenck P, Shahabi N (1980) Biol Reprod 22:817
Bongiovanni A, Santulli R, Wallach E (1983) Fertil Steril Suppl 397 (Abstr)
Bousquet D, Leveille MC, Roberts KD, Chapdelaine A, Blaeu G (1981) J Exp Zool 215:215
Bowring N, Earthly M, Mangan FR (1975) J Endocrinol 64:11P
Braden AWH (1952) Aust J Sci Res Ser B 5:460
Braden AWH, Austin CR, David HA (1954) Aust J Biol Sci 7:391
Brambell FWR (1956) In: Parkes AS (ed) Marshall's physiology of reproduction, 3rd edn. Longmans, Green, London, pp 397–542
Brand A, Jong De WHR (1973) J Reprod Fertil 33:431
Brandau H (1970) In: Butt WR, Crooke AC, Ryle M (eds) Gonadotrophins and ovarian development. Livingstone, Edinburgh London, pp 307–311
Braw RH, Bar-Ami S, Tsafriri A (1981) Biol Reprod 25:989
Brietenecker G, Friedrich F, Kemeter P (1978) Fertil Steril 29:336
Brinkworth RI, Masters CJ (1978) Mech Age Dev 8:299
Brinster RL (1965) Biochim Biophys Acta 110:439
Brinster RL (1966) Biochem J 101:161
Brinster RL (1967a) J Reprod Fertil 13:643
Brinster RL (1967b) Exp Cell Res 48:643
Brinster RL (1968) J Reprod Fertil 17:139

Brinster RL (1971a) Experientia 27:371
Brinster RL (1971b) Wilhelm Roux Arch Entwickl-Mech Org 166:308
Bronson R, Bryant G, Balk W, Emanuele N (1979) Fertil Steril 31:205
Brower PT, Schultz RM (1982a) Dev Biol 90:144
Brower PT, Schultz RM (1982b) J Exp Zool 220:257
Brower PT, Gizang E, Boreen SM, Schultz RM (1981) Dev Biol 86:373
Brunning JL, Hay MF, Moor RM, Cran DC, Dott HM (1979) J Reprod Fertil 55:195
Burden HW (1972) Am J Anat 133:125
Burden HW (1973) J Morphol 140:467
Burden HW (1978) In: Jones RE (ed) The vertebrate ovary. Plenum Press, New York, pp 615–638
Burden HW, Lawrence IE Jr (1980) In: Motta PM, Hafez ESE (eds) Biology of the ovary. Nijhoff,
    The Hague, pp 99–105
Burgoyne PS, Borland RM, Biggers JD, Lechene CP (1979) J Reprod Fertil 57:575
Burkholder GD, ComingsDE, Okada IA (1971) Exp Cell Res 69:361
Burkl W (1962) Z Zellforsch 58:369
Burkl W, Schiechl H (1976) Z Mikrosk Anat Forsch 90:273
Burne JL, Psychoyos A (1972) J Reprod Fertil 30:489
Burr JST, Davies JI (1951) Anat Rec 111:273
Byskov AGS (1969) Z Zellforsch 100:285
Byskov AGS (1974) J Reprod Fertil 37:277
Byskov AGS (1978) In: Jones RE (ed) The vertebrate ovary. Plenum Press, New York, pp
    533–562
ByskovAGS (1979) In: Midgley AR Jr, Sadler WA (eds) Ovarian follicular development and
    function. Raven Press, New York, pp 41–57
Byskov AGS, Rasmussen G (1973) In: Peters H (ed) The development and maturation of the
    ovary and its functions. Int Congr Ser No 267. Excerpta Medica, Amsterdam, pp 55–62
Cahill LP, Mauleon P (1980) J Reprod Fertil 58:321
Cahill LP, Mauleon P (1981) J Reprod Fertil 61:201
Cajander S (1976) Cell Tiss Res 173:437
Cajander S, Bjërsing L (1973) Acta Pathol Microbiol Scand Sect A 81:866
Cajander S, Bjërsing L (1975) Cell Tiss Res 164:279
Cajander S, Bjërsing L (1976) Cell Tiss Res 169:129
Calarco PG, Donahue RP, Szöllösi D (1972) J Cell Sci 10:369
Cameron IL, Lum JB, Nations C, Asch RH, Silverman AY (1983) Biol Reprod 28:817
Campbell MD, Neymark MA, Hill PK, Rothkope MM, Dimino MJ (1980) Biol Reprod 23:231
Canipari R, Pietrolucci A, Mangia F (1979) J Reprod Fertil 57:405
Cargille CM, Vaitukaitis JL, Bermudez JA, Ross GT (1973) J Clin Endocrinol Metab 36:87
Carnegie JA, Tsang BK (1983a) In: Greenwald GS, Terranova PF (eds) Factors regulating
    ovarian function. Raven Press, New York, pp 375–380
Carnegie JA, Tsang BK (1983b) Am J Obstet Gynaecol 145:223
Carson RS, Findlay JK, Burger HG (1979) In: Channing CP, Marsh J, Sadler WA (eds) Ovarian
    follicular and Corpus Luteum function. Plenum Press, New York, pp 89–94
Carson RS, Richards JS, Kahn LE (1981) Endocrinology 109:1433
Catt KJ, Dufau ML (1977) Ann Rev Physiol 39:529
Caucig H, Friedrich F, Breitenecker G, Golob E (1972) Gynaecol Invest 3:215
Cavallotti C, DiDio LJA, Familiari G, Fumagalli G, Motta P (1975) Acta Histochem 52:253
Challis JRG, Erickson GF, Ryan KG (1974) Prostaglandins 7:183
Chang MC (1955) J Exp Zool 128:378
Chang SCS, Jones JD, Ellefson RD, Ryan RJ (1976) Biol Reprod 15:321
Chang SCS, Anderson W, Lewis JC, Ryan RJ, Kang YH (1977) Biol Reprod 16:349
Channing CP (1969) In: Mc Kerns KW (ed) The gonads. Appleton Century, New York, pp
    415–490
Channing CP (1970) Recent Progr Horm Res 26:589
Channing CP (1979a) In: Midgley AR Jr, Sadler WA (eds) Ovarian follicular development and
    function. Raven Press, New York, pp 59–64
Channing CP (1979b) In: Channing CP, Marsh J, Sadler WA (eds) Ovarian follicular and Corpus
    Luteum function. Plenum Press, New York, pp 327–343

Channing CP, Batta SK (1981) In: Coutts JRT (ed) Functional morphology of the human ovary. MTP Press, Lancaster, pp 73–84

Channing CP, Coudert SP (1976) Endocrinology 98:590

Channing CP, Kammerman S (1973) Endocrinology 92:531

Channing CP, Ledwitz-Rigby F (1975) In: Coutino EM, Fuchs F (eds) Physiology and genetics of reproduction. Plenum Press, New York, pp 353–370

Channing CP, Reichert LE Jr (1983) In: Greenwald GS, Terranova PF (eds) Factors regulating ovarian function. Raven Press, New York, pp 329–332

Channing CP, Segal SJ (eds) (1982) Intraovarian control mechanisms. Plenum Press, New York

Channing CP, Tsafriri A (1977) Metabolism 26:413

Channing CP, Tsafriri A (1978) In: Sadler WA, Segal S (eds) Advances in fertility regulation through basic research. Plenum Press, New York

Channing CP, Wentz AC, Jones GS (1978) Symp Int Endocrinol Ovaire, 6–8 October, France

Channing CP, Marsh J, Sadler WA (eds) (1979a) Ovarian follicular and *Corpus Luteum* function. Plenum Press, New York

Channing CP, Anderson LD, Hodgen GD (1979b) In: Channing CP, Marsh J, Sadler WA (eds) Ovarian follicular and *Corpus Luteum* function. Plenum Press, New York, pp 407–415

Channing CP, Schaerf FW, Anderson LP, Tsafriri A (1980) In: Greep RO (ed) Reproductive physiology III. International Review of Physiology, vol 22. Univ Park Press, Baltimore, pp 117–201

Channing CP, Batta SK, Condon W, Ganjam VK, Kenney RM (1981) In: Schwartz NB, Hunzicker-Dunn M (eds) Dynamics of ovarian function. Raven Press, New York, pp 73–78

Channing CP, Anderson LD, Hoover DJ, Kolena J, Osteen KG, Pomerantz SH, Tanabe K (1982) Recent Progr Horm Res 38:331

Channing CP, Garrett R, Kroman N, Conn T, Gospodarowicz D (1983a) In: Greenwald GS, Terranova PF (eds) Factors regulating ovarian function. Raven Press, New York, pp 215–220

Channing CP, Pomerantz SH, Bae I-H, Evans VW, Atlas SJ (1983b) In: Channing CP, Segal SJ (eds) Intraovarian control mechanisms. Plenum Press, New York, pp 189–210

Chappel SC, Acott T, Spies HG (1979) In: Channing CP, Marsh J, Sadler WA (eds) Follicular and *Corpus Luteum* function. Plenum Press, New York, pp 361–371

Charlesworth MC, Grady RR, Schwartz NB (1983) In: Greenwald GS, Terranova PF (eds) Factors regulating ovarian function. Raven Press, New York, pp 169–174

Chatterjee A, Greenwald GS (1971) Endocrinology 88:491

Chaudhuri G, Elder MG (1976) Prostaglandins 11:727

Cherney DD, Didio LJA, Motta P (1975) Fertil Steril 26:257

Chiras DD, Greenwald GS (1980) Am J Anat 157:309

Cho, WK, Kim MK, Chung SO (1971) Proc 7th Congr Fertil Steril Excerpta Medica. Int Congr Ser No 278, Amsterdam, pp 706–708

Cho WK, Stern S, Biggers JD (1974) J Exp Zool 187:383

Cholewa-Stewart J, Massaro E (1972) Biol Reprod 7:166

Christensen AK (1975) In: Greep RO, Astwood EB (eds) Handbook of physiology, section 7: endocrinology, vol V. Male reproductive system. Am Physiol Soc, Washington DC, pp 57–94

Christensen AK, Gillim SW (1969) In: McKerns KW (ed) The gonads. Appleton Century, New York, pp 415–488

Church RB, Schultz GA (1974) Curr Top Dev Biol 8:179

Cinader B, Weck De A (eds) (1976) Immunological response of the female reproductive tract. Scriptor, Copenhagen

Clark MR, Marsh JM, LeMaire WJ (1976) Prostaglandins 12:209

Clark MR, Triebwasser WF, Marsh JM, LeMaire WJ (1978a) Ann Biol Anim Biochem Biophys 18:427

Clark MR, Marsh JM, LeMaire WJ (1978b) Endocrinology 102:39

Clark JR, Dierschke DJ, Wolf RC (1979) In: Midgley AR Jr, Sadler WA (eds) Ovarian follicular development and function. Raven Press, New York, pp 71–74

Clark MR, Thibier C, Marsh JM, LeMaire WJ (1980) Endocrinology 107:17

Clark JR, Dierschke DJ, Wolf RC (1981) Biol Reprod 25:332

Colonna R, Mangia F (1983) Biol Reprod 28:797

Conti M, Harwood JP, Hsueh AJW, Dufau, ML, Catt KKJ (1976) J Biol Chem 251:7729
Cons LW, Espey LL (1977) J Cell Biol 74:321
Coulson PB (1979) In: Midgley AR Jr, Sadler WA (eds) Ovarian follicular development and
    function. Raven Press, New York, pp 385–395
Coutinho E, Maia H (1972) Nature (London) 235:94
Cran DG (1983) J Reprod Fertil 67:415
Cran DG, Hay MF, Moor RM (1979) Cell Tiss Res 202:439
Cran DG, Moor RM, Hay MF (1976) Acta Endocrinol 82:631
Cran DG, Moor RM, Crosby I (1981) Exp Cell Res 134:251
Crespigny de LC, Herlihy CO, Robinson HP (1981) Am J Obstet Gynaecol 139:636
Crisp TM, Dessouky DA, Denys FR (1970) Am J Anat 127:37
Cross PC, Brinster RL (1974) Exp Cell Res 86:43
Crozet N, Szöllösi D (1980) Biol Cell 38:163
Crozet N, Motlik J, Szöllösi D (1981) Biol Cell 41:35
Cuatrecasas P, Hollenberg MD, Chang K, Bennet V (1975) Recent Prog Horm Res 31:37
Daguet M-C (1980) Reprod Nutr Dev 20:673
D'Amato C, Bahr J, Stockert CF, Kesler D (1979) In: Channing CP, Marsh J, Sadler WA (eds)
    Ovarian follicular and *Corpus Luteum* function. Plenum Press, New York, pp 235–239
D'Amato C, Calvo FO, Stockert B, Bahr JM (1981) Biol Reprod 25:843
Darga NC, Reichert LE Jr (1978) Biol Reprod 19:235
Darga NC, Reichert LE Jr (1979) In: Channing CP, Marsh J, Sadler WA (eds) Ovarian follicular
    and *Corpus Luteum* function. Plenum Press, New York, pp 383–388
Daume E, Chari S, Hopkkinson CRN, Sturm G, Hirschhauser C (1978) Klin Wochenschr 56:369
Dave BK, Graves CN (1978) Indian J Anim Sci 48:257
Davidson EH (1976) Gene activity in early development, 2nd edn. Academic Press, London New
    York
Davidson EH, Hough BR (1972) In: Biggers JD, Schuetz AW (eds) Oogenesis. Park Press,
    Baltimore, pp 129–139
Davies J, Broadus GD (1968) Am J Anat 123:441
Davis JS, Clark MR (1983) In: Greenwald GS, Terranova PF (eds) Factors regulating ovarian
    function. Raven Press, New York, pp 281–287
Davis JS, Farese RV, Clark MR (1983) Endocrinology 112:2212
Day SL, Nalbandov AV (1977) Biol Reprod 16:486
Deane HW (1951) Anat Rec 111:504
Deane HW (1952) Am J Anat 91:363
Deane HW, Adrews JS (1953) J Histochem Cytochem 1:283
Deane HW, Lobel BL, Romney SL (1962) J Obstet Gynaecol 83:281
Dekel N, Beers WH (1978) Proc Natl Acad Sci USA 75:4369
Dekel N, Beers WH (1980) Dev Biol 75:247
Dekel N, Kraicer PF (1974) In: Symposium: Functional morphology of the ovary. Glasgow,
    September 11–14 (Abstr)
Dekel N, Kraicer PF (1977) Isr J Med Sci 13:621
Dekel N, Phillips DM (1979) Biol Reprod 21:9
Dekel N, Hultborn R, Hillensjö T, Hamberger L, Kraicer P (1976) Endocrinology 98:498
Dekel N, David MP, Yedwab GA, Kraicer PF (1977) Int J Fertil 22:24
Dekel N, Hillensjö T, Kraicer PF (1979) Biol Reprod 20:191
Dekel N, Lawrence TS, Gilula NB, Beers WH (1981) Dev Biol 86:356
DeLeon V, Bachvarova R (1978) J Cell Biol 79:F907
Delforge JP, Thomas K, Roux F, Carneiro de Siqueira J, Ferin J (1972) Fertil Steril 23:1
Denahue RP, Stern S (1968) J Reprod Fertil 17:395
Dennefors BL, Hamberger L, Nilsson L (1983) Fertil Steril 39:59
Dickmann Z (1969) Adv Reprod Physiol 4:187
Dickmann Z, Dziuk PJ (1964) J Exp Biol 41:603
DiDio LJA, Allen DJ, Correr S, Motta PM (1980) In: Motta PM, Hafez ESE (eds) Biology of the
    ovary. Nijhoff, The Hague, pp 106–118
Dimino MJ, Cambell MD (1980) In: Motta PM, Hafez ESE (eds) Biology of the ovary. Nijhoff,
    The Hague, pp 191–195

Dimino MJ, Elfont EA (1980) In: Motta PM, Hafez ESE (eds) Biology of the ovary. Nijhoff, The
    Hague, pp 196–201
Dimino MJ, DeLiale FE, Downing JR (1979) In: Midgley AR Jr, Sadler WA (eds) Ovarian
    follicular development and function. Raven Press, New York, pp 199–201
Dodson KS, Watson J (1979) In: Channing CP, Marsh J, Sadler WA (eds) Ovarian follicular and
    Corpus Luteum function. Plenum Press New York, pp 95–103
Donahue RP, Stern S (1968) J Reprod Fert 17:395
Dorrington JH, Armstrong DT (1979) In: Midgley AR Jr, Sadler WA (eds) Ovarian follicular
    development and function. Raven Press, New York, pp 199–201
Dorrington JH, Gore-Langton RE (1982) Endocrinology 110:1701
Downs SM, Longo FJ (1982) Am J Anat 164:265
Downs SM, Longo FJ (1983) Anat Rec 205:159
Dubreuil G (1957) Acta Anat 30:289
Dudkiewicz AB, Shivers CA, Williams WL (1973) Biol RReprod 14:175
Dudkiewicz AB, Noske IG, Shivers A (1975) Fertil Steril 26:686
Dudkiewicz AB, Shivers CA, Williams WL (1976) Biol Reprod 14:175
Dufau ML, Hayashi K, Sala G, Baukal A, Catt KJ (1979) In: Channing CP, Marsh J, Sadler WA
    (eds) Ovarian follicular and *Corpus Luteum* function. Plenum Press, New York, pp 45–51
Dufy-Barbe L, Franchimont P, Faure JMA (1973) Endocrinology 92:1321
Dunbar BS (1980) In: IX Proc Int Congr Reprod Anim Insem Artif, vol III. Madrid, Spain, pp
    191–199
Dunbar BS (1982) In: Wegmann T, Gill T (eds) International congress on reproduction and
    immunology. Oxford Univ Press, New York
Dunbar BS, Raynor BD (1980) Biol Reprod 22:941
Dunbar BS, Roberts S (1982) Fed Proc 41:1159
Dunbar BS, Shivers CA (1976) Immunol Commun 5:375
Dunbar BS, Wardrip NJ, Hedrick JL (1978a) J Cell Biol 79:163a
Dunbar BS, Wardrip NJ, Hedrick JL (1978b) Biol Reprod 18, Suppl 1:22a
Dunbar BS, Wardrip NJ, Hedrick JL (1980) Biochemistry 19:356
Dunbar BS, Liu C, Sammons DW (1981) Biol Reprod 24:1111
Dvořák M, Tesařík J (1980) In: Motta PM, Hafez ESE (eds) Biology of the ovary. Nijhoff, The
    Hague, pp 121–137
Eager DD, Johnson MH, Thurley KW (1976) J Cell Sci 22:345
Edwards RG (1962) Nature (London) 196:446
Edwards RG (1974) J Reprod Fertil 37:189
Edwards RG, Steptoe PC (1975) J Reprod Fertil 22 (Suppl):121
Edwards RG, Fowler RE, Gore-Langton RE, Gosden RG, Jones EC, Readhead C, Steptoe PC
    (1977) J Reprod Fertil 51:237
Edwards RG, Steptoe PC, Fowler RE, Baillie J (1980) Br J Obstet Gynaecol 87:769
Ekholm C, Hillensjö T, Isaksson O (1981a) Endocrinology 108:2022
Ekholm C, Clark MR, Magnusson C, Isaksson O, Le Maire WJ (1981b) Endocrinology (in press)
Ekholm C, Hillensjö T, LeMaire WJ, Magnusson C, Sheeela Rani CS (1982) J Reprod Fertil (in
    press)
Elfont EA, Roszka JP, Dimino MJ (1977) Biol Reprod 17:787
El-Fouly MA, Cook B, Nekola M, Nalbandov AV (1970) Endocrinology 87:288
Ellinwood WE, Nett TM, Niswender GD (1978) In: Jones RE (ed) The vertebrate ovary. Plenum
    Press, New York, pp 583–609
Ellinwood WE, McClellan MC, Brenner RM, Resko JA (1983) Biol Reprod 28:505
Enders AC (1971) In: Blandau RJ (ed) The biology of the blastocyst. Univ Chicago Press,
    Chicago, pp 71–94
Engel W, Franke W (1975) In: Cropp A, Benirschke K (eds) Developmental biology and
    pathology. Springer, Berlin Heidelberg New York
Engel W, Kreutz R (1973) Human genetic 19:253
Engel W, Zenzes MT (1976) In: Crosignani PG, Mishell DR (eds) Ovulation in the human.
    Academic Press, London New York, pp 57–70
Engel W, Franke W, Petzoldt U (1975) In: Market CL (ed) Oozymes III developmental biology.
    Academic Press, London New York, pp 67–81

Eppig JJ (1976) J Exp Zool 198:375
Eppig JJ (1977) Dev Biol 60:371
Eppig JJ (1979) Nature (London) 281:483
Eppig JJ (1980) Biol Reprod 23:545
Eppig JJ (1981 a) Endocrinology 108:1992
Eppig JJ (1981 b) Biol Reprod 25:599
Eppig JJ (1982) Dev Biol 89:269
Eppig JJ, Koide SL (1978) J Reprod Fertil 53:99
Eppig JJ, Ward-Bailey PF (1982) Gamete Res 6:145
Eppig JJ, Ward-Bailey PF, Potter JER, Schultz RM (1982) Biol Reprod 27:399
Epstein CJ (1975) Biol Reprod 12:82
Epstein ML, Beers WH, Gilula NB (1976) J Cell Biol 70:302 a
Erickson GF (1982) In: Flamigni G, Givens JR (eds) The gonadotropins: basic science and
     clinical aspects in females. Academic Press, London New York, pp 177–185
Erickson GF, Ryan KJ (1976) J Exp Zool 195:153
Erickson GF, Sorenson RA (1974) J Exp Zool 190:123
Erickson GF, Challins JRG, Ryan KJ (1977) J Reprod Fertil 49:133
Erickson GF, Wang C, Hsuch AJW (1979) Nature (London) 279:336
Erickson GF, Hofeditz C, Hsueh AJW (1983) In: Greenwald GS, Terranova PF (eds) Factors
     regulating ovarian function. Raven Press, New York, pp 257–262
Ermini M, Carenza L (1980) In: Motta PM , Hafez ESE (eds) Biology of the ovary. Nijhoff, The
     Hague, pp 254–265
Ernst LK, Sviridov BE, Galieva LD, Golubev AK, Yanushka AA, Pimenova MN (1980)
     Tsitologiya 22:475
Esch F, Ling N, Ying S-Y, Ginllemin R (1983) In: McCann SM, Dhindsa DS (eds) Role of
     peptides and proteins in control of reproduction. Eslevier/North Holland, Biomedical Press,
     New York, pp 275–292
Eshkol A, Misgav N, Lunenfeld B (1976) In: Crosignani PG, Mishell DR (eds) Ovulation in the
     human. Academic Press, London New York, pp 79–86
Espey LL (1967) Endocrinology 81:267
Espey LL (1969) Fed Proc 28:638
Espey LL (1971) Endocrinology 88:437
Espey LL (1974) Biol Reprod 10:216
Espey LL (1976) Biol Reprod 14:502
Espey LL (1978 a) In: Jones RE (ed) The vertebrate ovary. Plenum Press, New York, pp 503–532
Espey LL (1978 b) Biol Reprod 19:540
Espey LL (1980) Biol Reprod 22:73
Espey LL (1982 a) Prostaglandin 23:330
Espey LL (1982 b) In: Program and abstracts. Endocr Soc, 64th Annul Meet San Francisco
Espey LL, Coons PJ (1976) Biol Reprod 14:233
Espey LL, Lipner H (1963) Am J Physiol 205:1067
Espey LL, Lipner H (1965) Am J Physiol 208:208
Espey LL, Rondell P (1968) Am J Physiol 214:326
Espey LL, Rawson JMR (1979) In: Midgley AR Jr, Sadler WA (eds) Ovarian follicular
     development and function. Raven Press, New York, pp 155–158
Espey LL, Stacey S (1970) Fed Proc Fed Am Soc Exp Biol 29:833
Espey LL, Stutts, RH (1972) Biol Reprod 6:168
Espey LL, Stein VI, Dumitrescu J (1982) Fertil Steril 38:238
Espey LL, Coons PJ, Marsh JM, LeMaire WJ (1981) Endocrinology 108:1040
Evans G, Dobias M, King GJ, Armstrong DT (1983 a) Biol Reprod 28:322
Evans G, Wathes DC, King GJ, Armstrong DT, Portes DG (1983 b) J Reprod Fertil 69:677
Ezzell RM, Szego CM (1977) J Cell Biol 75, 2, (pt 2):1739
Ezzell RM, Szego CM (1979) J Cell Biol 82:264
Faddy MJ, Jones EC, Edwards RG (1976) J Exp Zool 197:173
Fajer AB (1983) J Reprod Fertil 69:101
Falck B (1959) Acta Physiol Scand 47 (Suppl): 163
Familiari G, Simongini E, Motta PM (1981) Acta Histochem 69:193

Farookhi R (1980) Endocrinology 106:1216

Farookhi R (1981) In: Schwartz NB, Hunzicker-Dunn M (eds) Dynamics of ovarian function. Raven Press, New York, pp 13–23

Fléchon JE (1970) J Microsc 9:221

Fléchon JE, Gwatkin RBL (1980) Gam Res 3:141

Fléchon JE, Huneau D, Solari A, Thibault C (1975) Ann Biol Anim Biochem Biophys 15:9

Fleming WN, Saacke RG (1972) J Reprod Fertil 29:203

Fleming AD, Khalil W, Armstrong DT (1983) J Reprod Fertil 69:665

Fletcher WH (1979) In: Midgley AR Jr, Sadler WA (eds) Ovarian follicular development and function. Raven Press, New York, pp 113–120

Foote WD, Thibault C (1969) Ann Biol Anim Biochem Biophys. 9:329

Foote WD, Mills CD, Phelps DA, Tibbitts FD (1978) Ann Biol Anim Biochem Biophys 18:435

Fortune JE, Armstrong DT (1978) Endocrinology 102:277

Fortune JE, Armstrong DT (1979) In: Midgley AR Jr, Sadler WA (eds) Ovarian follicular development and function. Raven Press, New York, pp 193–198

Fortune JE, Hansel W (1979) In: Channing CP, Marsh J, Sadler WA (eds) Ovarian follicular and Corpus Luteum function. Plenum Press, New York, pp 203–208

Fox CL, Shivers CA (1975) Fertil Steril 26:599

Franchimont P, Lecomte-Yerna M-J, Hendersen K, Verhoeven G, Hazee-Hagelstein M-T, Jaspar J-M, Charlet-Renard C, Dimoulin A (1983) In: McCann SM, Dhindsa DS (eds) Role of peptides and proteins in control of reproduction. Elsevier/North Holland Biomedical Press, New York, pp 237–256

Franks S, Murray MAE, Jequier AM, Steele SJ, Nabarro JDN, Jacobs HS (1975) Clin Endocrinol 4:597

Frantz AG, Kleinberg DL, Noel GL (1972) Recent Progr Horm Res 28:527

Fraser IS, Baird DT, Cockburn F (1973) J Reprod Fertil 33:111

Fraser LR, Dandekar PV, Gordon MK (1972) J Reprod Fertil 29:295

Freeman ME, Butcher RL, Fugo NW (1970) Biol Reprod 2:209

Friedrich F, Breitenecker G, Salzer H, Bolzner JH (1974) Acta Endocrinol 76:343

Friedrich F, Kemeter P, Salzer H, Breitenecker G (1975) Acta Endocrinol 78:332

Fujii T, Hoover DJ, Channing CP (1983) J Reprod Fertil 69:307

Fukui Y, Sakuma Y (1980) Biol Reprod 22:669

Fukui Y, Fukushima M, Terawaki Y, Ono H (1982) Theriogeneology 18:161

Fukumoto M, Yajima Y, Okamura H, Midorikawa O (1981) Fertil Steril 36:756

Fukushima M (1977) Int J Fertil 22:206

Fumagalli Z, Motta P, Calavieri S (1971) Experientia 27:682

Gachechiladze TsV, Labadze MV (1977) Soobshch Akad Nauk Gruz 87:485

Gachechiladze TsV, Togonidze BM (1976) Soobshch Akad Nauk Gruz 84:209

Garavagno AA, Posada J, Barros C, Shivers CA (1974) J Exp Zool 189:37

Garcia RB, Pereyra-Alfonso S, Soteolo JR (1979a) Differentiation 14:101

Garcia RB, Pereyra-Alfonso S, Sateolo JR (1979b) In: Coutinho EM, Fuchs F (eds) Physiology and genetics of reproduction, vol IV A. Plenum Press, New York, pp 307–321

Gebauer H, Lindner HR, Amsterdam A (1978) Biol Reprod 18:350

Gemzell C (1963) In: Eckstein P, Knowles F (eds) Techniques in endocrine research. Academic Press, London New York, pp 213–230

Geschwind II (1963) In: Cole HH (ed) Gonadotrophins, their chemical and biological properties and secretory control. Freeman, San Francisco, pp 1–39

Gillet JY, Maillet R, Gautier C (1980) In: Motta PM, Hafez ESE (eds) Biology of the ovary. Nijhoff, The Hague, pp 86–98

Gilula NB (1977) Int Cell Biol Symp H: 61–69

Gilula NB, Reeves OR, Steinbach A (1972) Nature (London) 235:262

Gilula NB, Epstein ML, Beers WH (1978) J Cell Biol 78:58

Gimbo A, Germana G, Zanghi A (1980) Arch Ital Anat Embriol 85:71

Giorgi EP (1969) J Endocrinol 45:37

Glass LE (1963) Am Zool 3:135

Glass LE, Hansen JE (1974) Fertil Steril 25:484

Goff AK, Armstrong DT (1977) Endocrinology 101:1461

Golbus MS (1976) J Exp Zool 198:337

Golbus MS, Stein MP (1976) J Exp Zool 198:237

Goldenberg RL, Vaitukaitis JL, Ross TG (1972) Endocrinology 90:1492

Gondos B (1969) Anat Rec 165:67

Gonzáles-Santander R, Clavero Nunéz JA (1973) Acta Anat 84:106

Goodenough D (1974) J Cell Biol 61:557

Goodman AL, Neill JD (1976) Endocrinology 99:852

Gore-Langton RE, Lacroix M, Dorrington JH (1981) Endocrinology 108:812

Gosden RG, Laing SC, Flurkey K, Finch CE (1983) J Reprod Fertil 69:453

Gospodarowitz D, Bialecki H (1979) Endocrinology 104:757

Gougeon R, Lefévre B (1983) J Reprod Fertil 69:497

Gould K, Zaneveld LJD, Srivastava PN, Williams WL (1971) Proc Soc Exp Biol Med 136:6

Grady RR, Charlesworth C, Schwartz NB (1982) Rec Progr Horm Res 38:409

Greef de WJ, Jong De FH, Koning De FH, Stenbergen J, Vaart Van der PDM (1983) J
    Endocrinol 97:327

Green SH, Zuckerman S (1951) J Endocrinol 10:284

Greenwald GS (1961) J Reprod Fertil 2:351

Greenwald GS (1962) Endocrinology 71:664

Greenwald GS (1974) In: Greiger SR (ed) Handbook of physiology, sect 7: Endocrinology, vol
    IV, pt 2. Am Physiol Soc, Washington DC, pp 293–325

Greenwald GS (1978) In: Jones RE (ed) The vertebrate ovary. Plenum Press, New York, pp
    639–688

Greenwald GS (1979) In: Channing CP, Marsh JM, Sadler WA (eds) Ovarian follicular and
    Corpus Luteum function. Plenum Press, New York, pp 3–8

Greenwald GS (1980) In: Motta PM, Hafez ESE (eds) Biology of the ovary. Nijhoff, The Hague,
    pp 244–253

Greenwald GS, Siegel HI (1982) Proc Soc Exp Biol Med 170:225

Greenwald GS, Terranova PE (eds) (1983) Factors regulating ovarian function. Raven Press,
    New York

Greve JM, Satzmann GS, Wasserman PM (1982) Cell 31:749

Grinwich DL, Kennedy TG, Armstrong DT (1972) Prostaglandins 1:89

Gulyas BJ (1973) Anat Rec 177:195

Gulyas BJ (1974) Am J Anat 140:577

Gulyas BJ (1976) Am J Anat 147:203

Gulyas BJ (1980) Int Rev Cytol 63:357

Guraya SS (1965) J Exp Zool 160:123

Guraya SS (1966a) Acta Morphol Neerl Scand 4:395

Guraya SS (1966b) Am J Obstet Gynaecol 96:907

Guraya SS (1967a) Nature (London) 214:614

Guraya SS (1967b) Am J Obstet Gynaecol 98:99

Guraya SS (1967c) Z Zellforsch 83:187

Guraya SS (1967d) Res Bull Panjab Univ 18:191

Guraya SS (1968a) Acta Anat 70:447

Guraya SS (1968b) VI Proc Int Congr Reprod Anim Insem Artif, Paris 1:141

Guraya SS (1968c) Acta Biol Hung 19:279

Guraya SS (1968d) Acta Anat 70:623

Guraya SS (1968e) Am J Obstet Gynaecol 101:448

Guraya SS (1968f) J Reprod Fertil 15:381

Guraya SS (1969a) Z Zellforsch 94:32

Guraya SS (1969b) Acta Morphol Neerl Scand 7:211

Guraya SS (1969c) Acta Anat 74:65

Guraya SS (1969d) Acta Vet Acad Sci Hung 19:351

Guraya SS (1970a) Acta Anat (Basel) 77:617

Guraya SS (1970b) Acta Embryol Exp 3:227

Guraya SS (1971a) Physiol Rev 51:785

Guraya SS (1971b) J Reprod Fertil 24:107

Guraya SS (1972a) Acta Anat 82:284

Guraya SS (1972b) Acta Anat 81:507
Guraya SS (1973a) Ann Biol Anim Biochem Biophys 13:229
Guraya SS (1973b) Proc Indian Natl Sci Acad 39B:311
Guraya SS (1973c) Acta Endocr Cepench Suppl 171:72
Guraya SS (1974a) Int Rev Cytol 37:121
Guraya SS (1974b) In: Moudgal NR (ed) Gonadotropins and gonadal function. Academic Press,
    London New York, pp 220–236
Guraya SS (1976a) Int Rev Cytol 37:121
Guraya SS (1976b) Int Rev Cytol 44:365
Guraya SS (1976c) Arch It Anat Embriol 81:189
Guraya SS (1977a) Int Rev Cytol 51:49
Guraya SS (1977b) Arch It Anat Embriol 92:21
Guraya SS (1978) Int Rev Cytol 55:171
Guraya SS (1979a) Indian J Anim Sci 49:423
Guraya SS (1979b) Arch It Anat Embriol 84:321
Guraya SS (1979c) In: Talwar GP (ed) Recent advances in reproduction and regulation of
    fertility. Elsevier/North Holland Biomedical Preess, New York Amsterdam, pp 99–106
Guraya SS (1979d) Int Rev Cytol 59:249
Guraya SS (1980a) In: Motta P, Hafez ESE (ed) Biology of the ovary. Nijhoff, The Hague, pp
    33–51
Guraya SS (1980b) Int Rev Cytol 62:187
Guraya SS (1981) Arch It Anat Embriol 86:71
Guraya SS (1982a) Int Rev Cytol 78:257
Guraya SS (1982b) Arch It Anat Embriol (in press)
Guraya SS (1982c) In: Proc Symp Cell Control Mech. Bhaba Atomic Research Centre, Trombay,
    Bombay, pp 61–71
Guraya SS (1985) The biology of mammalian spermatogenesis and spermatozoa (in preparation)
Guraya SS, Greenwald GS (1964a) Am J Anat 114:495
Guraya SS, Greenwald GS (1964b) Anat Rec 149:411
Guraya SS, Greenwald GS (1965) Am J Anat 116:257
Guraya SS, Gupta SK (1979) Z Mikrosk Anat Forsch (Leipzig) 93:959
Guraya SS, Motta PM (1980) In: Motta PM, Hafez ESE (eds) Biology of the ovary. Nijhoff, The
    Hague, pp 68–85
Guraya SS, Stegner HE, Pape C (1974) Cytobiologie 9:100
Gwatkin RBL (1963) J Reprod Fertil 6:235
Gwatkin RBL (1964) J Reprod Fertil 7:99
Gwatkin RBL (1967) J Reprod Fertil 13:577
Gwatkin RBL (1976) In: Post G, Nicolson GL (eds) The cell surface in animal embryogenesis and
    development. Elsevier/North Holland Biomedical Press, New York Amsterdam, pp 1–54
Gwatkin RBL (1977a) In: Fertilization mechanisms in man and mammals. Plenum Press, New
    York
Gwatkin RBL (ed) (1977b) Fertilization mechanisms in man and mammals. Plenum Press, New
    York
Gwatkin RBL (1978a) In: Lerner RA, Bergsma D (eds) The molecular basis of cell-cell
    interaction. Alan R. Liss for Natl Found March of Dimes. BD: O AS XIV (2) New York, pp
    363–376
Gwatkin RBL (1978b) Birth defects. Orig Artic Ser 14:363 Natl Found
Gwatkin RBL (1979) In: Talwar GP (ed) Recent advances in reproduction and regulation of
    fertility. Elsevier/North Holland Biomedical Press, New York Amsterdam, pp 115–122
Gwatkin, RBL (1980) In: Motta PM, Hafez ESE (eds) Biology of the ovary. Nijhoff, The Hague,
    pp 209–214
Gwatkin RBL (1982) In: Hafez ESE, Semm K (eds) In vitro fertilization and embryo transfer.
    MTP Press, Lancester UK, pp 3–11
Gwatkin RBL, Andersen OF (1976) Life Sci 19:527
Gwatkin RBL, Haidri AA (1974) J Reprod Fertil 37:127
Gwatkin RBL, Williams DT (1974) J Reprod Fertil 39:153
Gwatkin RBL, Williams DT (1977) J Reprod Fertil 49:55

Gwatkin RBL, Williams DT (1978a) Gamet Res 1:19
Gwatkin RBL, Williams DT (1978b) Gamet Res 1:259
Gwatkin RBL, Williams D, Hartmann JF, Kniazuk M (1973) J Reprod Fertil 3:259
Gwatkin RBL, Rasmusson GH, Williams DT (1976) J Reprod Fertil 47:299
Gwatkin RBL, Williams DT, Carlo DJ (1977) Fertil Steril 28:871
Gwatkin RBL, Anderson OF, Williams DT (1980) Gamet Res 3:217
Habibi B, Franchi LL (1978) J Cell Sci 34:209
Haddad A, Nagai MET (1977) Cell Tiss Res 177:347
Hadek R (1963a) J Ultrastruct Res 8:170
Hadek R (1963b) J Ultrastruct Res 9: 99
Hadek R (1965) Int Rev Cytol 18:29
Hadek R (1969). Mammalian fertilization. Academic Press, London New York
Haegen van der BA (1978) Rev Bras Pesquisas Med Biol 11:53
Hall BU (1935) Proc Soc Exp Biol Med 32:747
Hamada Y, Bronson RA, Wright KH, Wallace EE (1977) Biol Reprod 17:58
Hamberger L, Nordenström K, Rosberg S, Sjögren A (1978) Acta Endocrinol 88:567
Hamberger L, Nilsson L, Nordenström K, Sjögren A (1979) In: Channing CP, Marsh J, Sadler
    WA (eds) Ovarian follicular and *Corpus Luteum* function. Plenum Press, New York, pp
    105–112
Hammond JM, Veldhuis JD, Seale TW, Rechler MM (1982) In: Channing CP, Segal SJ (eds)
    Intraovarian control mechanisms. Plenum Press, New York, pp 341–356
Hammond JM, Yoshida K, Veldhuis JD, Rechler MM, Knight AB (1983) In: Greenwald GS,
    Terranova PF (eds) Factors regulating ovarian function. Raven Press, New York, pp
    197–202
Han SS (1979) In: Midgley AR Jr, Sadler WA (eds) Ovarian follicular development and function.
    Raven Press, New York, pp 107–111
Harman SM, Louvet JP, Ross CT (1975) Endocrinology 96:1145
Hart DM, Baillie AH, Calman KC, Ferguson MM (1966) J Anat 100:810
Hartmann JF, Gwatkin RBL (1971) Nature (London) 234:479
Hastings RA, Enders AC, Schlafke S (1972) Biol Reprod 7:288
Hay MF, Cran DG (1978) Ann Biol Anim Biochem Biophys 18:453
Hay MF, Moor RM (1975) J Reprod Fertil 43:313
Hay MF, Moor RM (1978) In: Lamming GE, Chrighton DB (eds) Control of ovulation.
    Butterworths, London, pp 177–196
Hay MF, Cran DG, Moor RM (1976) Cell Tiss Res 169:515
Hedberg E (1953) Acta Endocrinol 14 (Suppl 15):1
Hedrick JL, Fry GN (1980) J Cell Biol 87:FE 1025
Hedrick JL, Wardrip N (1980) J Reprod Fertil 54:215
Heller DT, Schultz RM (1980) J Exp Zool 214:355
Heller DT, Cahill DM, Schultz RM (1981) Dev Biol 84:455
Henderson KM, Franchimont P (1983) J Reprod Fertil 67:291
Hermans WP, Jong de FH, Welschen R (1981) In: Coutts JRT (ed) Functional morphology of the
    human ovary. MTP Press, Lancaster, 85–93
Hermans WP, Debets MHM, Leeuwen van ECM, Jong de FH (1982) J Endocrinol 92:425
Hertig AG (1967) In: Wynn RM (ed) Foetal homeostasis. Acad Sci, New York, pp 98–123
Hertig AT (1968) Amer J Anat 122:107
Hertig AT, Adams EC (1967) J Cell Biol 34:647
Hertig A, Barton BR (1973) In: Astwood EB, Greep R (eds) Handbook of physiology-
    endocrinology II. Am Physiol Soc, Washington, DC, pp 327–348
Hicks JJ, Pedron N, Rosado A (1972) Fertil Steril 23:886
Hillensjö T (1976) Acta Endocrinol 82:809
Hillensjö T (1977) Acta Physiol Scand 100:261
Hillensjö T, Channing CP (1980) Gamet Res 3:233
Hillensjö T, Le Maire WJ (1980) Nature (London) 287:145
Hillensjö T, Hamberger L, Ahrén K (1975) Acta Endocrinol 78:751
Hillensjö T, Dekel N, Ahrén K (1976) Acta Physiol Scand 96:558
Hillensjö T, Ekholm C, Ahrén K (1978) Acta Endocrinol 87:377

Hillensjö T, Kripner AS, Pomerantz SH, Channing CP (1979) In: Channing CP, Marsh J, Sadler WA (eds) Ovarian follicular and *Corpus Luteum* function. Plenum Press, New York, pp 283–290

Hillensjö T, Magnusson C, Svensson U, Thelander H (1980) Endocrinology 106:584

Hillensjö T, Magnusson C, Svensson U, Thelander H (1981) In: Schwartz NB, Hunjicker-Dunn M (eds) Dynamics of ovarian function. Raven Press, New York, pp 105–110

Hillier SG, Ross GT (1979) Biol Reprod 20:261

Himelstein-Braw R, Byskov AG, Peters H, Faber M (1976) J Reprod Fertil 46:55

Hirschel MD, Hunter AC (1981) In: Schwartz NB, Hunzicker-Dunn M (eds) Dynamics of ovarian function. Raven Press, New York, pp 117–122

Hirshfield AN (1979) In: Midgley AJ Jr, Sadler WA (eds) Ovarian follicular development and function. Raven Press, New York, pp 19–22

Hirshfield A, Midgley AR Jr (1978a) Biol Reprod 19:597

Hirshfield A, Midgley AR Jr (1978b) Biol Reprod 19:606

Hirshfield AN (1982) Biol Reprod 26 (Suppl P):90A

Hiura M, Fujita H (1977a) Arch Histol Jpn 40:95

Hiura M, Fujita H (1977b) Histochemistry 51:321

Hoage TR, Cameron IL (1976) Anat Rec 186:585

Holmberg SRM, Johnson MH (1979) J Reprod Fertil 56:223

Hope J (1965) J Ultrastruct Res 12:592

Hoppe PC, Whitten WK (1974) J Reprod Fertil 39:433

Hoyer PE (1980) In: Motta P, Hafez ESE (eds) Biology of the ovary. Nijhoff, The Hague, pp 52–67

Hsueh AJW, Erickson GF (1979) Science 204:854

Hsueh AJW, Jones PBC (1982) In: Channing CP, Segal SJ (eds) Intraovarian control mechanisms. Plenum Press, New York, pp 223–262

Hsueh AJW, Ling NC (1979) Life Sci 25:1223

Hsueh AJW, Wang C, Erickson GF (1980) Endocrinology 106:1697

Hsueh AJW, Jones PBC, Adashi EY, Wang C, Zhuang L-Z, Welsh TH Jr (1983) J Reprod Fertil 69:325

Hubbard CJ, Greenwald GS (1981a) In: Schwartz NB, Hunzicker-Dunn M (eds) Dynamics of ovarian function. Raven Press, New York, pp 25–28

Hubbard CJ, Greenwald GS (1981b) J Reprod Fertil 63:455

Hubbard CJ, Greenwald GS (1982) Biol Reprod 26:230

Hubbard CJ, Greenwald GS (1983) Biol Reprod 28:849

Hubbard CJ, Terranova PF (1982) Biol Reprod 26:628

Hunter RHF, Lawson RAS, Rowson LEA (1972) J Reprod Fertil 30:325

Hunter RHF, Cook B, Baker TG (1976) Nature (London) 260:150

Hunzicker-Dunn M, Birnbaumer L (1976a) Endocrinology 99:211

Hunzicker-Dunn M, Birnbaumer L (1976b) Endocrinology 99:185

Hunzicker-Dunn M, Jungmann R, Derda D, Birnbaumer L (1979a) In: Channing CP, Marsh J, Sadler WA (eds) Ovarian follicular and *Corpus Luteum* function. Plenum Press, New York, pp 27–44

Hunzicker-Dunn M, Jungman RA, Birnbaumer L (1979b) In: Midgley AR Jr, Sadler WA (eds) Ovarian follicular development and function. Raven Press, New York, pp 267–304

Hunzicker-Dunn M, Wang MA, Jungmann R (1979c) In: Channing CP, Marsh J, Sadler WA (eds) Ovarian follicular and *Corpus Luteum* function. Plenum Press, New York, pp 113–121

Hunzicker-Dunn M, Jungmann RA, Birnbaumer L (1980) In: Motta PM, Hafez ESE (eds) Biology of the ovary. Nijhoff, The Hague, pp 231–243

Ichikawa S, Ohta M, Morioka H, Murao S (1983a) J Reprod Fertil 68:17

Ichikawa S, Morioka H, Ohta M, Oda K, Murao S (1983b) J Reprod Fertil 68:407

Iesaka T, Sato T, Igarashi M (1975) Endocrinol Jpn 22:279

Inaba T, Imori T, Matsumoto K (1976) J Steroid Biochem 9:1105

Ingram DL (1962) In: Zuckerman S (ed) The ovary, vol I. Academic Press, London New York, pp 247–273

Inoue M (1973) Biol Reprod 9:80

Inoue M, Wolf DP (1974a) Biol Reprod 10:512

Inoue M, Wolf DP (1974b) Biol Reprod 11:558
Inoue M, Wolf DP (1975) Biol Reprod 12:535
Ireland JJ, Richards JS (1978) Endocrinology 102:1458
Iwamatsu T, Keino H, Murakami U (1979) Zool Mag (Tokyo) 88:233
Iyengar MR, Iyengar CWL, Chen HY, Brinster RL, Bornslaeger E, Schultz RM (1983) Dev Biol
    96:263
Jackowski S, Dumont JN (1979) Biol Reprod 20:150
Jacobowitz D, Wallach EE (1967) Endocrinology 81:1132
Jacoby F (1962) In: Zuckerman S (ed) The ovary. Academic Press, London New York, pp
    189–245
Jagiello G, Graffeo J, Ducayen M, Prosser R (1977) Fertil Steril 28:476
Jagiello G, Ducayen M, Goonan D, Downey S (1978) Proc Soc Exp Biol Med 157:550
Jahn CL, Baran MM, Bachvarova R (1976) J Exp Zool 197:161
Jamiesson JD, Palade GE (1967) J Cell Biol 34:577
Janson PO, LeMaire WJ, Kallfelt B, Holmes PV, Cajander S, Bjersing L, Wiqvist N, Ahrén K
    (1982) Biol Reprod 26:456
Jarrell, RB (1981) In: Schwartz NB, Hunzicker-Dunn M (eds) Dynamics of ovarian function.
    Raven Press, New York, pp 47–53
Jilek F, Pavlok A (1975) J Reprod Fertil 42:377
Johnson MH (1973) Adv Reprod Physiol 6:279
Johnson MH, Eager DD, Muggleton-Harris A, Grave HM (1975) Nature (London) 257:321
Jonassen JA, Richards JS (1980) Endocrinology 106:1786
Jonassen JA, Bose K, Richards JS (1982) Endocrinology 111:74
Jones EC (1970) Bibliogr Reprod 15:129–132, 245–247
Jones EC, Krohn PL (1961a) J Endocrinol 21:469
Jones EC, Krohn PL (1961b) J Endocrinol 21:497
Jones PBC, Hsueh AJW (1982a) Endocrinology 110:1663
Jones PBC, Hsueh AJW (1982b) Endocrinology 111:713
Jones PBC, Hsueh AJW (1983) In: Greenwald GS, Terranova PF (eds) Factors regulating ovarian
    function. Raven Press, New York, pp 275–280
Jones PBC, Welsh TH, Hsueh AJW (1983) In: Greenwald GS, Terranova PF (eds) Factors
    regulating ovarian function. Raven Press, New York, pp 203–208
Jones RE (ed) (1978) The vertebrate ovary. Plenum Press, New York
Jong de FH, Sharpe RM (1976) Nature (London) 263:71
Jong de FH, Jansen EHJM, Hermans WP, Molen van der HJ (1982) In: Channing CP, Segal SJ
    (eds) Intraovarian control mechanisms. Plenum Press, New York, pp 37–52
Jong de FH, Jansen EHJM, Steenbergen J, Dijk van, Molen van der HJ (1983) In: McCann SM,
    Dhindsa DS (eds) Role of peptides and proteins in control of reproduction. Elsevier/North
    Holland Biomedical Press, New York Amsterdam, pp 257–273
Jung G (1965) Forsch Geburtsh Gynaekol 23:77
Kaleta E, Polak Z (1978) Zwierzeta Lab 15:53
Kang YH (1974) Am J Anat 139:535
Kang YH, Anderson WA (1975) Anat Rec 182:175
Kang YH, Anderson WA, Chang SC, Ryan RJ (1979) In: Midgley AR Jr, Sadler WA (eds)
    Ovarian follicular development and function. Raven Press, New York, pp 121–135
Kanzaki H (1981) Acta Obstet Gynaecol Jpn 33:1925
Kanzaki H, Okamura H, Okuda Y, Takenaka A, Morimoto K, Nishimura T (1981) Acta Obstet
    Gynaecol Jpn 33:11
Kaplan G, Abreu SL, Bachvarova R (1982) J Exp Zool 220:361
Kaulla von KN, Aikawa JK, Pettigrew JD (1958) Nature (London) 182:1238
Kaur P, Guraya SS (1983) Am J Anat 166:469
Kaur P, Guraya SS (1984) (submitted)
Kenney RM, Condon W, Ganjam VK, Channing CP (1979) J Reprod Fertil Suppl 27:163
Kimura J, Katoh M, Taya K, Sasamoto S (1983) J Endocrinol 97:313
Kiesling DO, Warren JE Jr, Bolt DJ (1983) J Anim Sci 57 (Suppl 1) 349 (Abstr)
King BF, Tibbitts FD (1977) Anat Rec 189:263
Kirchner C (1972a) J Embryol Exp Morphol 28:177

Kirchner C (1972b) Fertil Steril 23:131

Kling OR, Ujita EL, Campeau JD, Zerega di GS (1983) In: Greenwald GS, Terranova PF (eds) Factors regulating ovarian function. Raven Press, New York, pp 175–178

Klinken SP, Stevenson PM (1977) Eur J Biochem 81:327

Knecht M, Ranta T, Naor Z, Catt KJ (1983a) In: Greenwald GS, Terranova PF (eds) Factors regulating ovarian function. Raven Press, New York, pp 225–244

Knecht M, Ranta T, Catt KJ (1983b) In: Greenwald GS, Terranova PF (eds) Factors regulating ovarian function. Raven Press, New York, pp 269–274

Knecht M, Renta T, Catt KJ (1983c) Endocrinology 113:949

Knox E, Lowry S, Beck L (1979) In: Midgley AR Jr, Sadler WA (eds) Ovarian follicular development and function. Raven Press, New York, pp 159–163

Knudsen JF, Litkowski LJ, Wilson TL, Guthrie HD, Batta SK (1978) J Endocrinol 79:249

Knudsen JF, Litkowski LJ, Wilson TL, Guthrie HD, Batta SK (1979) J Reprod Fertil 57:419

Kobayashi Y, Wright KH, Santulli R, Kitai H, Wallach EE (1983a) Biol Reprod 28:385

Kobayashi Y, Kitai H, Santulli R, Wright KH, Wallach EE (1983b) Fertil Steril 39:396 (Abstr k)

Koering MJ (1969) Am J Anat 126:73

Koering MJ, Goodman AL, Williams RF, Hodgen GD (1982) Fertil Steril 37:837

Kolena J, Channing CP (1971) Biochem Biophys Acta 252:601

Kolena J, Channing CP (1972a) Endocrinology 92:531

Kolena J, Channing CP (1972b) Endocrinology 90:1543

Kolpovski VM (1978) Zool Zh 57:1860

Komar A, Roszkowski II. (1979) Ginekolpol 50:521

Koninckx PR, Verhoeven G, Moor de P (1983) In: Greenwald GS, Terranova PF (eds) Factors regulating ovarian function. Raven Press, New York, pp 75–80

Koos RD, Le Maire WJ (1983) In: Greenwald GS, Terranova PF (eds) Factors regulating ovarian function. Raven Press, New York, pp 191–196

Korfsmeier K-H (1979) Histochemistry 63:123

Korolev NV (1979) Vest Zool 1:53

Korolev VA, Zavarzina GA (1976) Tsitologiya 18:1281

Kraiem Z, Lunenfeld B (1979) In: Midgley AR Jr, Sadler WA (eds) Ovarian follicular development and function. Raven Press, New York, pp 333–337

Krarup T (1969) Int J Cancer 4:61

Krarup T (1970) J Endocrinol 46:483

Krauskopf C (1968a) Z Zellforsch Mikrosk Anat 92:275

Krauskopf C (1968b) Z Zellforsch 92:296

Krishnan RS, Daniel JC (1967) Science 158:490

Kruip TAM, Cran DG, Beneden van TH, Dieleman SJ (1983) Gamet Res 8:29

Kumari GL, Kumar N, Duraiswami S, Datta JK, Vidyasagar IC, Vohra S, Roy S (1982) In: Channing CP, Segal SJ (eds) Intraovarian control mechanisms. Plenum Press, New York, pp 283–302

Kurilo LF (1981) Tsitologiya 23:894

Kvlividze VE (1976) Soobshch Akad Nauk Gruz SSR 82:569

Labrie F, Seguin C, Lefebvre F-A, Massicotte J, Pelletier G, Borgus J-P, Kelly P-A, Reeves JJ, Belanger A (1983) In: Channing CP, Segal SJ (eds) Intraovarian control mechanisms. Plenum Press, New York, pp 211–222

Lacassagne A, Duplan JF, Marcovitch H, Raynaud A (1962) In: Zuckerman S (ed) The ovary, vol II. Academic Press, London New York, pp 463–532

LaMarca MJ, Wassarman PM (1979) Dev Biol 73:103

Lamprecht SA, Zor U, Tsafriri A, Lindner HR (1973) J Endocrinol 57:217

Lamprecht SA, Zor U, Salomon Y, Koch Y, Ahrén K, Lindner HR (1977) J Cycl Nucl Res 3:69

Lamprecht SA, Zor U, Koch Y, Salomon Y, Lindner HR (1979) In: Midgley AR Jr, Sadler WA (eds) Ovarian follicular development and function. Raven Press, New York, pp 311–323

Langan TA (1973) In: Greengard P, Robison GA (eds) Advances in cyclic nucleotide research, vol III. Raven Press, New York, p 99

Larsen WJ, Tung H (1978) Tiss Cell 10:585

Lau IF, Saksena SK, Chang MC (1974) J Reprod Fertil 40:467

Lau IF, Hoogasian J, Wong SK, Saksena SK (1980) Prostaglandins and Medicine 4:121

Ledwiz-Rigby F (1979) In: Midgley AR Jr, Sadler WA (eds) Ovarian follicular development and
    function. Raven Press, New York, pp 79–84
Ledwitz-Rigby F (1983) Mol Cell Endocrinol 29:213
Ledwitz-Rigby F, Rigby BW (1979) In: Channing CP, Marsh J, Sadler WA (eds) Ovarian
    follicular and *Corpus Luteum* function. Plenum Press, New York, pp 347–359
Ledwitz-Rigby F, Stetson M, Channing CP (1973) Biol Reprod 9:94 (Abstr 85)
Ledwitz-Rigby F, Rigby BW, Gay VL, Young J, Stetson M, Channing CP (1977) J Endocrinol
    74:175
Ledwitz-Rigby F, Mc Conoughey, Maloney B (1981) Biol Reprod 24 (Suppl 1): 83 A
Ledwitz-Rigby F, Rigby BW, Ling SY, Stewart L, McLean M (1982) In: Channing CP, Segal SJ
    (eds) Intraovarian control mechanism. Plenum Press, New York, pp 331–340
Lee CY (1976) Endocrinology 99:42
Lee VWK (1983) In: Greenwald GS, Terranova PF (eds) Factors regulating ovarian function.
    Raven Press, New York, pp 157–162
Leibfried L, First NL (1979a) J Anim Sci 48:76
Leibfried L, First NL (1979b) J Exp Zool 210:575
Leibfried L, First NL (1980a) Biol Reprod 23:699
Leibfried L, First NL (1980b) Biol Reprod 23:705
Leibfried L, First NL (1982) Anat Rec 202:339
LeMaire WJ (1979) In: Midgley AR Jr, Sadler WA (eds) Ovarian follicular development and
    function. Raven Press, New York, pp 305–309
LeMaire WJ, Marsh JM (1977) In: Scholler R (ed) Endocrinology of the ovary. Fresnes
LeMaire WJ, Yang NST, Behrman HR, Marsh JM (1973) Prostaglandins 3:367
LeMaire WJ, Leidner R, Marsh JM (1975) Prostaglandins 9:221
LeMaire WJ, Clark MR, Marsh JM (1979) In: Hafez ESE (ed) Human ovulation: mechanisms,
    prediction, detection, and induction, vol III. Elsevier/NorthHolland Biomedical Press,
    Amsterdam New York, pp 155–195
LeMaire WJ, Clark MR, Chenny GBN, Marsh JM (1980) In: Tozzini RI, Reeves G, Pineda RL
    (eds) Endocrine physiopathology of the ovary. Elsevier/North Holland Biomedical Press,
    Amsterdam New York, pp 207–217
Lerner LJ, Oldani C, Vitale A (1978) Prostaglandins 15:525
Leung PCK, Tsang BK, Armstrong DT (1979) In: Channing CP, Marsh J, Sadler WA (eds)
    Ovarian follicular and *Corpus Luteum* function. Plenum Press, New York, pp 241–243
Lieberman ME, Tsafriri A, Bauminger S, Collins WP, Ahrén K, Lindner HA (1976) Acta
    Endocrinol 83:151
Lin C-T, Mukai K, Lee CY (1982) Cell Tiss Res 224:647
Lindner HR, Tsafriri A, Lieberman ME, Zor U, Koch Y, Bauminger S, Barnea A (1974) Recent
    Prog Hormone Res 30:79
Lindner HR, Amsterdam A, Salomon Y, Tsafriri A, Nimrod A, Lamprecht SA, Zor U, Koch Y
    (1977) J Reprod Fertil 51:215
Lindner HR, Bar-Ami S, Tsafriri A (1980) In: Serio M, Martini L (eds) Animal models in human
    reproduction. Raven Press, New York, pp 65–85
Lindsey AM, Channing CP (1979) In: Midgley AR Jr, Sadler WA (eds) Ovarian follicular
    development and function. Raven Press, New York, pp 343–345
Lindstedt G (1970) J Clin lab Invest 25:59
Lindvall O, Bjorklund A (1974) Histochemistry 39:97
Lintern-Moore S, Moore GPM (1979) Biol Reprod 20:773
Lipner H (1973) In: Astwood EB, Greep R (eds) Handbook of physiology-endocrinology, vol II,
    sect 7. Am Physiol Soc, Washington DC, pp 409–437
Lipner H, Maxwell B (1960) Science 131:1737
Liss RH (1964) Anat Rec 149:385a
Lobel BL, Levy E (1968) Acta Endocrinol 59 (Suppl 132):7
Lobel BL, Rosenbaum RM, Deane HW (1961) Endocrinology 68:232
Loeken MR, Channing CP, D'Eletto R, Weiss G (1983) Endocrinology 112:769
Lowenstein JE, Cohen AI (1964) J Embryol Exp Morphol 12:113
Löfman CO, Janson PO, Kallfelt BJ, Ahrén K, LeMaire WJ (1982) Biol Reprod 26:467
Longo FJ (1974a) Biol Reprod 11:22

Longo FJ (1974b) Anat Rec 179:27
Longo FJ (1981) Biol Reprod 25:399
Lopo AC, Calarco PG (1982) Gamet Res 5:283
Lorenzen JR, Schwartz NB (1979) In: Channing CP, Marsh J, Sadler WA (eds) Ovarian follicular
    and *Corpus Luteum* function. Plenum Press, New York, pp 375–380
Lorenzen JR, Schwartz NB, Channing CP (1978) Fed Proc 37:296
Louvet J-P, Vaitukaitis JL (1976) Endocrinology 99:758
Louvet J-P, Harman SM, Schreiber JR, Ross GT (1975) Endocrinology 97:366
Lucky AW, Schreiber J, Hillier S, Schulman J, Ross GT (1977) Endocrinology 100:128
Lui CW, Cornett LE, Meizel S (1977) Biol Reprod 17:34
Magnusson C, Hillensjö T (1977) J Exp Zool 201:139
Magnusson C, Hillensjö T, Tsafriri A, Hultborn R, Ahrén K (1977) Biol Reprod 17:9
Magnusson C, Ami SB, Braw R, Tsafriri A (1983) J Reprod Fertil 68:97
Magoffin DA, Erickson GF (1981) In: Schwartz NB, Hunzicker-Dunn M (eds) Dynamics of
    ovarian function. Raven Press, New York, pp 55–60
Mahajan DK, Little AB (1978) Biol Reprod 17:834
Mahi CA, Yanagimachi R (1979) Fol Morphol (Praha) 20:73
Makinoda S (1980) Hokkaido J Med Sci 55:521
Makris A, Ryan KJ (1975) Endocrinology 96:964
Makris A, Ryan KJ (1977) Endocrinol Res Commun 4:233
Makris A, Ryan KJ (1980) Steroids 35:53
Manarang-Pangan S, Menge AC (1971) Fertil Steril 22:367
Mancini RE, Vilar O, Heinrich JJ, Davidson OW, Alvarez B (1962) J Histochem Cytochem 11:80
Mandelbaum J, Plachot M, Thibault C (1977) Ann Biol Anim Biochem Biophys 17:389
Mandl AM (1959) Proc R Soc London Ser B 150:53
Mandl AM (1964) Biol Rev 39:288
Mandl AM, Zuckerman S (1951) J Endocrinol 7:112
Manes C (1969) J Exp Zool 172:303
Mangia F, Canipari R (1978) In: Johnson MH (ed) Development in mammals, vol II. Elsevier,
    Amsterdam, pp 1–29
Mangia F, Epstein CJ (1975) Dev Biol 45:211
Mangia F, Erickson RP, Epstein CJ (1976) Dev Biol 54:146
Marion CB, Gier HT, Choudhury JB (1968) J Anim Sci 27:451
Marsh JM (1975) In: Greengard P, Robison GA (eds) Advances in cyclic nucleotides research, vol
    VI. Raven Press, New York, p 137
Marsh JM (1976) Biol Reprod 14:30
Marsh JM, Mills JM, LeMaire WJ (1973) Biochem Biophys Acta 304:197
Marsh JM, Yang NST, LeMaire WJ (1974) Prostaglandins 7:269
Martin GG, Talbot P (1981a) J Exp Zool 216:469
Martin GG, Talbot P (1981b) J Exp Zool 216:483
Martin GG, Miller-Walker C (1983) J Exp Zool 225:311
Martinat N, Combarnous Y (1983) Endocrinology 113:433
Masui Y, Clarke HJ (1979) Int Rev Cytol 57:186
Mattison DR, Nightingale R (1981) In: Schwartz NB, Hunzicker-Dunn M (eds) Dynamics of
    ovarian function. Raven Press, New York, pp 89–94
Mauléon P (1978) In: Crighton DB, Haynes NB, Foxcroft GR, Lamming GE (eds) Control of
    ovulation. Butterworth, London, pp 141–158
May JV, Schomberg DW (1981) Biol Reprod 25:421
May, JV, McCarty K Jr, Reichert LE Jr, Schomberg DW (1980) Endocrinology 107:1041
McClellan MC, Dickman MA, Able JA Jr, Niswender GD (1975) Cell Tiss Res 164:291
McCullagh DR (1932) Science 76:19–20
McGaughey RW (1977a) Exp Cell Res 109:25
McGaughey RW (1977b) Endocrinology 100:39
McGaughey RW (1983a) In: Oxford reviews of reproduction. (in press)
McGaughey RW (1983b) In: Blerkom van J, Motta P (eds) Ultrastructure of reproduction,
    gametogenesis, fertilization and embryogenesis. Nijhoff, The Hague
McGaughey RW, Blerkom van J (1977) Dev Biol 56:3241

McGaughey RW, Montogomery DH, Richter JDA (1979) J Exp Zool 209:239
McLaren A (1966) Proc R Soc London Ser B 166:316
McLaren A (1969) Adv Reprod Fertil 4:207
McLaren A (1970) J Embryol Exp Morphol 23:1
McNatty KP (1978) In: Jones RE (ed) The vertebrate ovary. Plenum Press, New York, pp 215–259
McNatty K (1979) In: Channinng CP, Marsh J, Sadler WA (eds) Ovarian follicular and *Corpus Luteum* function. Plenum Press, New York, pp 465–482
McNatty KP (1980) J Clin Endocrinol Metab 49:851
McNatty KP, Baird DT (1978) J Endocrinol 76:527
McNatty KP, McNeilly AS, Sawers RS (1974) Nature (London) 240:653
McNatty KP, Hunter WM, McNeilly AS, Sawers RS (1975) J Endocrinol 64:555
McNatty KP, Neal P, Baker TG (1976) J Reprod Fertil 47:155
McNatty KP, Smith DM, Makris A, Osathanondh R, Ryan KJ (1979) J Clin Endocrinol Metab 49:851
McNatty KP, Makris A, Osathanondh R, Ryan KJ (1980) Steroids 36:53
McReynold H, Siraki C, Branson P, Pollock R Jr (1973) Z Zellforsch 140:1
Meinecke B (1977) Zentralbl Vet Aermed R A 24:422
Meinecke B, Meinecke-Tillmann S (1978) Anat Histol Embryol 7:58
Ménézo Y (1976) C R Acad Sci 282:1967
Ménézo Y, Gérard M (1977) Ann Biol Anim Biochem Biophys 17:313
Ménézo Y, Gérard M, Thibault C (1976) C R Acad Sci 283:1309
Ménézo Y, Gérard M, Szollosi D, Thibault C (1978) Ann Biol Anim Biochem Biophys 18:471
Menge AC (1970) J Reprod Fertil (Suppl):171
Menino AR Jr, Wright RW Jr (1979) Proc Soc Exp Biol Med 160:449
Merk FB (1971) In: Arcenaux CJ (ed) Proc 29th Ann Meet Electr Microsc Soc Am. Claitor's, Baton Rouge, pp 554–555
Merk FB, Botticelli CR, Albright JT (1972) Endocrinology 90:992
Merk FB, Albright JT, Botticelli CR (1973) Anat Rec 175:107
Merker H, Zimmermann B (1970) Z Zellforsch Microsc Anat 111:364
Merwe van der M (1979) Afr J Zool 14:111
Mestwerdt W, Müller O, Brandau H (1977a) Arch Gynaekol 222:45
Mestwerdt W, Müller O, Brandau H (1977b) Arch Gynaekol 222:115
Mestwerdt W, Müller O, Brandau H (1979) In: Channing CP, Marsh J, Sadler WA (eds) Ovarian follicular and *Corpus Luteum* function. Plenum Press, New York, pp 123–128
Metz CP (1973) Fed Proc Fed Am Soc Exp Biol 32:2057
Metz CP (1978) Current Top Dev Biol 12:107
Meyenhofer M, Andersen OF, Marx BS, Gwatkin RBL (1977) Scanning Elect Microsc 2:343
Midgley AR (1973) Adv Exp Med Biol 36:365
Midgley AR Jr, Richards JS (1976) In: Crosignani PG, Mishell DR (eds) Ovulation in the human. Academic Press, London New York, pp 87–94
Midgley AR Jr, Sadler WA (eds) (1979) Ovarian follicular development and function. Raven Press, New York
Miller KF, Crister JK, Ginther OJ (1979) In: Channing CP, Marsh J, Sadler WA (eds) Ovarian follicular and *Corpus Luteum* function. Plenum Press, New York, pp 417–421
Mills TM (1975) Proc Soc Exp Biol Med 148:995
Mills TM (1979) In: Midgley AR Jr, Sadler WA (eds) Ovarian follicular development and function. Raven Press, New York, pp 187–191
Minato Y, Toyoda Y (1982) Jpn J Zootech Sci 53:480
Mintz B (1962) Science 139:5
Mintz B (1964) J Exp Zool 157:85
Mintz B (1967) Proc Natl Acad Sci USA 58:344
Mirre C, Stahl A (1978) J Cell Sci 31:79
Miyamoto H, Ishibashi T, Utsumi K (1980) Jpn J Zootech Sci 51:582
Miyamoto H, Ishibashi T, Utsumi K (1981) Jpn J Zootech Sci 52:58
Mizuno O, Gtani T, Shirota M, Sasamoto S (1983) J Endocrinol 97:113

Mondschein JS, May JF, Gunn EB, Schomberg DW (1981) In: Schwartz NB, Hunzicker-Dunn M (eds) Dynamics of ovarian function. Raven Press, New York, pp 83–88

Moon YS, Zamecnik J, Armstrong DT (1974) Life Sci 15:1731

Moor RM (1977) J Endocrinol 73:143

Moor RM (1978) Ann Biol Anim Biochem Biophys 18:477

Moor RM, Cran DG (1980) In: Johnson MH (ed) Development in mammals, vol IV. Elsevier/North Holland Biomedical Press, Amsterdam New York, pp 3–37

Moor RM, Heslop JP (1981) J Exp Zool 216:205

Moor RM, Smith MW (1979) Exp Cell Res 119:333

Moor RM, Trounson AO (1977) J Reprod Fertil 49:101

Moor RM, Warnes GM (1977) In: Lamming GE, Crighton DB (eds) Control of ovulation. Butterworths, London, pp 159–176

Moor RM, Hay MF, Seamark RF (1975) J Reprod Fertil 45:595

Moor RM, Hay MF, Dott HM, Cran DG (1978) J Endocrinol 77:309

Moor RM, Smith MW, Dawson RMC (1980) Exp Cell Res 126:15

Moor RM, Osborn JC, Cran DG, Walters DG (1981) J Embryol Exp Morphol 61:347

Moore GPM, Lintern-Moore S (1974) J Reprod Fertil 39:163

Moore GPM, Lintern-Moore S (1978) Biol Reprod 18:865

Moore GPM, Lintern-Moore S, Peters H, Faber M (1974) J Cell Biol 60:416

Moore GPM, Lintern-Moore S, Peters H, Byskov AG, Andersen M, Faber M (1975) J Cell Physiol 86:31

Moore RM, Crayle RG (1971) J Reprod Fertil 27:401

Morales TI, Woessner JF, Marsh JM (1983) Biochim Biophys Acta 756:119

Mori T, Uchida TA (1981) J Reprod Fertil 63:391

Mori T, Nishimoto T, Kitagawa M, Noda Y, Nishimura T, Oikawa T (1978) Experientia 34:797

Mori T, Nishimoto T, Kohada H, Takai Z, Nishimura T, Oikawa T (1979) Fertil Steril 32:67

Morikawa H, Okamura H, Takenaka A, Morimoto K, Nishimura T (1981) Int J Fertil 26:283

Morimoto K, Okamura H, Kanzaki H, Okuda Y, Takenaka A, Nishimura T (1981) Int J Fertil 26:14

Moscarini M, Amenta F (1980) Cell Mol Biol 26:275

Mossman HW, Duke KL (1973) Comparative morphology of the mammalian ovary. Univ Wisconson Press, Madison

Mossman HW, Koering MJ, Ferry D (1964) Am J Anat 115:235

Motlik J, Fulka J (1976) J exp Zool 198:155

Motlik J, Koefoed HN, Fulka J (1978a) J Exp Zool 205:377

Motlik J, Kopency V, Pivko J (1978b) Ann Biol Anim Biochem Biophys 18:735

Motta P (1965) Z Zellforsch 68: 308

Motta P, Blerkom van J (1974) J Submicrosc Cytol 6:297

Motta P, Blerkom van J (1975) Am J Anat 143:241

Motta PM, Blerkom van J (1980) In: Motta PM, Hafez ESE (eds) Biology of the ovary. Nijhoff, The Hague, pp 162–178

Motta P, DiDio LJA (1974) J Submicrosc Cytol 6:15

Motta PM, Familiari G (1981) Acta Anat 109:103

Motta PM, Hafez ESE (eds) (1980) Biology of the ovary. Nijhoff, The Hague

Motta P, Cherney DD, DiDio LJA (1971) J Submicrosc Cytol 3:85

Mueller PL, Schreiber JR, Luckey JR, Schulman JD, Rodbard D, Ross GT (1978) Endocrinology 102:824

Mumford RA, Hartmann JF, Ashe BM, Zimmerman M (1981) Dev Biol 81:332

Murdoch WJ, Dunn TG (1982) Biol Reprod 27:300

Murdoch WJ, Nix, KJ, Dunn TG (1983) Biol Reprod 28:1001

Nakajo S, Sato K, Fujima M (1982) Fertil Steril 27:118

Nakamura M, Yasumasu I, Okinaga S, Arai K (1982) Dev Growth Differ 24:265

Nandedkar TD, Munshi SR (1981) J Reprod Fertil 62:2

Neal P, Baker TG (1973) J Reprod Fertil 37:399

Neal P, Baker RG (1975) J Endocrinol 65:27

Neal P, Baker TG, McNatty KP, Scaramuzzir J (1975) J Endocrinol 65:27

Neal P, Baker TG, McNatty KP (1976) J Reprod Fertil 47:157

Nekola MV, Nalbandov AV (1971) Biol Reprod 4:154
Nekola MV, Smith DM (1974) Eur J Obstet Gynaecol Reprod Biol 4:S-125
Nekola MV, Smith DM (1975) J Exp Zool 194:529
Neymark MA, Dimino, MJ (1983) Biol Reprod 28:142
Nicholsen GL, Yanagimachi R, Yanagimachi H (1975) J Cell Biol 66:263
Nicosia SV (1980a) In: Tozzini RI, Reeves G, Pineda RL (eds) Endocrine physiopathology of the
    ovary. Elsevier/North Holland Biomedical Press, Amsterdam New York, pp 43–62
Nicosia SV (1980b) In: Tozzini RI, Reeves G, Pineda RL (eds) Endocrine physiopathology of the
    ovary. Elsevier/North Holland Biomedical Press, Amsterdam New York, pp 101–119
Nicosia SV, Mikhail G (1975) Fertil Steril 26:427
Nicosia SV, Wolf DP, Inoue M (1977) Dev Biol 57:56
Nicosia SV, Wolf DP, Mastroianni L (1978) Gamete Res 1:145
Nigi H (1977a) J Reprod Fertil 50:387
Nigi H (1977b) Primates 18:243
Nilsson L (1974) Acta Endocrinol 77:540
Nilsson L (1980) Gamete Res 3:369
Nilsson L, Rosberg S, Ahrén K (1974) Acta Endocrinol 77:559
Nilsson L, Rosberg S, Hillensjo T, Ahrén K (1975) Life Sci 16:517
Nilsson L, Munshi SF (1973) J Submicr Cytol 5:1
Nilsson L, Hillensjö T, Ekholm C (1977a) Acta Endocrinol 86:384
Nilsson L, Hamberger L, Hillensjö T, Ekholm C (1977b) Acta Endocrinol 85:39
Nimrod A, Lindner HR (1979) In: Midgley AR Jr, Sadler WA (eds) Ovarian follicular
    development and function. Raven Press, New York, pp 211–213
Nimrod A, Erickson GF, Ryan KJ (1976) Endocrinology 98:56
Nimrod A, Bedrak E, Lamprecht SA (1977) Biochim Biophys Res Commun 78:977
Nimura S, Ishida K (1981) Jpn J Fertil Steril 26:11
Nishimoto T, Mori T, Yamada E, Nishimura T (1980) Fertil Steril 34:352
Niwa K, Chang MC (1975) J Reprod Fertil 43:435
Noda Y, Mori T, Takai I, Kohda H, Nishimura T (1981) J Reprod Immunol 3:147
Norberg HS (1972) Z Zellforsch 131:497
Norberg HS (1973) Z Zellforsch 141:103
Norman RL, Greenwald GS (1972) Anat Rec 173:95
Norrevang A (1969) Int Rev Cytol 23:113
Nowacki R (1977) Anat Histol Embryol 6:217
Noworyta B, Szoltys M (1975) Bull Acad Sci Ser Sci Biol Cl V XXIII:65
Numazawa A, Kawashima S (1982) Annat Zool Jpn 55:82
Oakberg EF (1967) Arch Anat Microsc Morphol Exp 56 (Suppl 3–4)
Oakberg EF (1968) Mutat Res 6:155
Oakberg EF, Tyrrell PD (1975) Biol Reprod 12:477
Ochs D, McConkey EH, Sammons D (1981) Electrophoresis
Odor DL, Blandau RJ (1969) Am J Anat 125:177
Odor DL, Blandau RJ (1972) Peters H (ed) The development and maturation of the reproductive
    organs and functions in the female. Workshop Meeting, July 4–5, Copenhagen. Excerpta
    Medica, ICS No 267
Oh Y-K, Brackett BG (1975) Fertil Steril 26:665
Oikawa T, Yanagimachi R (1975) J reprod Fertil 45:487
Oikawa T, Yanagimachi R, Nicolson GL (1973) Nature (London) 241:256
Oikawa T, Nicolson GL, Yanagimachi R (1974) Exp Cell Res 83:239
Oikawa T, Nicolson GL, Yanagimachi R (1975a) J Reprod Fertil 43:133
Oikawa T, Yanagimachi R, Nicolson GL (1975b) J Reprod Fertil 43:137
Okamura H, Virutamasen P, Wright KH, Wallach EE (1972) Am J Obstet Gynaecol 112:183
Okamura H, Takenaka A, Yajima Y, Nishimuka T (1980) J Reprod Fertil 58:153
Okamura H, Hibeharukanzaki Y, Takenaka A, Morimoto K, Nishimura T (1981) Acta Obstet
    Gynaecol Jpn 33:215
Okuda Y, Okamura N, Kanzaki H, Takenaka A, Morimoto K, Nishimura T (1980) Acta Obstet
    Gynaecol Jpn 32:859
Olds PJ, Stern S, Biggers JD (1973) J Exp Zool 186:39

O'Rand MG (1981) Biol Reprod 25:621
Osborn JC, Moor RM (1981) Cell Biol Int Ret Suppl. A 5:131
Osborn JC, Moor RM (1983) J Steroid Biochem 19:133
O'Shea JD (1970a) J Anat 106:191
O'Shea JD (1970b) Anat Rec 167:127
O'Shea JD (1973) Res Vet Sci 14:273
Osteen KG, Channing CP (1983) In: Greenwald GS, Terranova PF (eds) Factors regulating
    ovarian function. Raven Press, New York, pp 185–190
Osvaldo-Decima L (1970) J Ultrastruct Res 29:218
Owman CH, Sjoberg N, Svennson K, Walles B (1975) J Reprod Fertil 45:553
Owman CH, Sjoberg N, Wallach E, Walles B, Wright K (1979) In: Hafez ESE (ed) Human
    ovulation. Elsevier/North Holland Biomedical Press, Amsterdam New York, pp 57–100
Owman CH, Edvinsson L, Sjöberg JO, Spörrong B, Stefenson A, Walles B (1980) In: Motta PM,
    Hafez ESE (eds) Biology of the ovary. Nijhoff, The Hague, pp 202–208
Ownby CL, Shivers CA (1972) Biol Reprod 6:310
Oxberry BA, Greenwald GS (1982) Biol Reprod 27:505
Oxberry BA, Greenwald GS (1983) Biol Reprod 29:1255
Palm VS, Sacco AG, Syner FN, Subramanian MG (1979) Biol Reprod 21:709
Palombi F, Viron A (1977) J Ultrastruct Res 61:10
Paracchia C (1973) J Cell Biol 57:66
Paracchia C (1977) J Cell Biol 72:628
Parr EL (1974a) Biol Reprod 11:504
Parr EL (1974b) Biol Reprod 11:509
Parr EL (1974c) Biol Reprod 11:483
Parshad RK, Guraya SS (1984a) Unpublished observation
Parshad RK, Guraya SS (1984b) Indian J Exp Biol 22:635
Parshad VR, Guraya SS (1983) Proc Indian Acad Sci (Anim Sci) 92:121
Parshad VR, Guraya SS (1984) Indian J Exp Biol 22:185
Pascu T, Dema A, Mihai D, Lunca H, Salageanu G (1968) VI Int Congr Reprod Anim Insem
    Artif (Paris) 1:785
Pascu T, Tudorascu R, Stancioiu N, Lumca H (1971) Rec Med Vet 147:979
Payer AF (1975) Am J Anat 142:295
Payne RW, Runser RH (1958) Endocrinology 62:313
Payne RW, Helbaum AR, Owens JN (1956) Endocrinology 59:306
Peacock T, Leon AR, Browdder W, Schultz GA (1978) Can J Genet Cytol 20:291
Pedersen T (1972) In: Biggers JD, Schuetz AW (eds) Oogenesis. Univ Park Press, Baltimore, pp
    261–276
Pedersen H, Seidel G (1972) J Ultrastruct Res 39:540
Peluso JJ (1979) In: Midgley AR Jr, Sadler WA (eds) Ovarian follicular development and
    function. Raven Press, New York, pp 85–88
Peluso JJ, Butcher RL (1974) Fertil Steril 26:665
Peluso JJ, Hutz R (1980) Cell Tiss Res 213:29
Peluso JJ, Steger RW (1978) J Reprod Fertil 54:275
Peluso JJ, Steger RW, Hafez ESE (1977a) J Reprod Fertil 49:215
Peluso JJ, Steger RW, Hafez ESE (1977b) Biol Reprod 16:600
Peluso JJ, Bolender DL, Perri A (1979) Biol Reprod 20:423
Peluso JJ, England-Charlesworth C, Bolender DL, Steger RW (1980) Cell Tiss Res 211:105
Pendergrass PB, Talbot PT (1979) Biol Reprod 20:205
Peters H (1969) In: McLaren A (ed) Advances in reproduction physiology, vol IV. Academic
    Press, London New York, pp 149–185
Peters H (1979) In: Midgley AR, Sadler WA (eds) Ovarian follicular development and function.
    Raven Press, New York, pp 41–57
Peters H, Levy E (1966) J Reprod Fertil 11:227
Peters H, McNatty KP (1980) The ovary. Granada, London
Peters H, Byskov AG, Faber M (1973) In: Peters H (ed) The development and maturation of the
    ovary and its functions. Int Congr Ser No 267.Excerpta Medica, Amsterdam, pp 559–566
Peters H, Byskov AG, Himelstein-Braw R, Faber M (1975) J Reprod Fertil 45:559

Peters T (1975) In: Putnam F (ed) The plasma proteins, vol I. Academic Press, London New York, pp 133–181

Peterson RN, Russell LD, Bundman D, Conway M, Freund M (1981) Dev Biol 84:144

Petzoldt U, Hoppe PC, Illmensee K (1980) Whilt Roux's Arch Dev Biol 189:215

Phillips DM, Shalgi RM (1980) J Exp Zool 213:1

Phillips DM, Shalgia R, Kraicer P, Segal SJ (1978) Scanning Electr Microsc 1978:113

Pikó L (1969) In: Metz CB, Monroy A (eds) Fertilization. Academic Press, London New York, pp 325–404

Pikó L, Clegg KB (1982) Dev Biol 89:362

Pincus G, Enzmann EV (1935) J Exp Med 62:655

Pivko J, Motlik J, Kopecyn V, Flechon JE (1982) Reprod Nutr Dev 22:93

Plunkett ER, Moon YS, Zamecnik J, Armstrong DT (1975) Am J. Obstet Gynaecol 123:391

Pool WR, Lipner H (1968) Endrocrinology 79:858

Porter CW, Highfill D, Winovich R (1970a) Int J Fertil 15:171

Porter CW, Highfill D, Winovich R (1970b) Int J Fertil 15:415

Potter E (1963) In: Grady HG, Smith DE (eds) The ovary II. Williams & Wilkins, Baltimore, pp 11–23

Powers RD (1982) J Exp Zool 221:365

Poznakhirkina NA, Serov PL, Korochkin LT (1975) Biochem Genet 14:65

Pratt HPM, Bolton VN, Gudgeon KA (1983) In: Molecular biology of egg maturation. CIBA Sym 98, Pitman

Printz RH, Greenwald GS (1970) Endocrinology 86:290

Racowsky C, McGaughey RW (1983) J Exp Zool 224:103

Rajaniemi HJ, Jääskeläinen K (1979) In: Channing CP, Marsh J, Sadler WA (eds) Ovarian follicular and Corpus Luteum function. Plenum Press, New York, pp 129–135

Rajaniemi HJ, Rönnberg L, Kauppila A, Ylostalo P, Jalkanen N, Saastamoinen J, Selander K, Pystynen P, Vinko R (1980) J Clin End Metab 52:307

Rajendran KG, Menge AC, Menon KMJ (1983) In: Greenwald GS, Terranova PF (eds) Factors regulating ovarian function. Raven Press, New York, pp 129–134

Rapola J, Koskimies O (1967) Science 157:1311

Rao MC (1979) In: Midgley AR Jr, Sadler WA (eds) Ovarian follicular development and function. Raven Press, New York, pp 325–331

Ranta T, Knecht M, Catt KJ (1983)In: Greenwald GS, Terranova PF (eds) Factors regulating ovarian functions. Raven Press, New York, pp 263–268

Rawson JMR, Espey LL (1977) Biol Reprod 17:561

Readhead C, Kaufman MH, Schuetz AW, Abraham GE (1979) In: Channing CP, Marsh J, Sadler WA (eds) Ovarian follicular and Corpus Luteum function. Plenum Press, New York, pp 293–300

Reamer GR (1963) Ph D Thes, Boston Univ

Reed MY, Burton FA, Diest van PA (1979) J Anat 128:195

Reichert LE Jr (1972) In: Saxena BB, Beling CG, Gandy HM (eds) Gonadotrophins. Wiley Interscience, New York, pp 107–119

Reichert LE Jr, Sanzo MA, Fletcher PW, Dias JA, Lee CY (1982) In: Channing CP, Segal SJ (eds) Intraovarian control mechanisms. Plenum Press, New York, pp 135–144

Reimers TJ, Gluss PM, Seidel GE Jr (1977) Proc 10th Am Meet Soc Stud Reprod, Austin, Texas, p 72

Renaud RL, Macler J, Dervain I, Ehret MC, Aron C, Plas-Roser S, Spira A, Pollack H (1980) Fertil Steril 33:272

Repin VS, Akimova IM (1976) Biokhim USSR 41:50

Rice C, McGaughey RW (1981) J Reprod Fertil 62:245

Richards BD (1966) Masters Thes, Fla State Univ, Tallahasee

Richards JS (1975) Endocrinology 97:1174

Richards JS (1978) In: Jones RE (ed) The vertebrate ovary. Plenum Press, New York, pp 331–360

Richards JS (1979) In: Midgley AR Jr, Sadler WA (eds) Ovarian follicular development and function. Raven Press, New York, pp 225–242

Richards JS (1980) Physiol Rev 60:51

Richards JS, Bogovich K (1980) In: Mahesh VB, Muldoon TG, Saxena RB, Sadler WH (eds) Functional correlates of hormone receptors in reproduction. Elsevier, Amsterdam New York, pp 223–244

Richards JS, Kersey KA (1979) Biol Reprod 21:1185

Richards JS, Midgley AR Jr (1976) Biol Reprod 14:82

Richards JS, Williams JJ (1976) Endocrinology 99:1562

Richards JS, Ireland JJ, Rao MC, Bernath GA, Midgley AR Jr, Reichert LE Jr (1976) Endocrinology 99:1562

Richards JS, Rao PC, Ireland JJ (1978) In: Crighton DB, Haynes NB, Foxcroft GR, Lamming GE (eds) Control of ovulation. Butterworths, London, pp 197–216

Richards JS, Uilenbroek J Th J, Jonassen JA (1979) In: Channing CP, Marsh J, Sadler WA (eds) Ovarian follicular and *Corpus Luteum* function. Plenum Press, New York, pp 11–26

Richter JD, McGaughey RW (1979) J Exp Zool 209:81

Richter JD, McGaughey RW (1982) Dev Biol 83:188

Rigby BW, Ling SY, Ledwitz-Rigby F (1983) In: Greenwald GS, Terranova PF (eds) Factors regulating ovarian function. Raven Press, New York, pp 179–184

Robertson JE, Baker RD (1969a) Proc Soc Stud Reprod Davis, Cal (Abstr) 29

Robertson JE, Baker RD (1969b) Proc Soc Stud Reprod Davis, Cal (Abstr) 57:27

Roca R, Garofalo E, Piriz H, Martino I, Rieppi G, Sala M (1976) Biol Reprod 15:464

Rocereto T, Jacobowitz D, Wallach E (1969) Endocrinology 84:1336

Rodman TC, Bachvarova R (1976) J Cell Biol 70:251

Rolland R, Hammond J (1975) Acta endocr (Kbh) Suppl 199:101

Rondell P (1964) Am J Physiol 207:590

Rondell P (1970) Biol Reprod (Suppl) 2:64

Rondell P (1974) Biol Reprod 10:199

Rosado A, Velázquez A, Lara-Ricalde R (1973) Fertil Steril 24:349

Rosenbaur KA, Janse B, Lindauer S, Schloesser HW (1977) Folia Morphol (Prague) 25:53

Rosenfeld MG, Joshi MS (1981) J Reprod Fertil 62:199

Ross GT, Hillier SG, Zeleznik AJ, Knazek RA (1979) In: Klopper A, Lerner L, Molen vander AJ, Sciarra F (eds) Research on steroids. Academic Press, London New York, pp 185–192

Rubin LB, Deane HW, Hamilton JA (1963) Endocrinology 73:748

Ryan RJ (1981) In: Schwartz NB, Hunzicker-Dunn M (eds) Dynamics of ovarian function. Raven Press, New York, pp 1–11

Sacco AG (1977a) Biol Reprod 16:158

Sacco AG (1977b) Biol Reprod 16:164

Sacco AG (1978) J Exp Zool 204:181

Sacco AG (1979) J Reprod Fertil 56:533

Sacco AG (1981) Obst Gynaecol Ann 10:1

Sacco AG, Moghiss KS (1979) Fertil Steril 31:503

Sacco AG, Palm VS (1977) J Reprod Fertil 51:165

Sacco AG, Shivers CA (1973a) J Reprod Fertil 32:421

Sacco AG, Shivers CA (1973b) J Reprod Fertil 32:415

Sacco AG, Shivers CA (1973c) Biol Reprod 8:481

Sacco AG, Yuregicz EC Subramanian MG, DeMayo FJ (1981a) Biol Reprod 25:997

Sacco AG, Subramanian MG, Yurewicz EC (1981b) Proc Soc Exp Biol Med 167:318

Sacco AG, Yurewicz EC, Zhang S (1983) J Reprod Fertil 68:21

Saidapur SK, Greenwald GS (1979a) Biol Reprod 18:401

Saidapur SK, Greenwald GS (1979b) Biol Reprod 20:226

Saidapur SK, Greenwald GS (1979c) Endocrinology 105:1432

Sairam MR, Kato K, Manjunath P, Ramasharma K, Miller WM, Haung ESR, Madhwa Raj HG (1982) In: Channing CP, Segal SJ (eds) Intraovarian control mechanisms. Plenum Press, New York, pp 79–88

Saling PM (1981) Proc Natl Acad Sci USA 78:6231

Salustri A, Martinozzi M (1980) Boll Soc Ital Biol Sperm 56:826

Salustri A, Siracusa G (1983) J Reprod Fertil 68:335

Sanders MM, Midgley AR Jr (1982) Endocrinology 111:614

Sanders MM, Midgley AR Jr (1983) Endocrinology 112:1382

Santhananthan AH, Trounson AO (1982) Gamete Res 5:191

Sanyal MK (1979) In: Channing CP Jr, Marsh J, Sadler WA (eds) Ovarian follicular and *Corpus Luteum* function. Plenum Press, New York, pp 321–324

Sasaki H, Nimura S, Ishida K (1979) Jpn J Zootech Sci 50:88

Sato E, Ishibashi T (1977) J Zootech Sci 48:22

Sato E, Iritani A, Nishikawa Y (1979) Jpn J Anim Reprod 25:95

Sato E, Ishibashi T, Iritani A (1982) In: Channing CP, Segal SJ (eds) Intraovarian control mechanism Plenum Press, New York, pp 161–174

Schaar H (1976) Acta Anat 94:238

Schlafke S, Enders AC (1973) Anat Rec 175:539

Schmell ED, Gulyas BJ (1980) Biol Reprod 23:1075

Schmidtler W (1980) Histochemistry 70:77

Schomberg DW (1979) In: Channing CP, Marsh J, Sadler WA (eds) Ovarian follicular and *Corpus Luteum* function. Plenum Press, New York, pp 155–168

Schomberg DW, May JV, Mondschein JS (1983) In: Greenwald GS, Terranova PF (eds) Factors regulating ovarian function. Raven Press, New York, pp 221–224

Schorderet-Slatkine S, Schorderet M, Baulieu EE (1982) Proc Natl Acad Sci USA 79:850

Schreiber JR (1979) In: Midgley AR Jr, Sadler WA (eds) Ovarian follicular development and function. Raven Press, New York, pp 243–253

Schreiber JR, Nakamura K, Erickson G (1981) In: Schwartz NB, Hunzicker-Dunn M (eds) Dynamics of ovarian function. Raven Press, New York, pp 67–72

Schreiber JR, Nakamura K, Weinstein DB (1983) In: Greenwald GS, Terranova PF (eds) Factors regulating ovarian function. Raven Press, New York, pp 311–316

Schroeder PC, Talbot P (1982) J Exp Zool 224:417

Schuetz AW (1969) In: McLaren A (ed) Advances in reproductive physiology, vol IV. Academic Press, London New York, pp 99–148

Schuetz AW (1977) Res Reprod (IPPF London) 9:3

Schuetz AW (1979) In: Channing CP, Marsh J, Sadler WA (eds) Ovarian follicular and *Corpus Luteum* function. Plenum Press, New York, pp 307–313

Schultz GA (1975) Dev Biol 44:270

Schultz RM, Wassarman PM (1977a) Proc Natl Acad Sci USA 74:538

Schultz RM, Wassarman PM (1977b) J Cell Sci 24:167

Schultz RM, Lamarca J, Wassarman PM (1978a) Proc Natl Acad Sci USA 75:4166

Schultz RM, Letourneau GE, Wassarman PM (1978b) J Cell Sci 30:251

Schultz RM, Letourneau GE, Wassarman PM (1979) Dev Biol 23:120

Schultz RM, Montgomery RR, Belanoff JR (1983) Dev Biol 97:264

Schwall RH, Erickson GF (1981) In: Schwartz NB, Hunzicker-Dunn M (eds) Dynamics of ovarian function. Raven Press, New York, pp 29–34

Schwall RH, Erickson GF (1983) In: Greenwald GS, Terranova PF (eds) Factors regulating ovarian function. Raven Press, New York, pp 351–356

Schwartz NB (1982a) Res Fronts Fertil Reg 2 No 2:1–9

Schwartz NB (1982b) In: Channing CP, Segal SJ (eds) Intraovarian control mechanisms. Plenum Press, New York, pp 15–36

Schwartz NB (1983) In: McCann SM, Dhindsa DS (eds) Role of peptides and proteins in control of reproduction. Elsevier/North Holland Biomedical Press, Amsterdam New York, pp 193–214

Schwartz NB, Channing CP (1977) Proc Natl Acad Sci USA 74:5721

Schwartz NB, Hunzicker-Dunn M (eds) (1981) Dynamics of ovarian function Raven Press, New York

Schwartz P (1974) Biol Reprod 10:236

Schwartz-Kripner A, Channing CP (1979) In: Channing CP, Marsh J, Sadler WA (eds) Ovarian follicular and *Corpus Luteum* function. Plenum Press, New York, pp 137–143

Schwarzel WC, Kruggel WG, Broodie HJ (1973) Endocrinology 92:866

Sellens MH, Jenkinson EJ (1975) J Reprod Fertil 42:153

Selman K (1974) J Cell Biol 63:308a (Abstr)

Selman K, Anderson E (1975) J Morphol 147:251

Selstam G (1975) Ph D Thes, Univ Göteborg, Sweden

Senger PL, Saacke RG (1970) J Cell Biol 46:405
Shaha C, Greenwald GS (1982) J Reprod Fertil 66:197
Shahabi N, Bahr J, Dial OK, Glenn S (1979) In: Channing CP, Marsh J, Sadler WA (eds) Ovarian
    follicular and *Corpus Luteum* function. Plenum Press, New York, pp 179–184
Shlagi R, Kraicer PF, Soferman N (1972) J Reprod Fertil 28:335
Shalgi R, Kraicer PF, Rimon A, Pinto M, Soferman N (1973) Fertil Steril 24:429
Shalgi R, Kaplan R, Kraicer PF (1977) Biol Reprod 17:333
Shander D, Anderson LD, Barraclough CP, Channing CP (1979) In: Channing CP, Marsh J,
    Sadler WA (eds) Ovarian follicular and *Corpus Luteum* function. Plenum Press, New York,
    pp 423–428
Shea BF, Baker RD, Latour JPA (1976a) Can J Anim Sci 56:377
Shea BF,Latour JPA, Bedirian KN, Baker RD (1976b) J Anim Sci 43:809
Sheela Rani CS, Ekholm C, Billig H, Magnusson C, Hillensjö T (1983) Biol Reprod 28:591
Shemesh M (1979) In: Channing CP, Marsh J, Sadler WA (eds) Ovarian follicular and *Corpus
    Luteum* function. Plenum Press, New York, pp 149–152
Shemesh M, Hansel W (1975) Biol Reprod 13:448
Sheridan JD (1971) Dev Biol 26:627
Shimada H, Oskamura H, Noda Y, Suzuki A, Tojo S, Takada A (1983) J Endocrinol 97:201
Shinohara H, Matsuda T (1982) Experientia 38:274
Shivers CA (1974) In: Diczfalusy E (ed) Karolinska symposia on research methods in
    reproductive endocrinology, immunological approaches to fertility control. Karolinska Inst,
    Stockholm, pp 223–244
Shivers CA (1975) Acta Endocrinol (Suppl) 78:223
Shivers, CA (1976) Development of vaccines for fertility regulation. Scripter, Copenhagen, pp
    81–92
Shivers CA (1977) In: Boetteher B (ed) Immunological influences on human fertility. Academic
    Press, London New York, pp 13–19
Shivers CA (1979) In: Alexander N (ed)Animal models for research on contraception and
    fertility. Harper & Row, Hagerstown, Maryland, pp 314–325
Shivers, CA, Dudkiewicz AB, Franklin LE, Fussell EN (1972) Science 178:1211
Shivers CA, Dunbar BS (1977) Science 197:1082
Shivers CA, Sieg PM (1980) In: Dhindsa DS, Schumacher GFB (eds) Immunological aspects of
    infertility and fertility regulation. Elsevier/North Holland, New York, pp 173–182
Shivers CA, Gengozian N, Franklin S, McLaughlin CL (1978) J Med Primatol 7:242
Shivers CA, Sieg PM, Kitchen H (1981) J Am Anim Hosp Assoc 17:832
Short RV (1962) J Endocrinol 24:39
Short RV (1964) Recent Prog Horm Res 20:303
Silavin SL, Greenwald GS (1981) In: 14th Annu Meet Soc Study Reprod. Corvallis, Oregon,
    Abstr No 49
Silavin SL, Greenwald GS (1982) J Reprod Fertil 66:291
Silberzahn P, Dehennin L, Zwain IH, Leymarie P (1983) J Endocrinol 97:51
Silver A (1978) Histochem J 10:79
Silverman A, Asch RH, Lum JB, Cameron IL (1982) Fertil Steril 38:269
Siracusa G (1973) Exp Cell Res 78:460
Siracusa G, Vivarelli E (1975) J Reprod Fertil 43:567
Sjöberg N-O, Owman C, Walles B (1979) In: Talwar GP (ed) Recent avances in reproduction and
    regulation of fertility. Elsevier/North Holland Biomedical Press, Amsterdam New York, pp
    107–113
Smith DM, Tenney DY (1978) J Reprod Fertil 54:401
Smith DM, Tenney DY (1979) J Reprod Fertil 55:415
Smith DM, Tyler JPP, Erickson GF (1978a) J Reprod Fertil 54:393
Smith DM, Conaway CH, Kerber WT (1978b) J Reprod Fertil 54:91
Solano AR, Garcia-Vela A, Catt KJ, Dufau ML (1981) In: Schwartz NB, Hunzicker-Dunn M
    (eds) Dynamics of ovarian function. Raven Press, New York, pp 123–128
Sorensen RA, Wassarman PM (1976) Dev Biol 50:531
Sorina VYa, Snetkova MG, Nikitin A, Pimenova MN (1979a) Ontogenez 10:298
Sorina VYa, Snetkova MG, Kitaev EM, Pimenova MN (1979b) Ontogenez 10:649

Soupart P (1973) In: Hafez ESE, Thibault C (eds) Biology of spermatozoon, transport, survival
    and fertilizing ability, vol 26. Coll INSERM, Paris, pp 407–434
Soupart P (1975) In: Thibault C (ed) La fécondation. Masson, Paris, pp 81–93
Soupart P, Noyes RW (1964) J Reprod Fertil 8:251
Soupart P, Strong PA (1974) Fertil Steril 25:11
Stadnicka A, Stoklosowa S (1976) Z Mikrosk Anat Forsch (Leipzig) 90:458
Staehelin LA (1974) Int Rev Cytol 39:191
Stahler E, Spaetling L, Dauhe E, Buchholz R (1977) Arch Gynaecol 223:41
Stahl A, Luciani JM, Devictor M, Capordano AM, Gagne R (1975) Humangenetik 26:315
Stastna J (1974) Scripta Med 47:527
Stefanini M, Oura C, Zamboni L (1969) J Submicrosc Cytol 1:1
Stefenson A, Owman C, Sjöberg NO, Sporrong B, Walles B (1981) Cell Tiss Res 215:47
Stegner HE (1967) Erg Anat Entwicklungsgesch 39:7
Stegner HE, Onken M (1971) Cytobiologie 3:240
Stegner HE, Wartenberg H (1961) Z Zellforsch 53:702
Stern S, Rayyis A, Kennedy JF (1972) Biol Reprod 7:341
Stewart LE, Rigby BW, Ledwitz-Rigby F (1982) Biol Reprod 27:54
Stoklosowa S, Nalbandov AV (1972) Endocrinology 91:25
Stoklosowa S, Szöltys M (1978) Ann Biol Anim Biochem Biophys 18:503
Stoklosowa S, Bahr J, Gregoraszckzuk E (1979) In: Channing CP, Marsh J, Sadler WA (eds)
    Ovarian follicular and Corpus Luteum function. Plenum Press, New York, pp 145–148
Stoklosowa S, Gregoraszczuk E, Channing CP (1982) Biol Reprod 26:943
Stone SL, Pomerantz SH, Schwartz-Kripner, A, Channing CP (1978) Biol Reprod 19:585
Strickland S, Beers WH (1979) In: Midgley AR Jr, Sadler WA (eds) Ovarian follicular
    development and function. Raven Press, New York, pp 143–153
Sturgis SH (1961) In: Villee CL (ed) Control of ovulation. Pergamon Press, Oxford New York, pp
    213–218
Suzuki F, Yanagimachi R (1983) Cell Tiss Res 231:265
Suzuki S, Kitai H, Tojo R, Seki K, Oba M, Fujiwara T, Iizuka R (1981) Fertil Steril 35:142
Suzuki S, Endo Y, Fujiwara T, Tanaka S, Iizuka R (1983) Fertil Steril 39:683
Swartz WJ, Schuetz AW (1980) Histochemistry 68:39
Swenson CE, Dunbar BS (1982) J Exp Zool 219:97
Symanski JA, Schomberg DW (1980) Endocrinology 106 (Suppl):226
Szöllösi D (1962) J Reprod Fertil 4:223
Szöllösi D (1967) Anat Rec 159:431
Szöllösi D (1970) J Cell Biol 44:192
Szöllösi D (1972) In: Biggers J, Schuetz A (eds) Oogenesis. University Park Press, Baltimore, pp
    47–64
Szöllösi D (1975a) In: Thibault C (ed) La fécondation. Masson, Paris, pp 13–35
Szöllösi D (1975b) In: Blandau RJ (ed) Aging gametes. Karger, Basel, pp 98–121
Szöllösi D (1976) In: Grundman E, Kirsten WH (eds) Current topics in pathology. Springer,
    Berlin Heidelberg New York, pp 9–27
Szöllösi D (1978) Res Reprod 10:3
Szöllösi D, Gérard M, Ménézo Y, Thibault C (1978) Ann Anim Biochem Biophys 18:511
Szoltys M (1976) J Reprod Fertil 48:397
Tadano Y, Yamada K (1978) Histochemistry 57:203
Tadano Y, Yamada K (1979) Histochemistry 60:125
Takai I, Mori T, Nishimuro T (1980) J Reprod Fertil 61:19
Talbot P (1983) J Exp Zool 226:129
Talbot P, Chacon RS (1982) J Exp Zool 224:409
Talbot P, Schroeder PC (1982) J Exp Zool 224:427
Tardini A, Vitali-Mazza L, Mansani FE (1961) Arch De Vecchi Anat Patol Med Clin 35:25
Taya K, Greenwald GS (1982a) Biol Reprod 27:1090
Taya K, Greenwald GS (1982b) Endocrinol Jpn 29:453
Taya K, Saidapur SK, Greenwald GS (1980) Biol Reprod 22:307
Terranova PF (1980) Biol Reprod 23:92

Terranova PF (1981 a) In: Schwartz NB, Hunzicker-Dunn M (eds) Dynamics of ovarian function. Raven Press, New York, pp 35–40
Terranova PF (1981 b) Endocrinology 108:1890
Terranova PF, Ascanio LE (1982) Biol Reprod 26:129
Terranova PF, Garza F (1983) Biol Reprod 29:630
Terranova PF, Martin NC, Chien S (1982) Biol Reprod 26:721
Terranova PF, Loker D, Garza F, Martin N (1983) In: Greenwald GS, Terranova PF (eds) Factors regulating ovarian function. Raven Press, New York, pp 39–44
Tesařík J, Dvořák M (1978 a) In: Klika E (ed) XXI Coll Sci Fac Med Univ Carol XIX Congr Morphol Symp, Prague, p 533
Tesařík J, Dvořák M (1978 b) Arch Anat Hist Embriol 75:11
Tesoriero JV (1981) J Morphol 168:171
Testart J (1983) J Reprod Fertil 68:413
Thanki KH, Channing CP (1979) In: Midgley AR Jr, Sadler WA (eds) Ovarian follicular development and function. Raven Press, New york, pp 215–222
Thebault A, Lefevre B, Testart J (1983) J Reprod Fertil 68:419
Thibault C (1949) Ann Sci Nat Zool 11:133
Thibault C (1977) J Reprod Fertil 51:1
Thibault C, Gérard M, Menezo Y (1975) Ann Biol Anim Biochem Biophys. 15:705
Thibault C, Gérard M, Menezo Y (1976) Prog Reprod Biol 1:233
Torre JC de la, Surgeon JW (1976) Histochemistry 49:81
Toshimori K, Oura C (1982) Cell Tiss Res 224:383
Travnik P (1977 a) Folia Morphol (Prague) 25:21
Travnik P (1977 b) Folia Morphol (Prague) 25:15
Trounson AO, Moore NW (1974) J Reprod Fertil 41:97
Trounson AO, Shivers CA, McMaster K, Lpata A (1980) Arch Androl 4:29
Tsafriri A (1978 a) In: Jones RE (ed) The vertebrate ovary. Plenum Press, New York, pp 409–442
Tsafriri A (1978 b) Ann Biol Anim Biochem Biophys 18:523
Tsafriri A (1979) In: Channing CP, Marsh J, Sadler WA (eds) Ovarian follicular and *Corpus Luteum* function. Plenum Press, New York, pp 269–281
Tsafriri A (1983) In: Metz CB, Monroy A (eds) Biology of fertilization. Academic Press, London New York (in press)
Tsafriri A, Bar-Ami S (1978) J Exp Zool 205:293
Tsafriri A, Bar-Ami S (1982) In: Channing CP, Segal SJ (eds) Intraovarian control mechanisms. Plenum Press, New York, pp 145–160
Tsafriri A, Channing CP (1975 a) Endrocrinology 96:922
Tsafriri A, Channing CP (1975 b) J Reprod Fertil 43:149
Tsafriri A, Lindner HR, Zor U, Lamprecht SA (1972) J Reprod Fertil 31:39
Tasfriri A, Koch Y, Lindner HR (1973 a) Prostaglandins 3:461
Tsafriri A, Lieberman ME, Barnea A, Bauminger S, Lindner HR (1973 b) Endocrinology 93:1378
Tsafriri A, Pomerantz SH, Channing CP (1976 a) Biol Reprod 14:511
Tsafriri A, Pomerantz SH, Channing CP (1976 b) In: Crosignani PG, Mishell DR (eds) Ovulation in the human. Academic Press, London New York, pp 31–39
Tsafriri A, Lieberman ME, Ahrén K, Lindner HR (1976 c) Acta Endocrinol 83:151
Tsafriri A, Liberman ME, Koch Y, Bauminger S, Chobsieng P, Zor U, Lindner HR (1976 d) Endocrinology 98:655
Tsafriri A, Bar-Ami S, Dekel N (1982 a) In: Fujji T, Channing CP (eds) Advances in the biosciences: non-steroidal regulations in reproductive biology and medicine, vol 34. Pergamon Press, Oxford New York, pp 147–156
Tsafriri A, Dekel N, Bar-Ami S (1982 b) J Reprod Fertil 64:541
Tsang BK, Carnegie JA (1983) In: Greenwald GS, Terranova PF (eds) Factors regulating ovarian functions. Raven Press, New York, pp 369–374
Tsonis CG, Quigg HE, Trounson AO, Findlay JK (1983 a) In: Greenwald GS, Terranova PF (eds) Factors regulating ovarian function. Raven Press, New York, pp 163–168
Tsonis CG, Quigg H, Lee VWK, Leversha L, Trounson AO, Findlay JK (1983 b) J Reprod Fertil 67:83
Tsujimoto D, Katayama K, Tojo S, Mizoguti H (1982) Acta Obstet Gynaecol Scand 61:269

Tsunoda Y (1977) J Reprod Fertil 50:353
Tsunoda Y, Chang MC (1976a) Biol Reprod 14:354
Tsunoda Y, Chang MC (1976b) J Exp Zool 195:409
Tsunoda Y, Chang MC (1976c) Biol Reprod 15:361
Tsunoda Y, Chang MC (1976d) J Reprod Fertil 46:379
Tsunoda Y, Chang MC (1977) Int J Fertil 22:129
Tsunoda Y, Chang MC (1978a) Biol Reprod 18:468
TsunodaY, Chang MC (1978b) J Reprod Fertil 54:233
Tsunoda Y, Sugie T (1977) Jpn J Zootech Sci 48:784
Tsunoda Y, Sugie T (1979a) In: Talwar GP (ed) Recent advances in reproduction and regulation
    of fertility. Elsevier/North Holland Biomedical Press, Amsterdam New York, pp 123–133
Tsunoda Y, Sugie T (1979b) Jpn J Zootech Sci 50:493
Tsunoda Y, Sugie T (1982) Gamet Res 6:73
Tsunoda Y, Whittingham DG (1982) J Reprod Fertil (in press)
Tsunoda Y, Soma T, Sugie T (1980) Arch Androl 5:58
Tsunoda Y, Soma T, Jinbu M, Tachiura K, Sugie T (1981a) Gamet Res 4:231
Tsunoda Y, Soma T, Sugie T (1981b) Gamet Res 4:133
Tsunoda Y, Sugie T, Mori J, Isozima S, Koyama K (1981c) J Exp Zool 217:103
Tsunoda Y, Soma T, Sugie T (1982) In: Hafez ESE, Semn K (eds) In vitro fertilization and embryo
    transfer. MTP Press, Lancester, pp 117–126
Turnbull KE, Braden AWH, Mattner PE (1977) Aust J Biol Sci 30:229
Uilenbroek J Th, Richards J (1979) Biol Reprod 20:1159
Uilenbroek J, Th, Arendsen Th, Wolff-Exalto de E, Welschen R (1976) Ann Biol Anim Biochem
    Biophys 16:297
Uilenbroek J Th J, Wourtersen PJA, Schoot van der P (1980) Biol Reprod 23:219
Uilenbroek J Th J, Schoot van der P, Wourtersen PJA (1981) In: Schwartz NB, Hunzicker-Dunn
    M (eds) Dynamics of ovarian function. Raven Press, New York, pp 41–46
Ullmann SL (1979) J Anat 128:619
Vaitukaitis JL, Albertson BD (1979) In: Midgley AR Jr, Sadler WA (eds) Ovarian follicular
    development and function. Raven Press, New York, pp 247–253
Vazquez-Nin GH, Echeverria OM (1976) Acta Anat 96:218
Vazquez-Nin GH, Sotelo JR (1967) Z Zellforsch. 81:91
Velazquez A, Reyes A, Chargoy J, Rosado A (1977) Fertil Steril 28:96
Veldhuis JD, Hammond JM (1980) Nature (London) 284:262
Veldhuis JD, Klase P, Hammond JM (1981) In: Schwartz NB, Hunzicker-Dunn M (eds)
    Dynamics of ovarian function. Raven Press, New York, pp 111–116
Veldhuis JD, Klase PA, Strauss JF, Hammond JM (1982) Endocrinology 111:144
Veldhuis JD, Klase PA, Strauss JF, Hammond JM (1983) In: Greenwald GS, Terranova PF (eds)
    Factors regulating ovarian function. Raven Press, New York, pp 323–328
Verbitskii MSh, Papazov IP, Shoshev VM, Mikhailov AT, Babaev VR (1980) Ontogenez 11:583
Vermeiden JP, Zeilmaker PW (1974) Endocrinology 85:341
Vernon MW, Dierschke DJ, Sholl SA, Wolf RX (1983) Biol Reprod 28:342
Vigerski RA, Loriaux DL (1976) J Clin Endocrinol Metab 43:817
Virutasmasen P, Wright K, Wallach E (1972) Obstet Gynaecol 39:225
Virutasmasen P, Smitasiri Y, Fuchs A (1976) Fertil Steril 27:188
Vivarelli E, Siracusa G, Mangia F (1976) J Reprod Fertil 47:149
Volkova OV, Melnikova LM, Perkarskii, Povalli TM (1980a) Byull Eksp Biol Med 89:229
Volkova OV, Alkadarskaya IM, Milovidova NS (1980b) Arkh Anat Gistol Embriol 79:5
Wabik-Sliz B (1979) Biol Reprod 21:89
Wada Y (1978) Bull Azabu Vet Coll 3:313
Wallach EE, Wright KH, Hamada Y (1978) Am J Obstet Gynaecol 132:728
Wallach EE, Bronson RA, Kobayashi Y, Hamada A, Wright KH (1980) In: Tozzini RI, Reeves
    G, Pineda RI (eds) Endocrine physiopathology of the ovary. Elsevier/North Holland
    Biomedical Press, Amsterdam New York, pp 153–163
Walles B, Edvinsson L, Nybell G, Owman C, Sjöberg N (1974) Fertil Steril 26:602
Walles B, Edvinsson L, Falck B, Owman C, Sjöberg N, Svennson K (1975a) Biol Reprod
    12:239

Walles B, Edvinsson L, Falck B, Owman C, Sjöberg N, Svennson K (1975b) J Pharmacol Exp Theor 193:460
Walles B, Edvinsson L, Owman CO, Sjöberg NO, Sporrong B (1976) Biol Reprod 15:565
Walles B, Flack B, Owman C, Sjöberg NO, Svennson K (1977) Biol Reprod 17:423
Walles B, Groshcel-Stewart U, Owman C, Sjöberg N, Unsicker K (1978) J Reprod Fertil 52:175
Wang C (1983) Endocrinology 12:1130
Wang C, Chan V (1982) Endocrinology 110:1085
Wang C, Leung A (1983) Endocrinology 112:1201
Wang C, Hsueh AJW, Erickson GF (1979) J Biol Chem 254:11330
Wang C, Hsueh AJW, Erickson GF (1980) Mol Cell Endocrinol 20:135
Wang C, Hsueh AJW, Erickson GF (1981) In: Schwartz NB, Hunzicker-Dunn M (eds) Dynamics of ovarian function. Raven Press, New York, pp 61–66
Ward DN, Liu W-K, Glenn SD, Channing CP (1983a) In: Channing CP, Segal SJ (eds) Intraovarian control mechanisms. Plenum Press, New York, pp 263–282
Ward DN, Glenn SD, Liu W-K, Gordon WL (1983b) In: Greenwald GS, Terranova PF (eds) Factors regulating ovarian function. Raven Press, New York, pp 141–156
Ware VC (1982) J Exp Zool 222:155
Warnes GM, Moor RM, Johnson MH (1977) J Reprod Fertil 49:331
Wassarman PM, Josefowicz J (1978) J Morphol 156:209
Wassarman PM, Letourneau GE (1976a) Nature (London) 261:73
Wassarman PM, Letourneau GE (1976b) J Cell Sci 20:549
Wassarman PM, Mrozak SC (1981) Dev Biol 84:364
Wassarman PM, Josefowicz WJ, Letourneau GE (1976) J Cell Sci 22:531
Wassarman PM, Ukena TE, Josefowicz WJ, Letourneau GE, Karnovsky MJ (1977) J Cell Sci 6:323
Wassarman PM, Schultze RM, Letourneau GE (1979a) Dev Biol 69:94
Wassarman PM, Schutz RM, Letourneau MJ, LaMarca MJ, Josefowicz WJ, Bleil JD (1979b) In: Channing CP, Marsh J, Sadler WA (eds) Ovarian follicular and Corpus Luteum Function. Plenum Press, New York, pp 251–268
Watzka M (1957) Das Ovarium in Handbuch der mikr. Anatomie des Menschen, vol VII, 3. Springer, Berlin Göttingen Heidelberg
Weakley BS (1966) J Anat 100:503
Weakley BS (1967) J Anat 101:435
Weakley BS (1976) Cell Tiss Res 168:531
Weakley BS (1977) Cell Tiss Res 180:515
Weakley BS, James JL (1982) Cell Tiss Res 223:127
Weakley BS, Webb P, James JL (1981) Cell Tiss Res 220:349
Weiner S, Wright KH, Wallach EE (1977) Am J Obstet Gynaecol 128:154
Weiss GK, Dail WG, Ratner A (1982) J Reprod Fertil 65:507
Weiss TJ, Armstrong DT (1977) ASRB Meet Austr
Weiss TJ, Seamark RF, McIntosh JEA, Moor RM (1976) J Reprod Fertil 46:347
Weitlauf H, Greenwald GS (1965) J Reprod Fertil 10:203
Weitlauf H, Greenwald GS (1967) Anat Rec 159:249
Welschen R, Dullaart J (1976) J Endocrinol 70:301
Welsh TH, Zhuang L-Z, Hsueh AJW (1983) Endocrinology 112:1916
Weymarn von N, Guggenheim R, Mueller H (1980a) Anat Embryol 161:19
Weymarn von N, Guggenheim R, Mueller H (1980b) Anat Embryol 161:9
Wiebel ER, Bolender RP (1973) In: Hayatt MA (ed) Principles and techniques of electron microscopy, vol III. Van Nostrand-Rheinhold, New York, p 237
Wiel van de DFM, Bar-Ami S, Tsafriri A, Jong de FH (1983) J Reprod Fertil 68:247
Wildt DE, Levinson CJ, Saeger SWJ (1977) Anat Rec 189:443
Wilkinson RF Jr, Byskov AG, Anderson E (1979) In: Midgley AR Jr, Sadler WA (eds) Ovarian follicular development and function. Raven Press, New York, pp 65–70
Williams PC (1956) In: Wolstenholme GEM, Miller ECP (eds) Ciba foundation colloquia on ageing, vol II. Churchill, London, p59
Williams AT, Lipner H (1982) In: Channing CP, Segal SJ (eds) Intraovarian control mechanisms. Plenum Press, New York, pp 99–116

Williams AT, Rush ME, Lipner H (1979) In: Channing CP, Marsh J, Sadler WA (eds) Ovarian follicular and *Corpus Luteum* function. Plenum Press, New York, pp 429–435

Wise PM, Paolo de LV, Anderson LD, Channing CP, Barraclough CA (1979) In: Channing CP, Marsh J, Sadler WA (eds) Ovarian follicular and *Corpus Luteum* function. Plenum Press, New York, pp 437–447

Wolf, DP, Hamada M (1977) Biol Reprod 17:350

Wolf DP, Nicosia SV (1978) J Cell Biol 79:160a (Abstr)

Wolf U, Engel W (1972) Humangenetik 15:99

Wolgemuth DJ (1983) In: Hartmann JF (ed) Mechanisms and control of animal fertilization. Academic Press, London New York, pp 415–452

Wolgemuth DJ, Jagiello GM (1979) In: Midgley AR Jr, Sadler WA (eds) Ovarian follicular development and function. Raven Press, New York, pp 379–383

Wolgemuth DJ, Jagiello GM, Henderson AS (1978) Exp Cell Res 118:181

Wolgemuth DJ, Jagiello GM, Henderson AS (1980) Dev Biol 78:598

Wood DM, Dunbar BS (1981) J Exp Zool 217:423

Wood DM, Liu C, Dunbar BS (1981) Biol Reprod 25:439

Wyche JH, Noteboom WD (1977) Exp Cell Res 110:135

Yajima Y (1980) Acta Obstet Gynaecol Jpn 32:1103

Yanagimachi R (1969) J Exp Zool 170:26

Yanagimachi R (1974) J Reprod Fertil 38:485

Yanagimachi R (1977) In: Edidin M, Johnson MH (eds) Immunobiology of gametes. Cambridge Univ Press, Cambridge, pp 187–207

Yanagimachi R (1978) Curr Top Dev Biol 12:83

Yanagimachi R, Chang M (1961) J Exp Zool 148:185

Yanagimachi R, Nicolson GL (1976) Exp Cell Res 100:249

Yanagimachi R, Winkelhake JL, Nicolson GL (1976) Proc Natl Acad Sci USA 73:2405

Yanagishita M, Hascall VC (1979) J Biol Chem 254:344

Yanagishita M, Rodbard D, Hascall C (1979) J Biol Chem 254:911

Yanagishita M, Hascall VC, Rodbard D (1981) Endocrinology 109:1641

Yang NST, Marsh JM, LeMaire WJ (1973) Prostaglandins 4:395

Yang NST, Marsh JM, LeMaire WJ (1974) Prostaglandins 6:37

Yang WH, Papkoff H (1973) Fertil Steril 24:633

Yen SSC (1978) In: Yen SSC, Jaffe RB (eds) Reproductive endocrinology: physiology, pathophysiology and clinical management. Saunder, Phil, pp 152–170

Ying S-Y, Guillemin R (1979) C R Acad Sci 289:943

Ying S-Y, Guillemin R (1980) 62nd Annu Meet Endocrinol Sco Abstr 158

Yin S-Y, Ling N, Bohlon P, Guillemin R (1981) Endocrinology 108:1206

Ying S-Y, Ling NC, Esch FS, Guillemin R, Watkins WB (1982) In: Channing CP, Segal SJ (eds) Intraovarian control mechanisms. Plenum Press, New York, pp 117–134

Yoon YD, Cho WK (1976) Korean J Zool 19:1

Yoshinaga K (1978) In: Jones RE (ed) The vertebrate ovary. Plenum Press, New York, pp 691–729

Young RJ (1977) Biochem Biophys Res Commun 78:32

Young RL, Stull GB, Brinster RL (1973) J Cell Biol 59:372a

Young RJ, Sweeny K, Bedford JM (1978) J Embriol Exp Morphol 44:133

Young Lai EV (1972) J Reprod Fertil 30:157

Young Lai EV (1973) Endocrinology 91:1267

Young Lai EV (1979) In: Midgley AR Jr, Sadler WA (eds) Ovarian follicular development and function. Raven Press, New York, pp 183–185

Yurewicz EC, Sacco AG, Subramanian MG (1983) Biol Reprod 29:511

Zachariae F (1958) Acta Endocrinol 27:339

Zachariae F, Jensen CE (1958) Acta Endocrinol 27:343

Zamboni L (1970) Biol Reprod Suppl 2:44

Zamboni L (1972) In: Biggers JD, Schuetz HW (eds) Oogenesis. Univ Park Press, Baltimore, pp 5–45

Zamboni L (1974) Biol Reprod 10:125

Zamboni L (1975) Am J Anat 144:525

Zamboni L (1976) In: Crosignani PG, Mishell DR (eds) Ovulation in the human. Academic Press, London New York, pp 1–30
Zamboni L (1980) In: Tozzini RI, Reeves G, Pineda RI (eds) Endocrine physiopathology of the ovary. Elsevier/North Holland Biomedical Press, Amsterdam New York, pp 62–99
Zamboni L, Mastroianni L Jr (1966a) J Ultrastruct Res 14:95
Zamboni L, Mastroianni L Jr (1966b) J Ultrastruct Res 14:118
Zamboni L, Thompson RS (1976) J Ultrastruct Res 55 (Abstr)
Zamboni L, Mishell DR Jr, Bell JH, Baca M (1966) J Cell Biol 30:579
Zamboni L, Thompson RS, Moore Smith D (1972) Biol Reprod 7:425
Zamboni L, Patterson H, Jones M (1976) Am J Anat 147:95
Zeilmaker GH (1978) Ann Biol Anim Biochem Biophys 18:529
Zeilmaker GH, Verhamme CMPM (1974) Biol Reprod 11:145
Zeilmaker GH, Verhamme CMPM (1976) Eur J Obst Gynaecol Reprod Biol 6:35
Zeilmaker GH, Verhamme CMPM (1977) Acta Endocrinol 86:380
Zeilmaker GH, Hulsmann WC, Wensinck F, Verhamme C (1972) J Reprod Fertil 29:151
Zeleznik AJ, Midgley AR Jr, Reichert LE Jr (1974a) Endocrinology 95:818
Zeleznik AJ, Desjardins C, Midgley AR Jr, Reichert LE Jr (1974b) Fed Proc 33:213
Zeleznik AJ, Schuler HM, Reichert LE (1981) Endocrinology 109:356
Zenzes MT, Engel W (1976) Cytobios 16:53
Zenzes MT, Engel W (1977) Cytobios 18:151
Zimniski SJ, Rorke EA, Helmy S, Vaitukaitis (1982) Endocrinology 111:635
Zor U, Lamprecht SA, Misulovin Z, Koch Y, Lindner HR (1976) Biochim Biophys Acta 428:761
Zor U, Strulovici G, Lindner HR (1977) Biochem Biophys Res Commun 76:1086
Zor U, Strulovicia B, Braw R, Lindner HR, Tsafriri A (1983) J Endocrinol 97:43
Zuckerman S (1960) In: Austin CR (ed) Sex differentiation and development. Mem Soc Endocrinol 7:63
Zuckerman S, Baker TG (1977) In: Zuckerman S, Weir BJ (eds) The ovary, 2nd edn, vol I. Academic Press, London New York, pp 41–112
Zuckerman S, Weir BJ (eds) (1977) The ovary, vol I, II, III. Academic Press, London New York
Zybina EV, Grishchenko TA (1977) Tsitologiya 19:1140
Zybina EV, Zybina TG, Dalma AR (1980) Tsitologiya 22:381

Subject Index

DATE DUE

DEC 16 1996			

DEMCO 38-297